P9-DGF-758

FIRE

—AND—

ICE

FIRE

— AND —

ICE

Soot, Solidarity, and Survival on the Roof of the World

Jonathan Mingle

St. Martin's Press
New York

 For information, address St. Martin's Press, 175 Fifth Avenue, New York, NY 10010.

www.stmartins.com

Library of Congress Cataloging-in-Publication Data

Mingle, Jonathan.
Fire and ice : soot, solidarity, and survival on the roof of the world / by Jonathan Mingle.—First edition.
pages cm
Includes bibliographical references and index.
ISBN 978-1-250-02950-8 (hardcover)
ISBN 978-1-250-02951-5 (e-book)
1. Carbon dioxide mitigation—India—Kumik. 2. Biomass stoves—India—Kumik. 3. Air—Pollution—India—Kumik. 4. Atmospheric carbon dioxide—Environmental aspects—India—Kumik. 5. Soot—Environmental aspects—India—Kumik. 6. Himalaya Mountains Region—Climate. 7. Kumik (India)—Climate. I. Title.
TD885.5.C3M56 2015
363.738—dc23 2014036955

St. Martin's Press books may be purchased for educational, business, or promotional use. For information on bulk purchases, please contact the Macmillan Corporate and Premium Sales Department at 1-800-221-7945, extension 5442, or write to specialmarkets@macmillan.com.

First Edition: March 2015

10 9 8 7 6 5 4 3 2 1

For my parents,

James and Barbara O. Mingle,

and for the Kumikpas

Contents

How wondrously strange, and how miraculous, this!
I haul water, I carry fuel.

—Layman P'ang, China, 8th century CE

FIRE

—AND—

ICE

Prologue:

A Hearth Tale

"When I was a boy we would go to other houses," Tashi Stobdan recalls with a smile, warming at the memory. "We all lie down in a line together, and we fall asleep as *Meme* tells stories."

Stobdan conjures a scene from a few decades ago in Zanskar, a remote valley high in the Himalayan mountains of northwest India, where temperatures plummet fast once the sun sets. With the day's chores done, everyone retreats to the *yokhang*, the slightly sunken winter kitchen on the ground floor, surrounded by stables and fodder storerooms for the sheep, goats, cows, and *dzo* (ponderous yak-cow hybrids prized for their draft power). The *yokhang* is always dark and smoke-filled, the warmest room in the house.

After dinner, the mother of the house rakes ashes over the embers in the earthen hearth, insulating them from the cold, so the fire can be coaxed back to life the next morning. *Meme* (Grandfather) settles into the place of honor near the fire. Neighbor kids crowd in to join the extended family, and everyone stretches out on the floor in a tightly packed row. Some have to squeeze into the next room, where *chang*, the local barley wine, is stored, but they can still hear the voices in the kitchen. Then someone calls out in the dark: "Who will tell a story?"

And *Meme* replies, "I will tell a story." And he begins: "Once upon a time . . ."

He spins tales of great Buddhist saints like Naropa and Marpa and Guru Rinpoche, who once visited Zanskar and performed fantastic feats to conquer the unruly nature spirits, founded great monasteries, and left mysterious signs of their passing in cliffside redoubts. Or perhaps he relates some fable, a morality tale about the *ridaks*—the wild goats and blue sheep that roam the mountains—and their hunters, from a canon not unlike that of Aesop or the even older Panchatantra stories of ancient India. Or he might tell part of the story of the legendary King Gesar of Ling—whose adventures, much like the fantastic journeys of the Odyssey, the Ramayana, and the Epic of Gilgamesh, took many days to recount in full. He describes the trials of ancient heroes and heroines or those distant moments of human folly responsible for certain singular features of the land around them—springs that gush improbably out of cliff faces and house-sized boulders that sit alone in the middle of the wide plain—reminders to his rapt listeners that the physical world is ceaselessly shaped by people's mental states and moral choices.

Meme might even pose a series of riddles and word games known as *tsot-tsot.* "There is a hollow stone," he'd say. "On the stone is the water. In the water swims a fish. What is this?" If the child gets the right answer—in this case, a local lamp, with the fish being the wick and the water the kerosene in which it swims—*Meme* announces, "You win Karsha Gompa!" and pretends to bestow this grand gift, or one of the other priceless monasteries of Zanskar, upon her.

The children soon grow drowsy, tucked under their woolens on the floor of the *yokhang*, until one by one they drift off to the soothing rhythms of his voice. And with everyone fast asleep, and the embers stowed safe and warm under their blanket of ashes (would they still be hot in the morning?), *Meme* ends his tale, perhaps to pick up where he left off the next evening. Soon he, too, nods off by the whispering coals.

A tale told around the fire: the cornerstone on which this young, collective venture of ours, roughly filed under the rubric of "civilization," has been built. In every culture, in every distant corner of the earth, you can find a scene that hasn't changed much since humans built their first hearth several hundred thousand years ago: people gathered around a fire, spinning yarns as they watch the flames gutter and dance, finding solace, joy, and mystery in their glow, and in the reflected warmth of each other's company. Since long before the time of Homer, the hearth has drawn out our deepest fears, hopes, longings—and stories that lend our lives a meaning as necessary as the air we breathe. In many ways, the story of fire *is* the story of humanity—we may only have become fully human, in both the biological and the social sense, once we learned to control the flames.

Think of this book as a set of linked tales told around the fire. We gaze at the flame, but most of us don't fully register what it is. Among other things, it is a catalyst for our tales, a metronome for their telling, a backdrop for other passions. But here our subject is the fire itself—and what comes out of it.

The particular hearth around which we are gathered, which illuminates both the space we all share and each other's faces—the fire that we are trying to really see for the first time—is in Kumik, Tashi Stobdan's village in Zanskar, ringed by high mountains on the edge of the roof of the world. This village is now in the final days of its more than thousand-year-long history. Its fire is dying down, its story is coming to an end. Tomorrow a new fire will be lit, from the embers still glowing under their blanket of ashes, and with it a new story will begin. But tonight there are many meanings to glean from the ashes themselves.

Like every tale, every fire has a beginning, a middle, and an end: fuel is gathered, kindling is lit, the flames take and roar for a while. Spent, they gutter and ebb, and all soon dies down to glowing coals. The wood and dung, turned into phantoms that have fled up the chimney pipe, leave a hazy, acrid rumor of their passing, in a warm room full of our imaginings. Through it all, smoke spreads beyond the firelight to settle in places unseen.

And for a time, dark particles fill the world.

PART ONE

The Question of Kumik

How from
a fire that never sinks
or sets,
would you escape?

—HERACLITUS

1

The Curse

In the garden we friends and equals
splash and play in the water.
Some who disapprove
tell us, "Don't frolic in the water!"
Why should we not enjoy the water?
How much life is left to us?
Why not enjoy the water?
How long will this life last?

—FROM *WAKHAI DRAKBU*, A LADAKHI FOLK SONG

The plane carries us westward, sunward, over the Gangetic Plain. To the north, anvil-shaped thunderheads hide the seam where the subcontinent accordions into foothills, then improbable mountains, as it plows into Asia. Blocked from advancing farther by the jagged wall of the Great Himalaya, which casts its "rain shadow" beyond onto the thirsty Tibetan Plateau, those clouds are parked over the hill towns of Uttarakhand, smothering them with rain.

Another cloud bank, murky brown, hides the plain below. This "atmospheric brown cloud," as the scientists call it, looks from this height like a vast shroud draped across the shoulders of South Asia. While the rain clouds beyond are bright, reflecting the sun's light, this cloud sponges it up greedily. From below this morning in the city of Lucknow it turned the sun into a vague, milky disc. Now, from above, it obscures the outlines of the towns that thrum along beneath us and turns the fertile fields into a patchwork of muted browns and beiges.

That soil has been plowed and planted almost continuously since the time of the Rig-Veda. Those ancient verses captured their authors' awe of, and gratitude for, the cyclic drama of flaring fire and flowing water that bounded and enabled their existence. "The waters which are from heaven," the Vedic peoples of north India sang over 3,000 years ago, "and which flow after being dug, and even those that spring by themselves, the bright pure waters which lead to the sea, may those divine waters protect me here."

They sang, too, of the fierce battle between the god Indra, lord of the thunderbolt, and Vritra, "the enveloper," the demiurge embodied in the clouds. The jealous Vritra had gathered all the waters of the world into himself, causing a terrible drought. Indra stormed Vritra's ninety-nine cloud fortresses and slew him, releasing the waters to flow back into the great rivers.

My fellow passengers and I on JetKonnect Flight S24233 to New Delhi are sojourning here in Indra's world, on brief parole from the earthbound realm of his brother Agni, the god of fire. Where the two realms meet, there is a kind of second horizon, a boundary line between brown cloud and blue sky above. As we climb higher, the brown cloud begins to resemble a puddle spreading across northern India, fed by some hidden leak—millions of them, actually. It is, in fact, the vast dark aftermath of the gift of Matarisvan, the Vedic Prometheus, who delivered Agni's sacred spark to mortals on the breeze.

Somewhere off to the northeast, behind the monsoon's cloud veil, pilgrims and tourists are climbing the long steps up Swayambhunath, the sacred hill on the west edge of Kathmandu, where candles are still lit daily on a small altar to Agni. On the eastern side of that city, in the shadow of Pashupatinath Temple, mourners are gathered on the banks of the Bagmati River—which flows on to join the Koshi, and then the Ganges to the south—watching the bodies of their loved ones turn into clouds of flame and ash, memories and lofted soot. Far below us, the Yamuna River, India's holiest after the Ganges, snakes its way to the southeast, swollen with rains and Himalayan silt, all heading for the distant Bay of Bengal. And thus the central drama of the ancient Rig-Veda plays out over and over again, through the window of seat 7C: Indra slays Vritra, the waters flow down the mountainsides and through the ancient grooves of the subcontinent, surge upward into grains of wheat and rice and stalks of sugarcane, sigh into the sea, to cycle back as more rains that feed glaciers and rivers and fields and farms and people and creatures and the clouds that feed the rains again and again, world everlasting, amen.

I press my nose against the glass and try to discern the outline of the foothills of the Himalaya and the upstream origins of the holy rivers. But all is hopelessly obscured by the two cloud veils: Agni's smoke and Indra's rains.

The monsoon is very late this year. A few days ago, on my way to visit a grassroots project to recharge dried-up springs in hillside villages of the Kumaon Himalaya, I was caught in heavy downpours in Uttarakhand. My socks are still damp. I can feel the first tickling of a cold in my throat. When I reached the villages of Sitapur District in Uttar Pradesh, in search of some of the earthly sources of those leaks feeding the great brown cloud, the rains had stopped, and the air became thick instead with smoke from all manner of fires: burning straw, burning wood, burning dung, burning kerosene lamps, burning diesel in the bellies of buses and trucks and generators that keep the lights on in roadside *dhabas* and shops. Out there in the impoverished countryside of Uttar Pradesh, where grid electricity has yet to reach many millions, where the occasional widow still practices the forbidden *sati*, jumping on her husband's funeral pyre, most meals are cooked on simple mud stoves, fed by dung and wood. Fires that in turn feed the dark cloud above.

Having witnessed those fires, from my vantage way up here, the brown cloud seems to me one enormous, emphatic sign of life, making even the cloud fortresses that loom high over the world's highest mountains seem small and under siege by comparison. The haze is our loudest smoke signal, an immense version of that wisp of smoke rising from a chimney on a distant hill—and one of the largest manmade objects visible from space. Most of the particles in that vast soup come from the home fires of Pakistan, India, Bangladesh, and Nepal. From one and a half billion daily acts of survival.

The newspaper in my lap reminds me that there are other such clouds, of varying sizes and hues, in other places. It is 2012, the end of summer, and almost 60 percent of the United States is in drought. The dryness and heat have fueled record-setting wildfires across the western part of the country, turning 9.3 million acres of forest into huge plumes of ash and soot, clearly discernible in NASA satellite images. On the front page, I find a story about other record-setting events: the Arctic sea ice is at an all-time low, covering just half the total area measured in 1980, and there is unprecedented melting over most of the ice sheet of Greenland.

It's a short flight from Lucknow to Delhi, and soon the plane banks and descends. As we approach India's capital—the city with the world's worst air quality—the smog thickens and visibility shrinks to a couple of miles. Before we dive down, transecting the sharp line dividing the realms of Indra and Agni, I get one last glimpse of the clouds off to the right, the ones that are still dumping ceaseless, apocalyptic sheets of rain on Uttarakhand, rain that is swallowing entire villages in landslides, washing away hydroelectric dams, plugging them up with silt and debris, devouring houses on the riverbanks, pushing rivers over their rims and into the downstream sugarcane fields and rice paddies of Uttar Pradesh.

Vritra's revenge: a binary world of droughts and floods.

A few hundred miles to the northwest, on the other side of all that havoc, tucked into a mountainous fold of the Himalaya, are the thirty-nine mud brick homes of Kumik, where the single stream has already been dry for over a month.

Soon after the stream dried up, the villagers had held a meeting. Sitting together in a tight cluster astride their empty main canal, near the prayer wheel and the long stone *mani* wall, they passed around a battered little booklet containing a Tibetan astrological calendar. For a while the Kumikpas, as the people of Kumik are known, debated its contents and discussed the current state of the cosmos. They were trying to fix the most important date of the year in Zanskar, India's highest inhabited valley: when to begin the barley harvest.

An elder declared that the twenty-eighth day of the Tibetan month, three days hence, would be the most auspicious date to begin, with the traditional prayer recitations and offerings.

"No, no, it's better to start on the first of the next month," another man shouted.

"But that's too late, the water is already finished," another voice called out.

A chorus of affirmative *halas*—"Isn't it so?"—rang out in response.

More than a week had passed since each household had watered its fields. The clock was ticking. Fodder would be sparse again this year. And if they waited too long, the harvest could be lost.

"Kumik's a bad place," said Tsewang Zangmo. She stuck out her tongue—her version of a wink—and gave a taut smile. "No water—what to do?"

From where the group huddled, the contours of their dilemma were wholly visible. At the top of a U-shaped valley stretching above the village loomed the mountain of Sultan Largo and its neighboring peak, capped by a small glacier and coated with patchy snowfields. Together these formed the source of all of Kumik's water for drinking, washing, and irrigating their staple crops and fodder grasses. One could just make out the place farther down the mountain where a stream just a couple of feet wide, the village's lifeline, emerged out of the rocky hillside. From there it flowed through concrete sections of canal until it entered the willow grove at the village's upper edge. And looking down below, toward the Zanskar River, one could see dozens of fields of golden barley and wheat and peas and potatoes, green and yellow patches stretching down to the long stone wall that rings the entire settlement. Inside the wall was another ring, a band of brown and sandy-colored ovals—all fields that had lain fallow these past several years. And beyond the wall was Marthang, the "red place," all sun-baked dust and russet rocks and stunted sage bushes, where unruly spirits lived, the older people said. This parched world dwarfed the oasis of Kumik, offering all-enveloping testament to the limits of life here at 12,000 feet, in the Himalaya's rain shadow.

From the same spot, one could also see Meme Ishay Paldan, standing motionless on the roof of his modest house on the edge of the village. Hunched over, hands clasped behind his back, in the worn woolen *goncha* robe dotted with a lifetime of accumulated patches that he wore every day, Kumik's eldest citizen was taking in the whole scene from a distance. "It was a warm winter," he had told me in passing just a few days before. "There was not much snow, so half of the fields weren't planted." For those who knew what a good year looked like, back in the old days, this fact, too, was visible on the canvas of Sultan Largo. The snowfields up there now looked moth-eaten and threadbare. This fact posed a much bigger problem than a few cloudy days in mid-August: when the sun came out again, it would have little left to melt.

As the debate went on, people came and went on their morning rounds. Meme Yunten came by with an empty *tsepo* basket on his back, on his way to some task out in the fields, and gave the prayer wheel a spin. Three little girls walked by on their way back home from the springwater pipe near the primary school, all leaning to the same side, laden down with one small water jug each. Young Stenzin Konchok walked by the group, ragged and looking for water to wash with, his hair sticking up at odd angles, a toothbrush dangling from his mouth. "Too much *chang* last night," he mumbled dolefully. "Now, no good."

His neighbors nodded and chuckled in sympathy; almost everyone in Kumik knows what it's like to have drunk a bit too much of the local barley wine. Konchok greeted the young men who leaned on the hood of a Tata Mobile pickup truck, who in turn were watching their elders, who in turn were shouting and arguing and laughing and jostling like one big raucous, wisecracking extended family, the burbling of their joined voices taking the place of the now-silent stream.

Soon the matter was settled. The advocates of the earlier date had prevailed. The preharvest prayers would take place on the twenty-eighth in the old *lhakhang* temple up on the rocky ridge. Discussion moved on to another topic of perennial interest: who owed money to whom. Small mountains of bread and lakes of *thukpa*, the local soup, would have to be made for the occasion, food would have to be purchased in the market town of Padum, oil brought for cooking and lighting the lamps in the temple. Everyone had to chip in. People pressed in on the account keeper. Some clutched receipts for what they had spent for the four village weddings that had taken place that summer; some handed over money that they owed; others took money from a neighbor so they could make some of the upcoming *puja* purchases in town that day. A name was called out—someone who was, the ledger keeper noted, "late with payment again," prompting a titter of knowing laughter.

Accounts settled, most of the villagers rose and drifted off in clusters, heading back to their homes and the full day of tasks ahead. A dozen men remained behind near the empty channel, arrayed in a loose circle on the gravel, to address a final order of business. A grievance.

A new metal pipe, paid for by the public works department of the district government, based in nearby Padum, was going to be laid in the ground. This would bring another drinking water tap to the village, fed from the same spring up the valley. Work was slated to begin that week with a diesel-powered excavating machine, explained Tsewang Norboo, the spiky-haired spark plug of a village headman. The pipe would be routed from high above the village toward the old temple up on the ridge, making access more convenient for several households of the main village and preventing further loss of scarce water through seepage. Another pipe would go to Pang Kumik, a hamlet of three households on the other side of the ridge.

But two men from Pang Kumik were unhappy. The pipe to reach their homes half a kilometer away wouldn't be installed until later. Meanwhile, their hand-pump-operated bore wells had gone dry. Their own small spring was just an unreliable trickle. They had tanks that stored some water, but by September those could be finished. Why should they, alone among the villagers, have to wait for months, walking several hundred feet uphill to haul water, when a bit more investment, some pipe, and gravity would bring it down to them now?

The discussion became heated. Norboo and another man made the case for patience: Pang Kumik would get its turn. It wasn't feasible to change the plan for work that was already in progress. The Pang Kumikpas shouted back at the injustice of it: "It *is* possible!"

The four men gesticulated, shouted over each other, shook their fingers, smacked their hands into their fists to punctuate their points. In the outer ring of onlookers, some murmured opinions on both sides of the matter. A couple of peacemakers among them periodically stepped in to calm the speakers down, but most stared down at the ground, shifting their weight from one foot to the other. Their pinched expressions reminded me, once again, that Zanskaris are some of the most conflict-averse people on the planet.

The argument soon spilled over into another, larger question that loomed in the background, a question about water and the future of the village, Pang Kumik included: Why all the fuss and further investment in a village that was already doomed?

Someone pointed out that the new canal the Kumikpas had been digging for a decade down in the barren land of Marthang was dry today, too. Plugged up by silt carried down from the eroding Himalaya, it needed to be dug out, with money and labor and sweat. There was a need to focus on the bigger water picture, on the future, to deploy their scarce resources judiciously. Others betrayed ambivalence, even after a decade of unstinting effort.

"But the canal is empty."

"Yes, but the water came last year!"

"Water is guaranteed below!"

This last point came, in a tone equal parts imploring and exasperated, with an emphatic gesture down toward the Zanskar River, from Tashi Stobdan, the primary school headmaster and longtime ringleader of the effort to dig the new canal.

The overlapping disputes carried on for several more minutes in a sequence of anguished explanation, appeals for patience, indignant outbursts, interspersed with long pregnant pauses. Finally, the men stood facing each other across the empty canal in frustrated silence.

The silence of the men, and of the stream, spoke eloquently of the narrow margins for error visible in the dry world encircling them, and in the waning snows above. Soon Tsewang Norboo and the Pang Kumikpas were gently pulled apart by their elbows, by two soldiers home on leave.

"Enough, enough, *acho* [older brother]. Come away, let's go."

And Tashi Stobdan, the visionary who saw a bright green future every time he looked down at the "red place" by the river—who believed that, if only his neighbors could stick together and *focus*, they too would see those same rich possibilities, and if they could see them, then they could work together to *create* them—walked home with a weary worried look weighing down his boyish features.

• • •

If you were making a movie about life in the Himalaya, seeking a setting that shouts pastoral harmony, at first glance you might be inclined to film it in Kumik. On the surface, at least, Kumik is a little Himalayan Arcadia, a comely oasis in the sparsely populated, arid mountain reaches of northwest India.

Its thirty-nine whitewashed mud homes cascade down a southwest-facing hillside that overlooks sun-kissed terrace fields of barley laced with intricate irrigation canals and interspersed with groves of swaying poplars and willows, which the Kumikpas coppice for saplings and ceiling materials. Several *ranthaks*, elegant water-powered grain mills, turn roasted barley into flour, the centerpiece of the Zanskari diet. A hanging glacier caps Sultan Largo, which towers above the *phu*, the high pastures where animals graze in the summer. Laughing children race up and down the narrow footpaths, past amiable grandfathers spinning prayer wheels and grandmothers doing clockwise *skora*s around the small *lhakhang* temple. Even the acrid smoke that wafts down the alleys has a cheering tang, conjuring the hidden warmth of dung-fired hearths. And if you crouch down on a summer evening among the ripening barley up on the ridge above the *lhakhang*, as the children skip and shout to greet the return of the *rarzepa*, the shepherd of the day, with every household's sheep and goats, and you listen to the stalks rustle and rub against each other, with a sound like spreading rumors—a shimmery whisper of snowmelt transmuted into life—well, all talk of crisis and catastrophe seems ridiculous. Crazy Chicken Little stuff. After all, Kumik is thought to be the oldest village in Zanskar, one of the highest, most remote, permanently inhabited places on the planet. The Kumikpas seem to have life in the rain shadow pretty well figured out.

Yet the Kumikpas are busily preparing to abandon it all.

Four years earlier and just a stone's throw from the scene of the argument, I had stood next to Tashi Stobdan outside of his stately, squat mud brick home, as the symmetry of these two facts struck him with full force:

"Kumik was the first village in Zanskar—and now it is the first to be destroyed!"

A look of wry wonder lit up his face, and he laughed heartily. Somewhere in the whole mess—the juxtaposition of his village's millennium-plus tenure on this rocky patch of north India, and its recently revealed fragility; the impending, drought-induced abandonment of the only home he has ever known—he discerned a pretty good joke. He turned to me, expectantly, with eyes that asked if I agreed: *Pretty funny, right?*

It didn't strike me as all that funny. Ironic, okay, sure. Freshly arrived in his village, and now thoroughly puzzled, I was just beginning to absorb the magnitude of Kumik's slow-motion disaster, which Stobdan and his wife, Tsewang Zangmo, and several neighbors had spent the afternoon explaining to me over bottomless cups of salty butter tea.

So I just offered him a quizzical smile in response. We were standing next to the hand-lettered metal sign that Stobdan had mounted by the gate to his musty sheep pen, a sort of cri de coeur made to inform the odd foreign trekker who might be inclined to help out, though tourists almost never come through Kumik. I read through it again:

> Due to failure of snowfall in the last 2 years the people couldn't harvest even a blade of grass & consequently had to sell their yak, cows, etc. at very nominal prize . . . the people of this village are now constructing a irrigation cannal fed by the LUNGNAK river to bring the virgin land of MARTHANG under cultivation.

I looked up from the sign. Stobdan was still laughing. It was not a rueful chuckle, not the sort of glum nod in the direction of fickle cosmic forces that you might expect. And not one of those "If you didn't laugh you'd have to cry" cathartic sort of exclamations, either. I studied him. An expression of defiance? Simple gallows humor?

The latter would have been understandable. Kumik's disastrous drought had been long foretold; the people have had some time to adjust to the seeming inevitability of abandoning this place. There was the visiting monk a century ago who predicted the village stream would one day run dry. There was the story of the *zbalu*, the vengeful fairy spirit who in ancient times placed a very specific curse on Kumik: "You will one day run out of water." And there was the proverb familiar to all Zanskaris, a wry Himalayan version of Murphy's law, uttered in a "wouldn't you know it" kind of tone: *Kha Kumik, chu Shila.* "The snow falls above Kumik, but the water goes to Shila!" (Shila being a neighboring, thoroughly watered, comparatively lush village on the other side of the mountain that people in Kumik refer to with a kind of jocular faux envy.)

All these gloomy prognoses seemed squarely contradicted by the cheerful scene that greeted me that afternoon.

"The older people think Kumik is the perfect village," notes Tsewang Rigzin, Stobdan's oldest friend and an officer in the government agriculture department, "because it is close to the mountains, for grazing for animals, for fuel collection." These—and adequate water, of course—used to be the only relevant criteria. It was a pragmatist's sort of perfection: convenient access to the bare essentials. And by this standard, for most of its thousand-year history, Kumik indeed had it all. Enough food, enough fuel, enough friends and neighbors, all near at hand—what else could one ask for?

But as the villagers pointed out to me on that first day, and over the course of many subsequent visits, a closer look revealed some ominous signs—such as the brown fields that have lain fallow for years down by the stone wall. Over the years, I began to notice some troubling trends: fewer animals, for example, as they were sold for lack of fodder. Kumik has become a village of women—fewer men are found there from year to year, as more go in search of income-generating nonagricultural work. The number of men in Kumik who self-identify primarily as farmers has dropped from more than two-thirds thirty years ago to less than 15 percent in 2005, even as the average has stayed well above 50 percent in other villages across Zanskar. Why? The lure of government and tourism jobs pulls the men out of Kumik. But the lack of water pushes them, too: with little water for irrigation, the need for their labor in the fields is much diminished.

And those who have lived their whole life in Kumik notice an even more dramatic transformation, the distal cause of all these other changes, hanging above the village like the sword of Damocles.

"When I was a child, there were no problems with water," Meme Ishay Paldan told me on that July afternoon in 2008 over a simple meal of *kholak*, a mixture of roasted barley flour and butter tea that we scooped out of cups with our fingers. "The glacier was much bigger. It used to snow more."

I asked him how old he was. He lifted his yellow woolen cap and scratched his scalp. He didn't know exactly. Around eighty, he guessed.

"The snow line then almost came down to the top of the village," he continued. With a nearly toothless smile he pointed through the window up the valley toward the glaciated peak. "Now look."

I glanced out his window, which perfectly framed the cirque and one of the peaks of the 18,300-foot mountain above. The snow line lay several miles distant and 4,000 feet up, just below the col separating Sultan Largo from its neighboring peak to the south. The glacier, once a blanket over the top of the valley that stretched well below the lip of the ridge, had shrunk into a small dome on the mountaintop.

But the bigger problem, he explained, is twofold. For one, the old proverb rings less and less true every year: the amount of snowfall varies naturally from year to year, but Zanskaris have observed a pronounced decline in recent decades. Kumikpas have watched the upward march of the snowfield's bottom edge, like a puddle evaporating in slow motion, all their lives. And second, springs are much warmer and arrive much earlier than before. This means more of the snow melts before the short growing season begins in June. Less snow and earlier melting together mean that the village often runs out of water by mid-August, sometimes sooner, in the critical weeks before the harvest.

This resulting change in the timing of water availability is their most serious threat. It has led to lost food and fodder, troubling knock-on effects and hard choices. Anticipating the scarcity of summer, the people plant fewer fields than they did in years past. Anticipating increasing difficulty in feeding their animals—the lynchpins of their agricultural system, which feed their keepers with milk and provide draft power in the fields—people sell more of their precious capital in the form of cows, sheep, goats, and *dzo*. Households must buy more food from the market, on those limited incomes. And there may come a time when each family can only plant one field. It has happened before and it could happen again, the people say. You can't survive on that.

Government rations of rice and flour and sugar help, and packaged food and vegetables trucked in from Kashmir and then bought from the Padum market may make up the difference for those who can afford it. But when the long, lone road into Zanskar, open only during the brief summer, is blocked by snow or by floods or by politics, supplies can get thin. It's a brittle safety net.

And all this raises some equally urgent questions of identity: some say that a person who doesn't plant his fields is no longer Zanskari.

So year after year, when they looked up at the mountain in late winter, the writing on the wall of Sultan Largo became quite clear to the people of Kumik. All signs pointed to intensifying drought, with no end in sight. And with it, an end to the world they had known.

Clearly there were plenty of good reasons for gallows humor. But you'll just have to trust me: Stobdan's laughter was not that of a man defeated. It was a tune sung in a key I had never quite heard before.

Granted, I could have been misreading the moment—the way early foreign travelers to the Tibetan region once mistook locals' handclapping greetings as applause, when it was actually a gesture deployed to drive out evil spirits. Maybe Stobdan's laugh was some culturally opaque signal of deep, deep despair. Zanskaris can be hard to read, even at the best of times.

And at that point I didn't know Stobdan all that well. We had met just a few days before on Padum's dusty main street. I had been visiting my old friend Urgain Dorjay, who told me that a man who was married to his cousin Tsewang Zangmo had been searching for me.

We arranged to meet, and when we finally shook hands and settled into a booth at a smoke-filled tea stall, Stobdan revealed the reason he had sought me out: he wanted to discuss new ways to heat his house through the long, intensely cold Zanskar winter.

Great, I thought, *this is exactly why I'm here*. Zanskaris burn vast quantities of yak, sheep, and cow dung to both cook and stay warm through the almost seven-month-long snowbound winter. I was in Zanskar that summer researching household energy technologies and collaborating with various nonprofit organizations on the design of buildings using passive solar heating in different communities.

Stobdan explained that he was building a new house and wanted it to be as warm and clean as possible, while requiring less fuel for heating. He shyly pushed across the table some drawings of a simple rectangular stone-and-mud-brick structure—four rooms and a glass-fronted space on the south. His new home, he explained, with a note of pride.

"Where did you learn about passive solar buildings?" I asked with surprise as I studied them. In my years of travel in the region, I had rarely encountered this kind of self-replication outside the main city of Leh: someone spending his own money, on his own initiative, to insulate and solar-heat his house with no subsidies or prodding from government agencies or nonprofit groups.

His neighbor was a skilled mason, Stobdan said, who had made a dark, south-facing wall, covered with glass windows—his interpretation of a technique known as a Trombe wall, developed in France decades ago. This man's kitchen was warmer than his neighbors and his stove demanded less fuel. These simple facts had piqued Stobdan's curiosity, making him wonder if he could make a system that was even better. But mostly, Stobdan said, he learned about the idea from Sonam Wangchuk. "I bought a video CD he made in Leh."

I laughed. Sonam Wangchuk, of course! The solar guru of the western Himalaya. I explained to Stobdan that we shared the same teacher. In past years I had worked with Wangchuk—one of the most famous, and enigmatic, men in the entire Himalayan region of Ladakh and Zanskar, an irrepressible engineer, educator, social reformer, and grassroots visionary, at the time in self-imposed exile in Nepal. In a way, Wangchuk was the reason I was there in Zanskar. My first exposure to solar buildings had come during a sojourn teaching and volunteering at the solar-powered, solar-heated school campus he and his colleagues had created near Leh, on a stark patch of desert overlooking a sharp bend in the Indus River.

"You know him?" Stobdan leaned forward and smiled with satisfaction at this coincidence. In the video, he explained, Wangchuk described the basic principles of solar building and solar water heating and other techniques to improve people's comfort and help them save on fuels like dung and kerosene. Heating and cooking in a place as remote and as cold as Zanskar was expensive, in terms of both money and time spent combing the hillsides for fuel. Dung's importance can be seen in a glance, in any village across the region: on the parapets of every house this dry flaky treasure is piled high, as a buffer against winter's bite. Wangchuk's idea of harnessing the free energy of the sun—which blazes down 315 days a year through the thin, high-altitude air—made so much sense that Stobdan felt he just had to try it.

"But why are you building a new house?" I asked. In Zanskar and Ladakh, eldest sons like Stobdan inherit and inhabit their ancestral homes, occasionally modifying and adding on to them, but rarely starting over entirely. (These traditions are changing fast in towns like Leh and Padum and Kargil, but new construction is still constrained both by a limited amount of arable land and water to support new households and by many farmers' meager incomes.)

"There is a drought in my village," Stobdan said simply.

He and his neighbors were building a new village from scratch, he explained, in the no man's land called Marthang, north of Padum. There they would find enough water for everyone. I was welcome to come see for myself.

So a few days later I accompanied him to the windswept plateau just above the confluence of the Lungnak and Stod rivers. The wind screamed and dust devils whirled about. Several hundred yards away, near the river, a handful of half-built, one-story mud brick homes seemed to cower under the hammer-like summer sun. At that point Stobdan's new house was merely a stone foundation, some heaps of earth to be fashioned into mud bricks, and a gleam in his eye. We walked around his few acres of property, and he pointed out with a pioneer's enthusiasm where he would plant more saplings and dig a storage tank for water to see him through lean times.

I looked around. There were few signs of life: none of the bulwark stands of willow or poplars found in a typical village, none of the Technicolor green fields and tidy, capillary-like canals. Just some hardy desert shrubs sprouting out of cracked earth and a few beleaguered-looking saplings that Stobdan had recently planted. Roaming sheep had chewed them down to pith.

"What do you think?" he asked, beaming.

"It looks good, Stobdan," I murmured unconvincingly, shielding my eyes from the dust and glare. Sand gritted my teeth and thickened in the corners of my eyes. "Plenty of sun here."

Actually it looked like he and his neighbors were colonizing Mars.

In a way, that's what they were doing. Marthang means "the red place"—a reference to the russet dust that coats the entire plain, borne on a southwesterly wind that slams into one's face every summer afternoon—as well as an implicit acknowledgment of its hostility to human endeavor.

But apparently Stobdan could see something in that dusty patch that I couldn't—a glorious green future. What Marthang had going for it was proximity to the river and its copious store of melted snow and ice: a steady, reliable source of irrigation water. The villagers, he explained, had mostly finished digging a seven-kilometer-long canal to bring that water from upriver to the site of the new village. That patient effort, combined with even more Herculean endeavors to amend the soil, build new hearths and homes, plant windbreaks, and reinvent their ancient social compacts to fit their new living arrangements, would make all things possible.

Then perhaps Marthang would cease to be that shade of parched red dust and become instead "the green place."

Still, for a place as forbidding as this to seem like anyone's best option, I assumed things must have gotten pretty bad. And indeed, I soon learned that the tipping point for the people of Kumik came in 2000. Very little snow fell that winter, and the drought that followed in the summer was severe, affecting villages across central Zanskar.

But Kumik suffered the most. The poorest households only cultivated a few "rooms" of a single field that year (each *nang*, or room, measures one to five square yards). The villagers faced serious shortages of food and fodder. Some bought grain and grasses from relatives in other communities. Meme Gelag, an elderly Kumikpa who has since passed away, told visiting researchers that in the spring of that year, when the stream would normally be overflowing with melted snow, there was "barely a trickle." Like Ishay Paldan and other members of Kumik's eldest generation, he had lived through attacks by invading militias from Pakistan just after Partition, and then alternating bouts of neglect and heavy-handedness by distant governments in Kashmir and New Delhi, through the harsh exigencies of life before there was any road linking Zanskar to the wider world and its modern comforts. So the researchers took note when this grizzled man said, as he walked through the parched village that summer, "I feel like crying."

Kumik had suffered from a short water supply for decades, but now the noose was getting too tight. There was no more margin for error. Traditions designed to manage these risks—to conserve and equitably share water, to pool labor among households—were no longer enough. They had made the community resilient, but they couldn't conjure up more water.

They couldn't go on like this. Something had to be done.

So the Kumikpas held a meeting to discuss their options. For years the situation had threatened to divide the village. One small faction had argued, leading up to the decisive drought, that they should put more effort into (and get government support for) making a better canal to bring more meltwater from the fast-receding snowfields—a kind of Hail Mary pass to save their ancestral homes. Another group of Kumikpas, led by Tashi Stobdan, had long made a forceful case for shifting their focus and energy to building new lives below in Marthang. The only hope, they said, lay in starting over. But whatever they chose to do, all agreed it must be done together.

The dispute had simmered for years, until the drought of 2000 forced the issue. This time, everyone acknowledged that the single, erratic stream coming down from the mountain could no longer be counted on to irrigate every household's fields. The decision was difficult, but it was quickly and unanimously made: after more than a thousand years of living together on this rocky spur, the Kumikpas would have to leave. Together.

There wasn't much else to discuss. "Without water, you have nothing," many Kumikpas would tell me with a resigned shrug.

They sent a delegation to the deputy commissioner (DC) of Kargil District, asking for some land two miles away, on a windswept bluff above the river. This part of Marthang historically belonged to other villages such as Pipiting and Tongde, which intermittently grazed their cattle on its sparse grasses. But the DC was moved by the desperate urgency of the Kumikpas' situation, and instructed the *tehsildar*, the government revenue officer for Zanskar, to grant them the land. (The *tehsildar* proved to be a strong advocate for Kumik, Stobdan recalls, as the official's own home village near Kargil was water-stressed too.) The villagers then held a lottery to allocate properties of a bit more than three acres to each household.

This entire process took several months and patient dialogue, but it would prove to be the easy part. The Kumikpas would then have to dig a new canal to bring river water from over four miles away to their new fields. Each household was slowly gathering materials and saving money to build a new home and prepare their virgin fields in Kumik Yogma (Lower Kumik). Some households had more labor and more income than others, so that process would take place at varying speeds.

"What will you do?" I asked Ishay Paldan at the end of my first day in Kumik. Meme Paldan, who had lived in the village his entire life, was living in a modest mud brick house at the upper edge of Kumik, where the path leads off through the willow thickets to the high pastures. His household consisted of himself, his ailing wife, and their unmarried daughter, Dolkar. The rest of their children lived in other villages. They had little money and just a few aged hands with which to build. "Will you move?"

Kumik's eldest citizen just laughed. He seemed remarkably upbeat. "I was gathering stones for our new house yesterday."

Tsering Motup, a stocky farmer in his sixties and teacher of *bodyik*, the written form of the local language, told me he was preparing to build a new home too. But he wasn't all that thrilled about it. He framed the parameters of the villagers' decision as a matter of stark necessity: "Without water, there is no life."

"Are you sad to be leaving?"

"Of course I'm sad! The village here is happy and green. It is not like this below." He gave a smile slightly more bitter than sweet. "I feel the same sadness a young girl does when she marries and has to leave her home."

In the years that followed the meeting, the decision seemed more and more prudent, and prescient. In 2003, there was so little snow that Kumikpas could plant only 10 percent of their fields.

But there by the twilit sign on that July evening in 2008 was Stobdan, still chuckling. So, deferring to his emotional proximity to the sharp edge of the unknown, I finally started laughing too. And then he laughed some more, and then we stopped laughing, and we both stared down at the sign for a while.

With no proper segue out of this awkward silence, we began to discuss design ideas for the new home he wanted to build in Marthang—this, of course, being the whole reason he had sought me out in the first place, not to tell cryptic jokes about doomed villages—and Stobdan crouched down and sketched a new floor plan in the middle of the dusty path.

"I will put the greenhouse here," he said, pointing with a willow twig. "Is it good?"

It was good. Other Kumikpas wanted advice on how to make solar-heated homes, too, he said. I was no expert, I explained, but I knew some people who were, including Sonam Wangchuk, the solar guru himself. So I promised to return. We shook hands and said good-bye.

As I walked off into the cool Himalayan night, past the giant prayer wheel that straddles the trickle of meltwater that runs through the tapestry of fields like a fraying thread, bewildered but strangely buoyed, I became fixated on this question: Why would a man laugh at the prospect of his age-old home, his birthright, crumbling into the dust, and at the thought of rebuilding his life, stone by stone, seed by seed, in an unfamiliar and unforgiving landscape?

The stars started to wink on one by one, and the village's tight cluster of mud brick houses, ridge-top temple, ovals of rustling barley, burbling canals, all together created the impression of a well-built lifeboat tossed on a dark sea of mountains and sky. I reached the *lhato*, a rectangular shrine of whitewashed stones topped with juniper boughs and ibex skulls, which marks the boundary between the green order of the human realm—Kumik proper—and the red, parched chaos beyond: Marthang. The *lhato*, in a sense, commemorates and embodies the founding of the community, when the spirits of the valley were bound in a reciprocal relationship with its human inhabitants. The people's acts of propitiation and respect for the land would renew in perpetuity the ancient compact with the *lha*, who would respond in kind by blessing their efforts with prosperity, fertility, abundant snow, and strong sunshine to melt it. A cycle much like the one celebrated in the awe-filled paeans of the Rig-Veda.

Now that contract had been abrogated.

As I walked past the *lhato*, I sensed, or imagined, an air of reproach. I felt like someone who had stumbled onto a crime scene on a bright day in the park. I walked out onto the hard plain below, reaching the main road, and waited to hitch a ride from a passing truck. The questions welled up like springwater.

How on earth would these people build new homes, dig new canals, sow new fields—the civilization-making work of dozens and dozens of generations squeezed into one—on that wind-blasted desert, with little outside help?

Then there was Stobdan's choice of words: Kumik was being "destroyed." It suggested a malevolent agent, an unseen hand at work. So what force, then, was capable of destroying the oldest, most resilient community in one of the world's most demanding environments?

And, seriously, why the hell was this guy laughing about it all?

Only time would answer the first question and, as it would turn out, the third.

As for the second, I assumed that any thorough forensic analysis of Kumik's sun-drenched crime scene would yield a straightforward answer, based on some clearly established physical facts. The explanation for their intensifying drought—the identity of the "destroyer"—should be blindingly obvious: burning was to blame. A whole lot of fossil fuel burning, to be specific.

You know the story. We've been combusting coal and petroleum products for a few centuries now, toward all sorts of useful ends, and cutting down forests to clear land and burning those, too. As a result, we've been adding carbon dioxide to the atmosphere at an unprecedented rate. That colorless, odorless gas, which traps the earth's outgoing infrared radiation, is largely responsible for keeping the temperature of our planet's atmosphere within a comfortable, livable range—without its warming blanket, average temperatures would be below zero. But now that we are adding over 30 gigatons of carbon dioxide to the atmosphere each year—and rising—we have trapped way too much heat.*

Now the whole world is warming, mostly due to increasing concentrations of globally mixed carbon dioxide in the atmosphere. Most of those emissions are coming from the burning of fossil fuels to power our vehicles, light and heat our homes and offices, and manufacture the goods we need and want, and from countless other fires, too. This planetwide transformation manifests in different ways in different places. In far-flung places like the Himalaya, this uptick in the global thermostat seems to be disrupting traditional snowfall patterns, speeding spring's arrival, gnawing at glaciers, killing harvests. Sure, there are microclimates doing funky things here and there, but the overall trend seems clear enough, and clearly ominous.

What's more, the molecules of carbon dioxide exiting my car's tailpipe bear equal responsibility for all the havoc as the ones streaming out of a coal-fired power plant's smokestack in China on the very same day. But the vast majority of historical greenhouse gas emissions have come from the fires lit by those of us who live in European countries, the United States, Canada, and other "developed" nations. Ergo, the overwhelming scientific consensus suggests that, plainly speaking, I and my frequent-flyer-mile-accruing, SUV-driving ilk back home in the United States bear much, if not most, of the blame for the displacement of people in Kumik. As we do for the plight of the people of Kiribati, the Polynesian nation slowly disappearing under the waves, and the distress

*Partly prompting some earth scientists and geologists to proclaim that we have exited the Holocene—that geological epoch characterized by stable climatic patterns highly conducive to the development of agriculture, and then technology, and then, you know, modern human civilization—and barreled right into the Anthropocene—a new era in which human influence has overridden the natural systems that underpin our survival and our relatively brief period of flourishing. Kind of like the global, chronological equivalent of leaving green, safe Kumik for red, wild Marthang, where all bets are off.

of the residents of the seaside Alaskan village of Kivalina, which is getting chewed up by and churned into the rising Chukchi Sea. I, for one, assumed that the carbon dioxide emissions from my own round-trip flights to Delhi from the United States were one more small but incremental upward twist of the thermostat dial, applying equally to Kumik as they did to southern California. Globally averaged climate change, I surmised, had come to Kumik.

All of which would make Kumik's story a sad but familiar tale of the early twenty-first century (albeit with a few intriguing twists), and the Kumikpas just the latest additions to the steadily growing ranks of canaries in the global warming coal mine (occasioning yet another short bout of hand-wringing, some fleeting sympathy for these blameless victims of our excesses halfway across the world, and maybe, if this sort of thing moves you, a click on that email petition to your congressperson calling for a carbon tax). In sum, it seemed clear enough that the Kumikpas were climate refugees, wrongfully evicted from their ancient home by all the greenhouse gases we greedy few have pumped into the atmosphere. Case closed.

Thus my thinking went, as I flew back to California (my home at the time), guiltily tabulating the tons of dangerous gases being added to my karmic tab.

But I was in for a few surprises. The first came when I discovered, on subsequent visits, that the Kumikpas mostly blame themselves for their problems. When I asked Tsering Motup, the teacher of *bodyik*, the Tibetan script, why the water and snow weren't coming anymore, he had a ready answer: "The *lha* are angry."

The *lha*—the gods of the place, of the village, of the mountain, of the sky—needed to be propitiated by prayer, by upright behavior, by daily signs of respect for their power and ability to influence human affairs, to bestow life and death.

"Now, people are not praying as much as before," Motup said. "They are becoming jealous, thinking only of money."

As he said this, I glanced around. The only sign that we were not in the eighteenth century was the four-wheel-drive jeep, covered in dust, parked by the central prayer wheel and owned by one of the seven brothers from the Gonpapa house, a now vital part of the village economy that ferries goods and people to and from the main town of Padum. Small government-subsidized solar panels dotted the earthen roofs, powering single bulbs, a radio, maybe a small TV here and there. (Grid electricity hadn't yet reached Kumik and wouldn't for another four years, after which it would still be offline more often than not.) If there was rampant consumption going on, it certainly wasn't of the conspicuous variety.

"Many prayers are done, and mantras, but still there is no water," Motup went on. "The *lha* are punishing us." Among older Zanskaris and Ladakhis, the *lha* are believed to punish selfish and careless people, polluters especially. (An acquaintance of mine once told me a story about his friend's son, who had urinated near a spring. A week later the boy drowned while swimming in a stream. "The *lha* were angry and killed him," the man, a highly educated NGO employee, concluded simply.)

Many other Kumikpas echo Motup's view. We've brought it on ourselves, they told me, and the only way to make it right is to change our behavior. They used to faithfully perform the prayers signaling respect and gratitude for the gifts of water, soil, fuel, sun—all those ancient, ultimate origins of wealth. But now, the refrain went, we are too distracted by money and greed, by incurring the respect and envy of our neighbors.

The people had since made efforts to shape up. They had made countless appeals to the angry *lha*, to relent and send more snow. They had prayed through many long nights in the village temple. They had brought an illustrious monk to chant in solitude in a tiny hut for fifteen days straight. They had gathered some funds to finance an epic recitation of the Kangyur, the bible of Tibetan Buddhism, in distant Dharamsala, home of His Holiness the Dalai Lama.

But none of it had worked. The *lha*, apparently, were no longer listening.

Another explanation for Kumik's woes is to be found in the oft-repeated proverb, *Kha Kumik, chu Shila*. A handful of Kumikpas, including Ishay Paldan, support this more prosaic diagnosis. The earliest settlers had probably looked up at the glacier carved bowl sloping generously down to their chosen spot and guessed that most of its stores of frozen water would flow their way. But they were victims of an optical illusion.

If you go up to the snowfields, you'll see that the topography near the summit drains most of the melt down the other sides of the mountain. What's more, the northwest-facing aspect of the snowfields that *do* lie in Kumik's drainage get less sun during the day.

Those who had climbed the mountain, like Ishay Paldan in his youth, knew this. For most of Kumik's thousand-plus years, whether those founders were right or wrong, whether the bulk went to Shila or Kumik or Shade or some other village, didn't really matter. There was plenty of snow, and, as the old stories attest, there was plenty of water to go around. There was the occasional drought or lean year, but the climatic shift didn't really start until the 1960s, when the intensifying trends of declining snowfall and warmer winters and earlier spring melting set in.

As the snow and ice have retreated, hard physical truths have been revealed. Kumikpas discovered that their relatively small slice was coming from a fast-shrinking hydrological pie. But what was now causing that pie to shrink?

Whereas many Kumikpas ascribe their misfortune to divine wrath, an ancient curse, or plain bad topographical luck, from the perspective of scientists studying environmental changes across the Himalaya and Tibetan Plateau, they are victims of a pronounced warming trend. The air temperature increase varies in different parts of the vast Himalayan region, but in almost every area the rate of warming is dramatically higher than the worldwide average. Over the northwest Indian Himalaya, for example, temperatures have risen by 1.6 degrees Celsius over the past century. During that same period, the global average temperature has risen by about half that amount. The Himalaya and Tibetan Plateau are warming faster than any other part of the planet except for the Arctic, which has warmed 1.5 degrees Celsius since the 1970s. And in recent decades, this trend has been accelerating.

These facts would seem to augur some obvious consequences for the 760,000 square kilometers of snow cover and 54,000 glaciers—including the one above Kumik—covering 60,000 square kilometers across the Himalaya and the Tibetan Plateau. These frozen alpine reservoirs are often referred to as Asia's "water towers." The glaciers of the Himalayan region hold more water than Lake Superior, about 12,000 cubic kilometers; collectively they amount to the biggest reservoir of freshwater outside of the polar regions. During the dry season, this snow- and icemelt provides critical flow in the Indus, Brahmaputra, Ganges, Indus, Yellow, Yangtze, Mekong, and Amu Darya rivers, among others—waters that help sustain somewhere between 1.3 to 2.5 billion people.

As temperatures continue to rise, these water towers will slowly dissolve. Scientists continue to debate how fast that might happen—one recent study says glacier melting in the Himalaya will peak around 2070—but they're largely agreed that it *will* happen.

"This is not rocket science," the veteran glaciologist Shakeel Romshoo, of the University of Kashmir, who has studied Zanskar's glaciers for over twenty years, told me with a self-deprecating smile. "When temperatures rise, ice melts!"

The process seems to be well under way. Though glaciers are complex and behave differently at various elevations and latitudes, the vast majority of new peer-reviewed studies continue to confirm what people who live on the roof of the world have been observing firsthand for decades: permanent snowfields are in retreat; most of the region's glaciers are shrinking and thinning; precipitation patterns are shifting. And the changes are happening fast. Perhaps of greatest concern is the regionwide shift toward warmer winters and earlier springs. This new regime upsets the familiar balance of snowfall and melt onset, altering patterns that the region's farmers and herders have come to depend on over the centuries.

All these transformations are most certainly due in part to the heat-trapping effects of carbon dioxide, the unhappy by-product of humanity's energy-intensive economic development, dating roughly from 1750 on. Those carbon dioxide emissions—along with methane and nitrous oxide and other greenhouse gases that we dump into the atmosphere every day—are without question playing a role in Kumik's unfolding drama, insofar as they contribute to the changes in regional climate that are the proximate cause for the Kumikpas' eviction.

But here's the problem. After I started looking into the matter, I discovered that carbon dioxide, while a very important part of the story, isn't the *whole* story.

My assumptions as I left Kumik that first evening were indeed partly correct: *fire* was the overwhelming distal cause of Kumik's plight. But it turns out there's something even more destructive to snow and ice than carbon dioxide. And it, too, is a by-product of so many of our daily fires.

Scientists are still trying to figure out why there are certain temperature anomalies around the globe. For example, the Arctic has warmed by 2 degrees Celsius since the 1980s, while the global average temperature went up by 0.6 degrees in the same period. The European Alps, which have lost half of their ice since 1850, are warming twice as fast as the global average. And so are the Himalayas; in some parts they are warming as much as five times as fast as the rest of the world. Climate models suggest that atmospheric carbon dioxide concentrations can't explain these regional differences.

In 2009, physicist Surabi Menon and her collaborators at Lawrence Berkeley National Laboratory conducted a computer simulation to try to explain why glaciers and snowfields in the Himalayan region were melting so fast. "Our simulations showed that greenhouse gases alone are not nearly enough to be responsible for the snowmelt," she told reporters. Several other scientists were coming to similar conclusions. "Based on the differences it's not difficult to conclude that greenhouse gases are not the sole agents of change in this region," William Lau, the head of atmospheric sciences at NASA's Goddard Space Flight Center, told reporters in the wake of his own study. "There's a localized phenomenon at play."

So what's going on here? If warming caused by greenhouse gases alone can't account for these changes, then what is the mysterious agent driving up the temperatures on the roof of the world, chewing away at ice and snow from Sultan Largo to Mount Everest to southwest China, accelerating the melting of the world's largest expanse of frozen freshwater outside of the poles? What is destroying Kumik?

A clue to this puzzle was recently found buried in ice thousands of miles away from Zanskar, when scientists uncovered the solution to another climate mystery.

Around 1865, without warning, the glaciers of the European Alps suddenly began retreating. Just a couple of decades prior, they had reached their maximum extent. Given the temperature and precipitation trends of that time, if anything, Alpine glaciers should have been *advancing*.

Instead, they mysteriously started shrinking. For decades, scientists were at a loss to explain this phenomenon, which they dubbed the Little Ice Age paradox. (The Little Ice Age refers to an unusual period of cooling across Europe in the medieval era, possibly due to cyclical changes in Earth's orbit around the Sun, affecting the amount of solar energy reaching the Northern Hemisphere. Or it could have had to do with the ocean's circulation patterns; researchers still debate the cause.)

A glacier will shrink in response to either declining input via precipitation, or rising temperatures, or both. These two factors cause the rate of melting at lower elevations to outstrip the rate at which snowfall adds mass, higher up the glacier in the accumulation zone. But records show that precipitation wasn't changing much in the second half of the nineteenth century, and that temperatures in the Alps region didn't start to rise until the early twentieth century. Alpine glaciers clearly started retreating a half century before the data suggest they should have. Until recently it was a kind of scientific locked-room mystery: if temperatures and snowfall remained the same, what could be causing this abrupt shift from growth to retreat?

Tom Painter studies the complex interactions between pollutants, dust, and snow as an expert hydrologist at NASA's Jet Propulsion Laboratory. Much of his research has focused on the complex dynamics of dust deposition on the mountain snows of his native Colorado. While working on a model of dust impacts on the Colorado River Basin, he began reading papers about the mystery of the Little Ice Age in Europe. None of the previously proffered hypotheses could explain the glaciers' abrupt reversal and retreat, a trend that continues to this day, and which seems to be speeding up in the past two to three decades. (This is also the reason why some luxury ski resorts in Switzerland have resorted to spreading giant reflective blankets across their glaciers.)

"I was looking at the best climate records we have in the Alps," he told me. "They quite clearly show that temperatures didn't start increasing until 1910, 1920. So it just dawned on me—God, what other forces could there be? I thought at that same time industrialization was going on, and certainly [there were] large emissions of black carbon."

"Black carbon" is scientists' term of art for those exceedingly small, exceedingly dark particles that are a product of incomplete combustion. Black carbon is the stuff that makes soot dark. And black carbon is produced wherever and whenever any carbon-based fuel—coal, biomass (organic matter such as wood, dung, and agricultural waste), petroleum distillates like diesel and kerosene, you name it—is burned, but not *completely* burned. As a constituent of soot (which can also contain other stuff, such as organic carbon) it is thus one measure of the inefficiency of a given fire. Those tiny particles, linked in little chains of dark spherules, and glommed onto by some of the other chemicals in soot and smoke, are simply fuel carbon that failed to make the complete transition to carbon dioxide, which is what happens in ideal combustion.

There was nothing "ideal" about the combustion that took place at the height of the Industrial Revolution: as anyone who's read Charles Dickens knows, the middle of the nineteenth century was a veritable orgy of soot production. Velvety black plumes spiraling up from the horizon were perhaps the surest sign of "progress" in those days, as they are today in some parts of rapidly developing China. During that period, western Europe became dotted with coal-fired factories and crisscrossed with coal-powered locomotives. Tom Painter's Austrian coauthor, Georg Kaser, of the University of Innsbruck, told him of "stories about how the maids in Innsbruck in the mid- to late 1800s had to start bringing their washed clothes in and drying them inside. They would get soiled by the particulate pollution. And that's sitting right there in the valley below the glaciers."

Painter and Kaser and their colleagues theorized that all those black particles, streaming out of smokestacks and carried by prevailing winds into the Alps, might be the solution to the Little Ice Age paradox. When it's deposited onto surfaces of snow and ice, black carbon dramatically reduces the amount of light reflected back into space. These dark particles absorb sunlight and turn it into heat, warming the atmosphere when they are aloft and melting the ice and snow when they wash out onto their surface. This enormous input of energy into the glacier can accelerate the rate at which it melts, helping to trigger other feedback mechanisms—for example, as a glacier retreats it exposes darker ground on its margin, which then absorbs more light and exerts more warming—that can pile on and cause runaway melting.

No one had tagged black carbon as a suspect in the caper of the Little Ice Age paradox before, but Painter sensed he was on the right trail. To test the theory, his team examined ice core records taken from Alpine glaciers, allowing them to analyze pollution and other anomalies found in successive layers of ice, dating back to different time periods. They found that black carbon concentrations rose sharply in the mid-nineteenth century and kept rising well into the twentieth century, a pattern that coincided closely with the industrialization of western Europe. The team then modeled the amount of radiative forcing those concentrations of black carbon would have contributed to the region's energy balance—in other words, how much sunlight those particles would absorb, turn into heat, and add to the system through various mechanisms. They concluded that "the associated increases in absorbed sunlight by black carbon in snow and snowmelt were of sufficient magnitude to cause this scale of glacier retreat."

"We're pretty damned confident," Painter told me a couple months after his study was published in the *Proceedings of the National Academy of Sciences*, "but it's not 100 percent because we have to infer the [black carbon] deposition down low from ice cores up high." Measuring black carbon on the lower part of the glaciers, closer to where the sources of the pollution would have been, was impossible, because those sections melted and vanished long ago. But there is, he argues, no other way to explain the phenomenon. "There is nothing else we know of in the climate record that can speak to this. From a general scientific standpoint, it's the first physically reasonable explanation, physically and climatically consistent."

The significance of the study goes beyond the solution of an obscure historical scientific puzzle. It suggests that people have been having a major impact on regional climates since long before the effects of increasing carbon dioxide emissions—with their long lag times, courtesy of their lengthy residence in the atmosphere—started showing up. Those tiny dark particles started administering thermal death blows to the Alpine glaciers well before the temperature rise associated with carbon dioxide and other greenhouse gases could even start to contribute in a significant way.

"It's a big deal," says Painter. "There's a human influence when we thought for sure there wasn't a human influence, and [at least] not at that magnitude."

Painter's study has the potential to reshape our understanding of the scale, timing, and extent of how human activities have transformed natural systems, thanks to something that we have long overlooked, tolerated, or just wrinkled our noses at: soot.

According to the latest estimates, we belch anywhere from 7.5 million to 17 million tons of black carbon into the air each year—a number that increases from year to year, as "emerging economies" such as China, India, and Kenya try to bootstrap their way to prosperity, with a little help from a whole lot of burning. If black carbon emitted during Europe's industrial revolution was responsible for the decline of Europe's great glaciers, what might today's dark particles be up to in other parts of the world that are currently following a similar playbook of development? And if black carbon's impact on the Alps was so potent, what, for example, could it now be doing to the snow and ice of the Himalaya? Or the Arctic, for that matter?

Consider China, which builds a new coal-fired power plant every week, where pollution in some cities gets so thick that it sometimes shuts down the airport because pilots can't see the runway, and where the lifeblood of both the new industrial and the traditional agricultural economies are two rivers, the Yangtze and the Yellow, that start as streams pouring off glaciers in Tibet.

Or, for that matter, consider California's Central Valley, which has the worst particulate air pollution in the United States, where half the nation's fruit and vegetables are grown, and where the majority of the irrigation water comes from the declining snows of the Sierra Nevada.

And what might black carbon's potent thermal punch mean for the melt-based microeconomy of the thirty-nine households in the oldest village in Zanskar, now in search of a new lifeline on the roof of the world?

There is another story, familiar to just about everyone in Zanskar, that purports to explain the root cause of Kumik's famous misfortune. It all goes back to a curse laid at the village's founding, you see. It all started with the *zbalu*.

"*Zbalu* is only small, like this." Stobdan puts his hands a couple of feet apart. The *zbalu*, he explains, are invisible spritelike spirits who dwell all around us, but mostly live out there on the edges of the village. They carry a cap and a stick. They are proud creatures, and can be generous, but are very dangerous when piqued or provoked.

"Old people say if we play dice alone—imagine that *zbalu* is here, is coming. So one person is playing and he catches the *zbalu*. Just imagine." This was way back in the beginning, soon after the founding of the village.

"And the *zbalu* is now afraid with the person, and he says, 'Please free me. Don't catch me like this. I will do everything you tell me.' And the person says, 'You should wall around this village.' "

"So the *zbalu* asks the question: 'You need this wall, or you need water?' He gives two options. At that time many snow here, and water here. So the person says, 'No, we have many water. We must need this wall.' And the *zbalu* does in one night."

The ancient stone wall that surrounds Kumik is three or four feet high and well over a mile long. It would be an enormous undertaking for a team of regular-sized workers with tractors and bulldozers. I am incredulous that the tiny *zbalu* could do this all by himself, let alone in one night, and I say so. Stobdan concedes that it's unlikely and offers clarification: "He brings many *zbalu*. In one night they wall all around Kumik. And the person didn't give food to *zbalu*."

Big mistake. "But there you see Pang Kumik"—the hamlet of three houses down at the base of the village—"there is one grandmother, and she gives some curd to *zbalu*, and *zbalu* is very happy and gives them a separate spring. And I think the *zbalu* is not happy with this other person, and the *zbalu* says, 'After this there is no water in Kumik.' "

Stobdan pauses, leans back, waggles his head. "So now it's finished. Very old story."

"Could the curse be reversed if the *zbalu* could be caught again? Where do the *zbalu* live?"

"Our old people say *zbalu* is everywhere. Right now *zbalu* is here." He waves a hand in front of the table. "If we play dice, the *zbalu* will come now." It's a high-stakes game, though, kind of like gambling with the devil. "But the person needs a very strong heart to win, to beat the *zbalu*. Otherwise they will die."

Later Stobdan tells me these are just stories, superstitions. "Many believe this is true, but I cannot believe."

"So why is Kumik running out of water?"

He shrugs. "I think weather is changing. I don't know why."

The original-sin connotation of the *zbalu* parable—with its twin warnings against shortsightedness and selfish behavior—seems to voice some unspoken anxiety lodged deep in the Zanskari psyche about their narrow hydrological tolerances, and about time: keep an eye on the clock, which was set at the beginning and which is now winding down. Or perhaps it speaks to a deeper understanding of a central fact about the universe: there's no such thing as a free lunch.

This idea seems to be enshrined in almost every culture's fire-giver myths: with the blessing comes a roughly commensurate curse. The Cherokee had the water spider, who bore the fire back from a hollow sycamore tree on a distant island in a bowl she had woven from her own silken thread. The hooting owl had tried, but the smoke hurt his eyes and made permanent rings of ash around them; the raven had also made an attempt, but as he flew over the fire, the flames singed his feathers forever black.

The Greeks, of course, had Prometheus, who, you may recall, pitied us benighted mortals, hunched out there in the damp cold night, ceaselessly hassled by callow gods, with no light or tools to pull ourselves up into organized societies that might have a fighting chance. So, hoping to buck us up, Prometheus spirited fire out of Olympus in a hollow fennel stalk. He gave us the flames and taught us how to control them. The lesson was simple: without fire we were more or less consigned to lives of darkness and misery.

But this gift came with a heavy price. Fire being beyond precious, its transfer to humans was expressly against the wishes of the Olympians. Zeus was livid when the stratagem of Prometheus was discovered. The father of the gods made an ominous pledge: "Son of Iapetos, deviser of crafts beyond all others, you are happy that you stole the fire, and outwitted my thinking; but it will be a great sorrow to you, and to men who come after. As the price of fire I will give them an evil, and all men shall fondle this, their evil, close to their hearts, and take delight in it."

So Zeus sent down Pandora, a girl shaped from clay, to Earth to beguile Prometheus's (whose name means "forethought") hapless brother Epimetheus ("afterthought"). She bore a gift, a jar loaded by the Olympians with "sad troubles for mankind." When she lifted its lid, pain and strife and all manner of diseases streamed out, to be carried by the winds across the face of the Earth, to plague the human race in myriad ways.

And thus we obtained the ultimate survival tool, the enabler of all civilization to follow—but with it, fire's dark alter ego. A plague borne by the winds across the face of the Earth, deadly, destructive, ubiquitous . . . and yet, for all that, still an "afterthought." Pandora, after lifting the lid more out of curiosity than any malevolence, quickly rushed to replace it. And "hope," the Greek poet Hesiod tells us, "was the only spirit that stayed there in the unbreakable closure of the jar, under its rim, and could not fly forth abroad."

The mystery of the rapid retreat of Himalayan snow and ice differs in important ways from that of the Little Ice Age paradox. Temperatures in the Himalayan region are rising and precipitation patterns are indeed changing. But the contours of the problem are similar. Greenhouse gases alone can't explain the decline in snow and ice, or why the temperature rise is two to three times as fast as the average global rate of increase. There's a missing *X* factor.

Not long after my first visit to Kumik, I discovered that a recent spate of research had solved this puzzle, too. The latest science suggests that the same dark agent that was behind the Little Ice Age paradox of 150 years ago is today driving the extraordinary regional warming on the roof of the world, destroying life-sustaining snow and ice from Kumik to the Khumbu Valley of Nepal.

Just before my first visit to Kumik in 2008, atmospheric scientists Veerabhadran Ramanathan and Greg Carmichael published a groundbreaking modeling study in *Nature Geoscience*. They concluded that "in the Himalayan region, solar heating from black carbon at high elevations may be just as important as carbon dioxide in the melting of snowpacks and glaciers." Black carbon, they argued, was likely responsible for at least half of observed glacial retreat and the associated rise in winter and spring temperatures.

A suite of other studies, conducted by agencies ranging from the Environmental Protection Agency to the World Bank, have come to a similar conclusion. Surabi Menon found that about 90 percent of the change in snow and ice cover area across the Tibetan Plateau and Himalayan region was attributable to aerosols, those long-overlooked pollutants (defined broadly as particles suspended in a gas) that make up much of the atmospheric brown clouds that form seasonally over different parts of the world, especially South and East Asia. Her team concluded that one aerosol alone—black carbon—accounts for 30 percent of observed decline and glacial thinning, and probably much more.

"We were underestimating it to some extent," Menon told me later. She pointed out that getting the right data on the amount of black carbon present in the atmosphere is a serious challenge because there's a big discrepancy between "bottom-up" inventories, estimating and adding up the amount of pollution from various sources based on measurements of economic activity (the amount of fuel burned in a power plant, for example), and "top-down" estimates, which use models that rely on direct measurement of pollutants in the atmosphere. (Direct observations perennially find greater concentrations of black carbon in the atmosphere than estimates using emission factors of different polluting activities would suggest, which means that black carbon likely has an even bigger impact than Menon's model indicated.) These modeling studies helped confirm the claim put forward a few years earlier by James Hansen, perhaps the foremost climate scientist of his generation, and his colleague Larissa Nazarenko, in a study in which they performed equilibrium climate simulations incorporating black carbon's calculated effect on snow and ice albedo: "We suggest that soot contributes to near worldwide melting of ice that is usually attributed solely to global warming."

The dark particles were melting snow and ice directly, they argued, and indirectly warming the atmosphere—"for a given forcing it is twice as effective as CO_2 in altering global surface air temperature." Part of the reason for this is that black carbon on snow ramps up snowmelt much more efficiently than does warming in the air. Light bounces around so much within a snowpack that there's a high likelihood it will be absorbed by black carbon particles before it exits. So it doesn't take much soot—just a few parts per billion—to make a serious dent in the reflectivity of snow.

Hansen and Nazarenko calculated that black carbon's impact on reducing snow albedo alone accounts for "more than a quarter of the warming observed in the past century." This effect hadn't been incorporated into most climate models at the time, including the assessment of the premier body studying climate change, the Intergovernmental Panel on Climate Change (IPCC).

"Soot is a more all-around 'bad actor' than has been appreciated," they concluded.

Another decade of research since that observation has borne out Hansen's suspicions, and then some. In fact, soot's troublemaking extends far beyond the frozen world of the high mountains and hits us all much closer to home.

Black carbon might be the oldest pollutant there is—and yet it is easily the most dangerous pollutant you've never heard of. And it has been right under, and in, our noses for a very long time.

Black carbon?

Perhaps you're thinking: Another thing to worry about, on top of flesh-eating viruses, MRSA, people texting while driving, political polarization, the spread of West Nile virus and Lyme disease (both spurred on, by the way, by warming temperatures). But the black carbon story is, on balance, a hopeful one. Like Kumik's, it gets worse before it gets better. But at the end of the day it offers an opening amid all the relentless gloom—a pall of crazy weather, rising inequality, intractable poverty—that seems to hover over the early twenty-first century. Even the darkest clouds admit some light.

Veerabhadran Ramanathan should know. And if anyone has a right to be pessimistic in the wake of the emerging revelations about black carbon, it's Ramanathan. He's an atmospheric chemist and expert on climate science, who has helped to break the story. He has spent over three decades studying the behavior of aerosols in the atmosphere, and how they affect climate and other earth systems. He was the first scientist to fly unmanned aerial vehicles (drones) into the haze over parts of Asia and measure its light absorption and other properties. (He was also one of the key investigators who discovered the atmospheric brown cloud over India during a huge 1999 multinational field campaign called INDOEX—the Indian Ocean Experiment.)

And after a long career spent rigorously measuring and trying to understand the properties and physical behavior of black carbon and other airborne pollutants around the world, from California to India, Ramanathan says he finds cause for rejoicing in our emerging knowledge of the full scope of black carbon's disastrous effects on both human health in the near term, and on the natural systems that underpin human health and welfare in the medium to long term. (Or in the case of the Kumikpas, right now.)

Like the redoubtable residents of Kumik, he exudes an optimism seemingly at odds with the tough spot we find ourselves in, and with the fine details of the story his own research has revealed.

"I've worked thirty-five years on short-lived climate pollutants," he told me, referring to black carbon and other warming agents such as methane and hydrofluorocarbons (which are used as refrigerants in air conditioners, etc.). "Every time I come back from a field campaign, you know, I have more bad news to report: 'These people are dying, that mountain is vanishing, melting.' "

We were meeting in a hotel lobby in Lucknow, the largest city in

Uttar Pradesh, one of India's poorest states, and its most populous. According to the World Health Organization, Lucknow is also the tenth most polluted city in the world in terms of particulate matter. Ramanathan had just come back from several days in rural villages that hadn't changed much in centuries, checking in on the progress of Project Surya, an ambitious experiment he launched to test whether providing clean cooking devices to households that rely on dung and wood fires could punch a "hole" in the haze of black carbon and other pollutants that seasonally blankets north India. (*Surya* means "sun" in Hindi and Sanskrit.)

The preliminary results were encouraging, and Ramanathan seemed energized as he described the ambitious next phases of the project: 10,000 households in Uttar Pradesh, then on to Africa and the world. In the midst of the steady drumbeat of bad news about all the intractable problems we face, and geopolitical inertia in response, the emerging story of the vast damage wrought by black carbon was, to him, actually cause for hope, not despair. Black carbon, he suggested, offered humanity a way to get a handle on some of its thorniest, most daunting, and interlocking problems—and start seeing some immediate results.

"Now I can come to you guys and say, now we have found ways to help people." His eyes lit up behind his thick glasses lenses as he leaned in across the table, and his soft, measured voice took on a note of passion.

"True happiness is there!"

2

Our Dark Materials

Be thou as chaste as ice, as pure as snow,
thou shalt not escape calumny.

—*HAMLET*

An old Tibetan proverb captures the isolation of mountain kingdoms like Zanskar in stark terms: "If the valley is reached by a high pass, only the best of friends and worst of enemies are its visitors." Or, as Tashi Stobdan sums up the situation: "In winter, leaving Zanskar is harder than jumping out of a wolf trap!"

Six months have passed since my first visit to Kumik, and I have jumped back into the "wolf trap," slipping and sliding, at times crawling and wading, for four days on the *chadar*, the "ice way."

Between November and May, the only way of getting in or out of Zanskar is by walking for several days on the frozen surface of the Zanskar River. For six weeks every January and February a crust forms that is thick enough to serve as a slow-motion highway connecting Zanskar to the rest of the world. For centuries it was the preferred path for carrying famously rich Zanskari butter to trade in Leh, which was an old hub on a spur trail of the Silk Route. (The precious cargo would melt on long summer journeys.) Today the ice way still carries a steady traffic of Zanskaris of all ages, in search of work, goods, or an education.

This time, I have walked in from Ladakh with a man named Lobzhang, from Pishu (a village famous for its *chadar* guides), and a man named Thinlas, from the village of Lingshed. We encountered seventy-year-old men hauling rugs back from Leh, and seven-year-old girls on their way to distant schools in Leh and Himachal Pradesh, in equally high spirits. We slept in caves, burning bonfires of scavenged driftwood, and spent our last night in an abandoned tin hut at the road builders' camp at Kilima, where my contact lens solution froze solid and I had to hold my boots over the stove for an hour the next morning so I could wedge my feet into them. The diesel fuel in their only truck had turned to gel; the road camp workers tried to get it going by building a flaming fire under the engine block. We watched their futile efforts for a few hours and then started walking down the road, eventually catching a ride on a jeep from Zangla to Padum.

And now, after thawing out for a couple days in the kitchen of my old friend Urgain Dorjay, I am climbing the long slope up to Kumik through knee-deep snow. Sunlight punctures gauzy high clouds and bounces off the snow that coats the neat stacks of dung outside each home. Urgain takes long, deliberate strides beside me. He wears an old dusty ski jacket over his immaculate woolen robe, the traditional dress known as the *goncha*.

Urgain, who looks a bit like a Zanskari Gary Cooper, has a dignified bearing and an expression on his long face that is equal parts kind and shrewd. He is the cousin of Tsewang Zangmo, Tashi Stobdan's tireless wife. Urgain is also the first Zanskari I met, back in 2003, on a trip to ski a glacier beneath the 24,000-foot mountain known as Nun, the sharp-tipped sentinel of western Zanskar. On that trip, I first discovered that Urgain is a Himalayan Renaissance man: farmer, tailor, metalworker, ski instructor, trekking guide, carpenter. On subsequent journeys, I learned that Urgain is also a human Swiss army knife, that guy you want around when the yak dung hits the fan, the first guy you would choose for your team in any dystopian Mad Max–type situation: endlessly resourceful, unflappable, quick-thinking, good-humored, good company. This is probably why the local authorities always call on him to lead rescues when disaster strikes—as when avalanches trapped dozens of *chadar* travelers on the ice in 2012 for several days, when a bus full of passengers crashed and tumbled down the Pensi-la pass the year before, and when a visiting *rinpoche* (a reincarnate lama) and his 500 followers were bogged down for days without food by freak summer snows at the foot of the 5,100-meter Shingo-la two years before that. Soon he will guide me on the *chadar* trek back to Leh, but today he has come along to visit his relatives and learn more about the slow-motion disaster unfolding in Kumik.*

We reach the empty reservoirs on the edge of the village. Two young girls sled past us, giggling and clinging to each other on a single wooden board, down past the *mani* wall of sacred carved stones. We reach Stobdan's house and duck our heads under the low threshold as we enter the dark ground-floor hallway. On the left are storerooms for the animals' fodder, on the right the winter kitchen where Stobdan and Zangmo spend most of their hours between November and May.

We find them there, bustling about, making tea and setting out biscuits. They greet us warmly and usher us over to sit near the stove, in the place reserved for honored guests. Zangmo asks after my family as she pours me one cup of hot salt tea, one cup of cool *chang*. Neighbors start to filter in and soon the kitchen is full of weather-bitten Kumik men of all ages, sitting cross-legged along the pitted mud walls. They slurp tea and fall into easy chatter.

Their talk soon turns to the difficult summer past. Soon after I departed, they tell me, in early August, the village ran out of water before the harvest. Again. Some of the crop was lost. The harvest of grasses for winter fodder was lean.

---

*Urgain knows disasters. Before Kumik, the last time an entire village was relocated in Zanskar was in 1985. At that time the cause wasn't a lack of snow, but too much. After a once-in-a-century storm, a massive avalanche crushed the village of Shagar Yogma in the middle of the night. My friend Thukje Tashi, then a young boy living across the valley in Phe village, remembers the aftermath. Rescuers from neighboring villages had assumed everyone had died but kept digging down for survivors—and found a nun weeks afterward, trapped in her house, who had survived by subsisting on fermented barley grains left over from making *chang*, the Zanskari barley wine. (Perhaps, he speculated, the chimney had saved her, delivering oxygen in lieu of smoke.) Shagar's seven houses were completely destroyed. Urgain's oldest sister was among the forty or more villagers who died that night, along with her husband, mother-in-law, and all their children. Urgain once calmly told me the story and recalled fondly how he had gone to Shagar when he was a boy of ten to study religious scriptures with his brother-in-law, who was an *amchi* (a traditional doctor) and had mastered the ancient art of painting *thangkas*, intricate scrolls with religious scenes. The whole village of Shagar was rebuilt a few kilometers away, at a safe distance from the steep slopes. "Avalanche awareness is very important," Urgain concluded his sad tale, succinctly.

"*Kha Kumik, chu Shila*," I murmur. The snow falls above Kumik, but the water goes to Shila.

The men laugh riotously—either at the old saw itself or, more likely, at the novelty of hearing it from the mouth of a *chigyalpa*, "one from outside the kingdom." They know well that their village's name has become a kind of shorthand throughout Zanskar for water scarcity. Stobdan was once in a village called Abran with some other Kumikpas, in more thickly glaciated western Zanskar, where water is more abundant. The villagers there were putting on a drama as part of a festival, and the title of one comedic sketch was *Kumik Chu Met* (Kumik Has No Water!). When the actors came out and began their show, people from Stongde (the large village neighboring Kumik that occasionally has lean water years as well) started throwing stones at them. But the Kumikpas just sat calmly and watched. They are used to this kind of thing. After all, there's even a famous song that used to play on the radio: "Kumik, the Waterless Village."

So why shouldn't even a *chigyalpa* like me, from halfway across the world, know of their plight, too?

But one of the men amends the old saying. "If you go up to the mountain," he says, "you see that most of the water actually goes to Shade, not Shila."

The others nod. The spirit of the proverb was accurate, says Urgain, who has trekked all over the mountains of central Zanskar: the better part of the snow- and icemelt above Kumik does indeed flow to neighboring villages. Kumik's piece of the water pie, always small, is now shrinking—along with the pie itself. Less snow than usual had fallen this winter, too, in keeping with the general trend of the past couple of decades. But perhaps because of the *chang*, perhaps because sun-filled winter days like this are full of leisure and free from worry—you've done what you can, you've made your bed during the brief intense summer, and now you lie in it—the men are in a jovial mood. Sure, Shila, or Shade, or even Pipcha, across the mountains, may get much of the water.

"But the water then flows into the Lungnak River," one man says. "And then it will reach lower Kumik eventually, in our new canal!"

Everyone laughs uproariously at the symmetry and justice of this—the idea of Kumik's purloined snows returning to their rightful home, courtesy of some shrewd planning and collective bursts of labor. Stobdan grins with pleasure at this prediction. As one of the masterminds and driving forces behind the new canal, and after years of harrying sometimes reluctant neighbors into action, he relishes the idea of cosmic irony redounding to Kumik's benefit for a change.

Stobdan clears his throat and explains that he has called this meeting to discuss building plans in the new village for the coming summer. Everyone present is interested in using solar heat for their new home in Marthang, so I have come to help them weigh their options. "Zanskari winters are long and cold, and the first need is warmth," one man announces. After my journey on the way in, I don't need much persuading on this point.

They debate the cost, whether bigger windows and insulated walls and other alterations to their traditional building patterns will be worth the expense. In old homes the animals are stabled in rooms on the ground floor, both for convenience and to take advantage of their body heat. Over the course of a summer, sheep and goats would trample their dried dung into the floor. This would later become insulation, covered by another layer of packed earth, in the traditional winter kitchen, such as the one where we sit.

But the men want their new homes to be cleaner, more light-filled, more modern—more *pukka*, as people say down in the plains of north India. A strong, durable house, like the ones you see on the TV shows beamed in from Delhi. Stobdan listens intently. He wants the same, but he has other ideas.

"They want to build a separate house for animals and for people," he says. "But I don't like that." He will build a solar house, and use modern materials like cement plaster and big glass windows, but keep his animals close, in stables on the north side of his building. "I need the culture and the modern too, combined."

A consensus soon emerges from the group: mostly, the toilet should be on the north side, and the *chotkhang* (the shrine room) on the east side, and the winter room on the south side, where the sun hits. "Solar room is very good, perfect, and useful," Urgain summarizes. "If somebody wants to build solar room, but he is not good financially, then he can manage from some relatives some money. They help each other. It is perfect for everyone. Everyone agrees."

Zangmo serves steamed dumplings called *momos*, and listens closely as we go on to discuss the basic principles of solar design: how the sun's light is admitted, absorbed, radiated as heat, and then trapped by the glass windows; how the heat is stored in the thermal mass of the mud brick walls and floors; and how dark surfaces are warmer than lighter-colored surfaces. These things have the ring of common sense to the Kumikpas, and they nod and go on to debate the finer points: stone versus mud, big windows versus small, spending a bit more money up front to heat with the sun versus time spent gathering scarce dung for fuel.

Since she and the other women of the village do most of the fuel collection and tending of the hearth fires, Zangmo has a greater stake in these debates than anyone else in the room. As she bustles about, she asks detailed questions about the kitchen design, and I recall a meeting from the prior summer about building a passive solar community hall in Shisherak, Zangmo's home village, where an older woman, fed up with all the ceaseless talking, had gotten to her feet and shouted: "It's *cold* here! Let's just *do* something!"

The Kumikpas voice some concern about *stanches*: how the house will look to others. In Zanskar and Ladakh, some homes and all monasteries are whitewashed with lime; lighter colors are considered auspicious and pleasing to the eye. But as everyone knows, darker colors are warmer to the touch, since they capture light and transfer more heat—and so darker-colored walls are part of many solar heating strategies. *Stanches* is important, the men agree, but in Zanskar, function trumps appearances.

One man concludes: "If it's warm, then it's beautiful!"

We sip our butter tea as Zangmo places another piece of dung in the firebox. Smoke escapes from corroded holes in the sides of the rusty stove and rushes past her into the room. Sunlight streams through the *yokhang*'s single high window, illuminating the lazy curls of dark particles above our heads.

One of the men looks up from his tea. He has just had a notion.

"The solar house is happening on the snow," he says.

The others laugh appreciatively, but it's hard to tell if he's just joking around or chewing on a sudden insight. He shrugs and adds, "Maybe this is causing the water shortage."

He was on to something.

In 1998, Urgain Dorjay was hired by an American traveler to guide him on a week-long trek across the Himalaya from Zanskar into Kishtwar. On the second day, under a cloudless sky, they climbed up onto the crevasse-riddled glacier that coats the 5,450-meter-high Kang-la pass.

Light bounced off the snow as they climbed, and they moved slowly over difficult terrain. Halfway to the pass, the client's sunglasses broke in two. So Urgain gave the man his own pair, figuring he would be able to deal with the trying conditions better than the foreigner, and continued to route-find up the glacier. Soon the client was flagging from the altitude and exertion, and the snow bridges covering the crevasses seemed increasingly thin. After a while, Urgain decided the route had become too treacherous, so they turned around to descend to Padum.

That night they made camp on a bluff high above the Lungnak River. Urgain cooked dinner and tended the open fire all evening. The following morning he awoke to a rude surprise.

"Next morning, not possible to open my eyes. Ooh la la!"

He laughs at the memory of the discomfort. Urgain had come down with a bad case of snow blindness; it feels like sandpaper is rubbing the surface of your eye every time you close your eyelids.

"We use *burtse*, which makes very strong smoke," he recalls, referring to a kind of wild sage that grows all over the hillsides. The particles streaming out of the fire only made the condition worse. "The smoke damaged my eyes. It feels like my eyes are burning."

But they still had ten miles of walking to get back to Padum. "So I just am putting water in my eyes, and I open eyes quickly and take few steps, walking like this," he says and gets up to demonstrate the halting steps of a blind man. He walked that way—opening his eyes briefly, taking a couple steps on a rocky, twisting path high above the raging river, then closing them again with his hands held out—all day long.

Snow blindness is simply a burn of the cornea, the clear fragile tissue covering the pupil of the eye. Like any burn, it causes painful dryness and inflammation. And it happens because snow is the most reflective naturally occurring surface on the planet. Snow has a very high albedo, reflecting about 80 percent of incident light, including ultraviolet. (Fresh, unbroken snow can have an albedo as high as 90 percent.) It is nature's best mirror.

In the thin air at high altitudes, with fewer gaseous molecules to deflect and attenuate their path, those rays are even more intense when they reach exposed flesh. That's why veteran mountaineers put sunblock on the underside of their noses, and why Zanskaris used to smear soot under their eyes—to reduce the light bounced into them.

Of course, tiny, gritty particles of partly burned *burtse* deposited *in*, rather than *around*, the eyes don't help that much. Like many Zanskaris, Urgain's eyes are now chronically red, thanks to daily exposure to soot, dust, and sun. He dates the problem to the twin hammer blows of smoke and snow that day on Kang-la. "From that time, my eyes are a little weak," he says with an uncomplaining shrug.

Zanskaris have an intimate and practical understanding of the principle of albedo that goes beyond the occasional bout of snow blindness and is derived from long experience. The fact that dirty snow is less reflective than fresh snow, and therefore melts faster, seems both intuitively obvious to most of us and not of much practical import. But Zanskaris used to exploit this knowledge every spring to squeeze an extra few days out of the short growing season.

"Many years before, in my parents' time," Urgain recalls, "in wintertime a lot of snow we have." Back then, it was quite common to have four to eight feet of snow piling up over the course of a winter and lasting into early April. (Locals told the French traveler Michel Peissel in the 1970s that sometimes they would round a corner in the eight-foot-high passageways dug in the snow between two villages and find a wolf prowling, unable to find its way out. It would try to escape by jumping into a deep snow bank, then get stuck, until the people stoned it.)

So they dug down six or seven feet through the snow into the soil. "We take out mud and put on the field to help for good melting, and quick. Many times we used like this."

The dark soil would absorb the sun's light, turn it into heat, speed the melting of the snow, which would soften the soil for plowing and hasten the arrival of the day when the fields were clear for planting. Without those critical few extra days, the barley might not have time to ripen before snows returned in September. Some families across Ladakh and Zanskar used to gather ashes from their hearths and spread them on their fields in early spring to the same effect.

But times have changed. "Now we don't need to do like this," Urgain says. "Very less snow, so it's automatically melting quick."

In an experiment much less famous than flying a kite in a thunderstorm, Benjamin Franklin figured all of this out almost 300 years ago. On a sun-filled winter day in the late 1720s, he gathered up a dozen different square patches of cloth and went out into a "bright Sunshiny Morning" on the streets of Philadelphia. Coming to a patch of clean snow, he stooped down and laid them all—black, deep blue, light blue, green, purple, red, yellow, white, and other shades in between—upon the surface. Then he went off, waited a while, and came back to see the results of his experiment.

"In a few Hours," he recalled many years later in a letter to a friend, "the Black, being warm'd most by the Sun, was sunk so low as to be below the Stroke of the Sun's Rays; the dark Blue almost as low, the lighter blue not quite so much as the dark, the other colours less as they were lighter; and the quite White remain'd on the Surface of the Snow, not having entred it all."

With his simple experiment Franklin confirmed what experience had long taught: materials of darker colors "imbibed," as he put it, more heat from the sun than did lighter-colored objects. This fact had some consequences: black cloths hung out wet will dry more quickly in the sun; and (perhaps most importantly) beer in a black mug set near a fire will warm much faster than beer in a "bright Silver Tankard." He might have been one of the earliest investigators to realize its wider practical applications, hinting at the broader implications future generations might deduce from this fundamental physical property of matter.

"What signifies Philosophy that does not apply to some Use?" Franklin wrote. "May we not learn from hence, that black Clothes are not so fit to wear in a hot Sunny Climate or Season, as white ones. . . . That Fruit-Walls black'd may receive so much Heat from the Sun in the Daytime, as to continue warm in some degree thro' the Night, and thereby preserve the Fruit from Frosts, or forward its Growth?—with sundry other particulars of less or greater Importance, that will occur from time to time to attentive Minds?"

This insight had been put to use long before Franklin's musings, of course: just witness the whitewashed homes of ancient Greek villages that cool living spaces by reflecting the noonday Mediterranean sun. The implications for snow and ice have been long understood, as well. Like Zanskaris darkening their fields, air forces around the world have long used the same trick to clear their runways of snow.

A few attentive minds in the Chinese Communist Party grasped this principle, too, but sought to test it on a much wider scale.

In 1959, a young, enterprising engineer named Shi Yafeng, later to become the father of Chinese glaciology, was dispatched to barren Gansu Province in northwestern China. His mission was to identify and test ways to increase irrigation flows to help feed his country's rapidly growing population; party officials were becoming quietly alarmed at the widening shortfalls of staple grains.

"The research team was required to put all its effort into work on trying to increase glacier melting," scholars Juichen Zhang and David Oldroyd wrote about the project. Guided by Soviet advisers, Shi's team settled on a bold solution. For four months, they recruited thousands of local peasants to load coal dust and soot, made from burnt dung and grass, onto donkeys, onto porters' backs, onto planes—and then spread, sprinkle, and dump vast quantities of the stuff over glaciers and snowfields. The glaciers of Gansu survived this dark onslaught, but not before the basic principle had been demonstrated. Thanks to the particles spread "over 19 glaciers in the Tienshan . . . the glacial meltwater was increased by 12.5 million cubic meters between May 10 and June 12."

The project was suspended in 1961, but it had proved that darkening snow and glacier surfaces with soot on a massive scale could rapidly accelerate melting.

Black carbon attacks snow and ice on multiple fronts, beyond its initial absorption and reradiation of the sun's rays. These force multiplier effects mean that it doesn't take much of the stuff sprinkled on top of a glacier to set a thawing feedback loop in motion—leading to accelerated melting, altered timing of runoff, and disruption to familiar farming patterns.

As Urgain's parents understood, the first impact is through direct reduction of the albedo of the snow—light at almost every wavelength is absorbed by those dark particles and converted to heat, which is then radiated outward in all directions by the particles, melting the snow grains around them. That part's pretty straightforward.

But then things get a bit more complicated. As the sun heats these impurities, which then heat the snowpack, the grains of snow get rounder and larger. Larger, coarser snow grains reflect less light than smaller, finer snow grains. So as black carbon raises the temperature of the snowpack, it creates another feedback loop: the grains that it melts grow in size, letting even more light even deeper into the snowpack, to bounce around and increase the rate of melting even further. This feedback process is one of the primary mechanisms by which black carbon hastens the arrival of spring melting. Because snow is incredibly efficient at scattering light, just a little bit of black carbon has an amplified impact on the total energy balance. The black carbon particle embedded in the snowpack is subjected to a barrage of incident radiation from all angles: up, down, right, left. And because it absorbs light more efficiently than any other aerosol, it converts all that incident light to heat faster than if it were just suspended in the air.

During late winter and early spring, when temperatures historically would still be below freezing in the lower reaches of a glacier or snowfield, black carbon can supply a decisive push toward the warm end of the scale. As snow melts, the concentration of black carbon on the surface increases, and these effects become even more potent, spurring additional melting. As vegetation and rock surfaces (with much lower albedos than snow) are exposed as a result of this melting, they further increase warming in the immediate area, driving melting of the remaining snow even faster. (This effect is less common on alpine glaciers or ice sheets, but more powerful on seasonal snowpack, where land is exposed beneath.) Researchers at the Pacific Northwest National Laboratory have conducted simulations that suggest black carbon reduces the reflectivity of snow on the Tibetan Plateau by 4 to 6 percent during the crucial spring period. It may not sound like much, but this change has a huge impact, warming the average surface air temperature by 1 degree Celsius. And all the while, black carbon suspended in the atmosphere soaks up light and heats the air directly over snowfields and glaciers, opening up yet another front of thermal assault.

All these various mechanisms add up to an effect somewhat akin to pointing a 1,500-watt hair dryer on a tray of ice cubes. The snow and ice don't have much of a chance.

These dark particles have an exquisitely diabolical sense of timing, too. Black carbon experts Mark Flanner and Charles Zender, and their colleagues have noted that "slight changes in solar absorption can alter snowmelt timing, and snow spatial coverage is tightly coupled to climate through snow albedo feedback." This makes black carbon a scale-tipper, a kind of airborne hair trigger for the world's frozen places. In spring, the snowpack coating glaciers and rock in the high mountains is already near its melt point, at its maximum extent of area at the end of the winter—meaning impurities in the snow are at their maximum, too—and the amount of solar energy reaching the snow surface is also at its maximum. At this moment, the snowpack is like a skier poised at the top of a steep slope, and black carbon gives it a decisive push.

Black carbon has thus been busily expanding its empire of melt both vertically—snow lines climbing up slopes and sooty particles burrowing down into snowpacks—and horizontally, as particles from the Gangetic Plain head north up into the highest Himalayan valleys, and residues from fires lit every day all over the Northern Hemisphere penetrate the high Arctic. It is quite literally the ascendance of the forces of darkness over the forces of light. John von Neumann, the polymath mathematician and strategic thinker who was said in his day to possess "the world's greatest mind," wrote in 1955 about the possibilities and perils of tinkering with complex earth systems on such vast scales.

> It is known that the persistence of large ice fields is due to the fact that ice both reflects sunlight energy and radiates away terrestrial energy at an even higher rate than ordinary soil. Microscopic layers of colored matter spread on an icy surface, or in the atmosphere above one, could inhibit the reflection-radiation process, melt the ice, and change the local climate. Measures that would effect such changes are technically possible. . . . The main difficulty lies in predicting in detail the effects of any such drastic intervention. But our knowledge of the dynamics and the controlling processes in the atmosphere is rapidly approaching a level that would permit such prediction. Probably intervention in atmospheric and climatic matters will come in a few decades, and will unfold on a scale difficult to imagine at present. . . . There is no need to detail what such things would mean to agriculture or, indeed, to all phases of human, animal, and plant ecology. What power over our environment, over all nature, is implied!

What power? What would happen, say, if the Chinese soot-on-snow deposition experiment were conducted for a century or more, on a global scale? How would the conversion of all that sunlight to heat affect the frozen water towers of Asia that feed the rivers on which a third of humanity depends? Or the Sierra Nevada snowpack that irrigates half of all U.S.-grown produce? Or the Greenland ice sheet, which contains 23.6 feet of potential sea level rise?

It turns out that we are in the middle of just such a natural experiment, being conducted across the snow- and ice-covered regions of the globe. Astonishingly, concentrations of black carbon in some high-altitude valleys of the Himalaya are comparable to those in midsized cities. In one study, Veerabhadran Ramanathan found that, in the Mount Everest region between altitudes of 3,000 and 6,000 meters, the levels of black carbon are as high as those measured in traffic-choked downtown Los Angeles. An analysis of ice cores drilled on the East Rongbuk glacier, on Everest's northeast ridge, showed that black carbon concentrations tripled in the period 1975 to 2000, relative to the preceding 115 years.

What power is implied? We're now in the process of finding out.

The preliminary results suggest the answer is exactly what you might expect: a whole lot of snow and ice is turning to water. In some places faster than others.

It's late May, near the end of the climbing season in the Nepalese Himalaya. I am high above base camp, at around 18,000 feet, a bit lost, but exactly where I want to be.

Ahead of me rears Qomolangma, the peak known to Tibetans as "Goddess Mother of the Universe" and best known to the rest of us by the surname of a nineteenth-century British surveyor: Everest.

I gaze at the iconic summit pyramid and imagine that if I squint hard enough, I can see the figures of the mountaineers struggling up it at this very moment, bent over under the invisible burdens of their determination to reach the top of the world. But all I can see are the morning clouds wreathing the waists of impossible peaks all around me. Later I will learn that men have been dying up there this week, swept away by avalanche, hollowed out by exposure. But at the moment I have come here to witness the death of a glacier.

I scan the river of ice that pours down from the foot of Everest's monstrous north face, caked with dull brown debris, serrated by the sun. The Rongbuk glacier stretches out a thousand feet below me, terminating just above the pastel-tent mosaic of the climbers' base camp, where it turns into a stream that flows down onto the roof of the world, to be swallowed by the vastness of Tibet. The Rongbuk looks like a damn mess.

I can say this with confidence because I know what it used to look like, thanks to David Breashears. In 1921 the British mountaineer George Mallory (who disappeared during a summit attempt a few years later) took a photograph of the Rongbuk just across the valley from where I stand. In 2007, Breashears, a filmmaker, mountain guide, and five-time Everest summiter, took a photo from the very same vantage point. Viewed side by side, they tell a startling tale. Mallory's black-and-white image shows a solid white river of ice pinnacles snaking northward. The same glacier in Breashears' photo is much diminished. It looks like the snow piled at the edge of my driveway at the end of

March: dirty, pockmarked, sinking in on itself, sad, and beleaguered. Testament to a long winter now become a torrid spring day.

Breashears is a world-class mountaineer who's climbed and guided most of the world's 8,000-meter peaks. But the day he trekked to Mallory's vantage point, studied the scene before him, and then compared it with the black-and-white photo in his hand, he took on a new mission. Breashears now travels with a team of veteran Sherpas, all of whom have made multiple Everest ascents and possess the technical skills to get to places most scientists would find impossible to reach. They travel fast and light, not with a summit as their goal, but some very specific locations, offering an unparalleled view of the past, and of a possible future.

Breashears runs a nonprofit called GlacierWorks, which documents the dramatic retreat of glaciers like Rongbuk throughout Nepal and Tibet over the past century. On every trip, Breashears and his team follow a logistically maddening method: they locate spots where early explorers such as Mallory made a photographic record of their journeys, in terrain that has often shifted dramatically under decades of Himalayan weathering. Then they take their own photos, recording with shocking clarity the extent to which these glaciers have diminished in the course of the twentieth century. In the absence of extensive, ground-truthed data about some of these remote, high-altitude glacial systems, the snapshots in time assembled by Breashears provide compelling anecdotal evidence of glacial retreat—at varying altitudes and aspects and in different climate regimes—in the eastern Himalaya.

"I went off and took a picture of the main Rongbuk glacier from the exact same spot as George Mallory took his picture in July 1921. I didn't even need to have the image printed," Breashears told me later. "There it was. It was very surprising, astonishing to see. I've been traveling to that region since 1981. One doesn't notice the changes when you're there every year. The change from 1921 to 2007 was really surprising, to see the amount of devastation and vertical ablation of the glacier, and the formation of supraglacial lakes near the terminus."

Breashears has noticed some other dramatic changes since his first trips in the region, in the air over the Khumbu Valley. "Now we're starting to see dust and black carbon aerosols brought into the mix," he told me. "[People in the region] are burning wood and kerosene because they're poor. Farmers burn huge piles of debris from their fields at the end of season.

"Since I first went to Nepal in 1979, you know, the soot now—we see this haze higher and higher. Sometimes I've been in Namche and had trouble getting a clear shot over to Tengboche Monastery," just a few miles away.

Breashears and his team sometimes use helicopters to do aerial mapping of certain areas. "We've had trouble in the dry season in winter. It's sometimes very hard at lower elevations, 14,000 feet, to navigate the helicopter. There are times when you're in a thick soup, and the pilots have to be careful. It's probably like being in Los Angeles in the 1970s. It's just incredible. You just can't see some few hundred yards or a quarter mile.

"There are images of this faraway land, in a lot of people's minds a very exotic place, the Himalayas, the seeming abode of the snows, seemingly immutable to change. It's fixed in our imagination forever as these great cathedrals, these ice and snow structures, with massive glaciers at their base." But Breashears has seen firsthand just how vulnerable to change these monoliths really are.

Measurements back up the photos' sobering tale. Since Mallory's time, the Rongbuk glacier has lost more than 330 vertical feet and retreated more than half a mile; it retreated 558 feet between 1966 and 1997. And it's not just the Rongbuk that is in trouble. In the past fifty years, glaciers on and around Mount Everest have decreased in area by 13 percent, and their termini have retreated by an average of 400 meters. Small glaciers are disappearing much faster: those less than one square kilometer have lost 43 percent of their area over the same time period. Meanwhile the average snow line on Everest itself has climbed 180 meters up the mountain since 1962.

Now, four years after Breashears' visit, I compare my front-row view of the Rongbuk to those two images. The glacier looks even more besieged now than it does in his 2007 portrait. The surface is dotted with luminous but telling turquoise pools; lonely ice pinnacles have been left stranded far from the parent glacier by its steady retreat back toward the mountain. Where its debris-covered surface parts to reveal fissures of white beneath, the snow is darkly streaked and undercut, as though a herd of thirsty creatures has been patiently chewing on the base of each icy hummock. The dark spots are soil, mud, stone—not soot itself, but possible testament to the battle it has waged with now vanished, vanquished layers of snow.

A rock lubricated by the morning sun—or is that calving ice?—crashes down toward the mud-colored lake, which spreads like spilled coffee from the Rongbuk's snout, in the direction of the base camp. I take my own photo of the besieged glacier, then turn to go. (It occurs to me that the People's Liberation Army, which staffs the checkpoint before the base camp, might not appreciate my unauthorized inspection. Mountaineers pay as much as $40,000 in fees to get this far; I paid the standard $30 tourist's fee at the entrance to Sagarmatha National Park and just wandered blithely up the trail.)

As I throw my pack over my shoulder I take one more glance at Qomolangma. The sun has broken through the clouds to reveal a teasing edge, around the corner of a ridge just ahead of me, of the East Rongbuk glacier, which spills down from the mountain's northeast ridge to collide in slow motion with the Main Rongbuk. (The East Rongbuk retreated 270 meters between 1966 and 1997.) And then I have a sudden vision—of a denuded Everest, of the Goddess Mother of the Universe waking up one spring morning in the not-too-distant future to find herself presiding over a pyramid of dark, naked rock, and below it a deep, trenchlike valley, not unlike certain parts of Zanskar, lined with ghostly moraines and house-sized boulders where a magnificent river of ice had once flowed.

The top of the world laid bare.

How much of the Rongbuk's swift decline is due to black carbon? It may not be possible to estimate an exact proportion of responsibility, but scientists have been busy looking deep inside glaciers for clues. Susan Kaspari of Central Washington University and her colleagues trekked in 2002 up the East Rongbuk glacier, a few miles past the point where I turned around, and drilled a 102-meter-deep core, containing layers of ice dating from 1860 to 2000. Then they analyzed the old soot trapped in those layers using a single-particle soot photometer, which uses a laser to incandesce and measure the mass of the individual black carbon particles in the soot.

The ice core held an invaluable record of climate and air quality dating back to the start of the second (steel-producing and railroad-building) wave of the Industrial Revolution. And its buried story hints at the hidden forces behind the transformations Breashears has witnessed and documented.

Kaspari and her colleagues found that black carbon concentrations in the period from 1975 to 2000 increased threefold relative to the 1860–1975 period, "indicating that BC from anthropogenic sources is being transported to high elevation regions of the Himalaya." They speculated that the black carbon buried in the East Rongbuk's belly comes from as far away as eastern Europe and the Middle East, to mix with particles from its own neighborhood, from the Gangetic Plain and mountain valleys of Nepal. The deposition on the glacier peaked during winter and spring, suggesting a link to the inefficient brick kilns in Kathmandu Valley that burn coal and biomass at full tilt from December through April, and to biomass emissions from the rest of South Asia, which also peak in April and May.

Chinese scientist Xu Baiqing and his team extracted their own ice core from the East Rongbuk glacier (along with four others across the Tibetan Plateau) and analyzed the results with NASA's Jim Hansen in 2009. They found evidence of increasing black carbon concentrations on all of these glaciers, enough to significantly reduce their surface reflectivity, noting that "the black soot amount has increased rapidly since the 1990s, coincidentally with the accelerating glacier retreat and increasing industrial activity in South and East Asia." They further found that "the spatial variability of glacier retreat is also consistent with a role for black soot, because the most rapid glacier retreat and the highest black soot concentrations are located around the margin of the plateau."

In other words, echoing Kaspari's finding, not all glaciers in the region behave the same, but where there's fast melting, you're very likely to find lots of black carbon. Its smudged, incriminating fingerprints are there if you care to look—and drill and dig—closely enough. Even the highest point on the globe bears these telltale traces of upwind burning: particles that rise from millions of cooking fires, brick kilns, diesel exhaust pipes, and coal-fired boilers and power plants across Asia and beyond.

If you travel through the region, it doesn't take long to sample all the sources of this haze. A farmer in Punjab burns the straw left over from his rice harvest in the field. A gaudily painted truck loaded with cylinders of propane destined for Kathmandu rumbles up a steep incline in central Nepal, coughing out puffs of dense black smoke. On the road to Sitapur outside of Lucknow, seasonal laborers fire a kiln loaded with coal chunks and freshly molded bricks, and its smokestack sends forth billions of black carbon chains into a hazy morning sky. More dark materials stream out of millions of stovepipes like the one jutting through the roof of Stobdan's house and millions more from the tailpipes of vehicles from New Delhi's Ring Road to Calcutta's Park Street.

Some of these particles settle out just a few feet from their point of origin, landing harmlessly on the already-dark surface of the bare ground. But some rise up and up in plumes several miles high to merge into the vast smog bank that squats over northern India every spring—the atmospheric brown cloud made up of sulfates, dust, lighter-colored organic carbon aerosols such as brown carbon, and huge amounts of black carbon.

Meanwhile invisible convection currents rise off the Gangetic Plain and climb up the flanks of the Himalaya and onto the Tibetan Plateau. The temperature difference formed by these mountain heights helps drive the Indian monsoon: as hot air rises over the plateau, warm moisture-laden air from the ocean is drawn in over the subcontinent to replace it. These warm winds grab the sooty products from all these fires, lift them up more than five miles above the earth, and carry them hundreds or even thousands of miles away, during the one to two weeks, on average, that they remain aloft.

Like a vast army of tiny braziers deployed across the roof of the world, the tiny particles intercept the light that would otherwise be reflected back into space and radiate heat into the air over the glaciers, into the snow coating the mountains, extending the melting season earlier and earlier, causing spring to nibble away at winter's edge.

As it turns out, and as the man in Stobdan's kitchen had speculated, all that lofted soot functions very much like a "solar house on the snow."

"It would be such a catastrophic, shameful thing to do to melt away Mount Everest," Veerabhadran Ramanathan, an atmospheric scientist, told me when I met him in Lucknow, in the heart of the brown cloud. "The retreat is very clear in the Rongbuk. That's why, if you clean up the Indo-Gangetic Plain, and from the Chinese side, the Tibetan Plateau . . ."

"You think that's possible?" I asked, a bit incredulous.

"I think so."

It's impossible, of course, to attribute direct causality, with 100 percent confidence, to any single factor in a system as complex as the Himalayan climate. There will always be some measure of uncertainty. Instead, science deals in probabilities, and its practitioners seek to "bound," or constrain, those uncertainties, providing not ironclad proof but enough information to guide our decisions in a complex, ever-changing world.

Unless and until someone installs some air-sampling equipment on the summit of Sultan Largo, or on Mount Everest, pinning Kumik's plight or the Rongbuk's accelerating demise on black carbon will be a matter of some conjecture. "No work has been done on [black carbon] in this part of the Himalaya, in Kashmir," the glaciologist Shakeel Romshoo told me. "I think definitely there is a need to segregate the contribution of black carbon [versus the] contribution of global warming at the local level to the melting of glaciers. I think there is a need for some integrated research."

Efforts are under way to more precisely differentiate the effects of black carbon emitted in a given region from the effects of globally averaged warming caused by long-lived greenhouse gases like carbon dioxide. On top of these initiatives, there are other factors requiring deeper study: microclimates, shifting precipitation patterns, dust deposition, cloud cover dynamics. But many scientists told me that Kumik's situation is representative of trends observed across the Himalaya and the Tibetan Plateau. The data coming in from a wide range of ice-core, air-monitoring, and climate-modeling studies all seem to point an accusing finger at the soot we're sending skyward every day—that civilizational by-product that we've been more or less tolerating or ignoring since humans huddled around the first hearth.

One moment in particular drove this message home. In March 2010, still in the thick of my hunt for Kumik's destroyer, I sat for three days in a hotel ballroom in Chapel Hill, North Carolina, full of atmospheric scientists, combustion experts, air pollution researchers, and government officials from a variety of agencies, at a conference hosted by the Environmental Protection Agency on "short-lived climate forcers" (pollutants like soot and methane and ozone that stay in the atmosphere on much shorter time frames than carbon dioxide but have a much greater warming impact, molecule for molecule). For three days, I listened as eminent scientists presented their findings—often the result of painstaking modeling efforts or multiyear empirical studies—about the impacts of black carbon, methane, and other SLCFs on human health, the climate, and the economy. These talks were rife with caveats. Deep uncertainty plagues research on climate change impacts, and serious scientists are understandably obsessed with quantifying and accounting for this uncertainty in their analyses, projections, forecasts, and conclusions. Thus, many of the discussions focused on the complex emerging story around black carbon and its fellow products of incomplete combustion: Sulfates and organic carbon co-emitted with black carbon have a cooling impact on climate. Their impacts on cloud formation are difficult to model and poorly understood. Chemical species coating black carbon change the resulting particles' optical properties . . . and on and on.

So at the end of the third day, during the session entitled "Where Do We Go From Here? Making Good Policy Choices in the Face of Current Scientific Certainties, Uncertainties, and Gaps," when the conference organizers invited task force leaders to summarize conclusions from the dozens of breakout sessions," I was fully prepared for a muddied, hedged picture to emerge. Along the lines of: "More study is needed before we can confidently say X, Y, or Z."

One of the organizers took the podium and said, "So we've made a lot of progress in identifying the areas that require further research. What can we say with confidence about SLCF impacts and mitigation opportunities?"

Then something totally unexpected happened. In a room full of professionally cautious people, as close to an ironclad conclusion as you're likely to get emerged: "It's safe to conclude that acting on black carbon in areas near snow and ice, like the Himalayas and the Arctic, is a clear winner. One thing we can say with very high confidence is that mitigating black carbon is worth doing now, for health reasons, and especially near snow and ice."

Which is a scientist's way of saying "It's a no-brainer."

Soot. Who knew?

It seemed relatively harmless, a minor irritant, a small price to pay for the fire's warmth and light: a hand wave in front of your nose, some dirty linens. As Bert the chimney sweep crowed in the film *Mary Poppins*: "It's just good clean soot, Michael!"

More than a handful of journal articles on black carbon ironically cite that cheerful line from the smiling, soot-smeared Bert. But most neglect to mention some other implausible details in that movie. For one, Bert the chimney sweep and his colleagues dancing on those London rooftops would have had a high probability of getting sick from their chronic soot exposure. English chimney sweeps in the eighteenth century wore loose-fitting clothing that let soot particles invade and coat their skin. The soot mixed with their sweat and ran down into crevices, where the tars and fine particles and polyaromatic hydrocarbons it contained festered. Many chimney sweeps developed scrotal cancer later in life ("soot wart," they called it). Sometimes they'd cut it out themselves with a knife. You know, just good, clean soot.

John Evelyn seems to have been one of the earliest observers to properly size up this microscopic dark adversary. In 1661 he wrote a pamphlet titled *Fumifugium: The Inconvenience of the Aer and Smoake of London Dissipated: Together with Some Remedies Humbly Proposed by J. E. Esq. to His Sacred Majestie, and to the Parliament Now*

Assembled. It was basically one long effort to persuade King Charles II of the myriad dangers of soot—much of it produced by brewers, lime burners, and soap boilers who burned dirty "sea coal" day and night—then flying around London. The treatise is an extended fit of pique:

> That this Glorious and Antient City . . . which commands the Proud Ocean to the Indies, and reaches to the farthest Antipodes, should wrap her stately head in Clouds of Smoake and Sulphur, so full of Stink and Darkness, I deplore with just indignation.

A few hundred years ahead of his time, Evelyn describes how the foul air "violates the Larynx and Epiglottis" and penetrates deep into the lungs, even mixing with the blood and then traveling through the entire body. At times he quivers with outrage at this black pall all around him, and inside him, ascribing almost malevolent intent to it:

> It is this horrid Smoake which . . . corrupts the Waters, so as the very Rain, and refreshing Dews which fall in the several Seasons, precipitate this impure vapour, which, with its black and tenacious quality, spots and contaminates whatsoever is expos'd to it. It is this which scatters and strews about those black and smutty *Atomes* upon all things where it comes, insinuating itself into our very secret *Cabinets*, and most precious *Repositories*.

Evelyn's call to arms would prove to be more prescient than he could possibly know. Even 350 years later his warning still hasn't quite sunk in.

He got a few things wrong, though. Evelyn thought planting more trees around the city would go a long way toward solving the problem, and he suggested that Londoners should burn more wood, instead of coal, arguing that wood smoke would be easier on their lungs. But in the words of household air pollution expert Kirk Smith, we now know that wood and biomass fires are, just like burning coal, a "toxic waste factory."

• • •

As plans for solar houses were taking shape that winter afternoon in Stobdan's kitchen, as millions of dark plumes from hearth fires all over Asia were being borne up and across the world's highest mountains, a parallel drama was taking place: Zangmo tended her fire. She dipped a piece of dried yak dung into a plastic bottle of kerosene, placed it in the stove's firebox on a bed of embers, and leaned in to blow on the tentative flames. She turned her head and shut her eyes tight, and dark particles surged outward and up, like wild creatures uncaged.

They curled and billowed around Zangmo's red handkerchief-covered head, like starling flocks in slow motion, through shafts of sunlight admitted by the room's only window. Inky, sweetly pungent, they filled our nostrils, watered our eyes, coated our tongues. Some particles seemed to search for chinks in the stone and mud brick wall, gaps in the willow-stick ceiling—like suspects fleeing the scene of the crime. Some pooled under the ceiling beams, where they obscured the frayed poster taped to the wall above the stove, the incongruous one of a cruise ship sailing on a sun-kissed ocean with the caption: "A Steady Approach to New Horizons."

Inside the stove's firebox, a molecular tug-of-war was taking place. The fire flared and then, after some time, cooled again. As the temperature dipped, both in the room and in the stove's combustion chamber, the draft weakened, so less fresh air flowed in and over the fuel. Deprived of enough oxygen and warm temperatures to sustain a vigorous reaction, the fire sagged.

In ideal combustion, mixing a carbon-based fuel with oxygen in the presence of sufficient heat would spark a reaction that looks like this:

$$CxHy + O_2 \rightarrow CO_2 + H_2O$$

Like the oxidation process of rusting metal, this reaction involves the combination of oxygen with some material—but it takes place much faster, releasing great bursts of energy. We put this thermal energy to all kind of uses, from making steam to drive a power plant's turbine to warming a *yokhang* on a bitterly cold January day. Meanwhile, the mass in the fuel and the air are transformed into gaseous carbon dioxide and water vapor, which escape, invisible, up the chimney, into the atmosphere.

But there is no such thing as ideal combustion, anywhere. And it's fair to describe conditions in Stobdan's kitchen as suboptimal. The thin high-altitude air of Kumik, streaming in around the room's leaky windows and the stove's door, the low-energy density and impurities of the yak dung, the embers that piled up around it and covered the dung's surface, choking off access to already scarce oxygen—all these factors conspired to ensure that Zangmo's fire converted fewer of the dung's carbon atoms into carbon dioxide (which, if liberated from renewably grown and sustainably harvested biomass fuels, is harmless to humans and the world at large) and much more into a host of other stuff. Thus the actual fire in Zangmo's stove looked something like this:

$$\text{Fuel (C, H, N, S, Cl, K, etc.)} + \text{air } (N_2 + O_2) \rightarrow CO_2 + H_2O + (CO, CH_4, NOx, SOx, VOCs, K_2SO_4, KCl, HCl, PM) + \text{ash} + \text{other tars} + \ldots$$

In other words, a nasty business. Of the carbon locked in the dung's flaky, dry matrix, one out of every five atoms escaped to the air around us in some toxic avatar, the "other stuff" between the parentheses.

And what is that other stuff? Scientists are still cataloging all the exotic substances found in biomass smoke. But a partial list, depending on the fuel type and environmental conditions of a given fire, would include:

- Toxic polyaromatic hydrocarbons (PAHs) such as hexane, benzene, and styrene
- Organics such as formaldehyde, methanol, methylene chloride, and dioxin
- Carbon monoxide, a toxic, odorless, sometimes deadly gas
- Other trace gases such as nitrogen oxides that are precursors to the formation of ozone, which, at ground level is a major respiratory health threat to humans and causes significant damage to agricultural crops

- Tars and other volatile organic compounds, which also contribute to ozone formation, and are dangerous to human health in their own right
- And last but far from least, exceedingly small solids called particulate matter, the "PM" in the equation above

Of all that "other stuff," experts regard these ultra-fine particles as the most dangerous. Very small particles (often categorized as PM2.5, for particulate matter under 2.5 microns in diameter) can include PAHs, sulfates, and organic carbon. The amount produced varies from fire to fire, depending on the type of fuel burned and a wide range of other factors. For example, fossil fuel burning in kerosene lamps and old diesel engines produces a higher ratio of black carbon to lighter-colored organic carbon and sulfates than does burning biomass.

But even in Zangmo's flaming dung-and-twig fire, a sizable fraction of the PM2.5 streaming out was our oldest nemesis and companion, black carbon.

Soot signals a lost opportunity. As the *unwanted* by-product of combustion, it is, essentially, waste. The quantity of soot produced by a given fire is thus a rough measure of its inefficiency.

But soot has had its uses down through the centuries. Its darkness (which is to say, its black carbon content) made it a valuable tool for communication, for making signs and images. It was likely our earliest medium of artistic expression. Prehistoric painters working in the caves of Arnhem Land, in Australia, used soot to make haunting images of giant birdlike creatures, long since extinct. The mysterious artists working in the caves of Chauvet, France, 30,000 years ago mixed it with mineral pigments to create the impression of shadow on the manes of a row of muscular horses, in what is probably the earliest known use of perspective. Five thousand years ago, the peoples of ancient China burned bones and resinous wood and pine pitch, or collected the velvety residue we now call "lampblack" from the containers that held the wet end of the wick in shallow, oil-burning lamps. They mixed this tarry substance with water and gum and fashioned it into sticks that could be easily carried and traded. Later, because the soot used to make this mixture often came from India (or perhaps because it was traded to Europe via India) it became famous around the world as "India ink," highly prized because of its stability and durability. Colonial Americans used soot to make house paint. Today we still use lampblack (or "carbon black") to glaze ceramics, in printer cartridges and copy toner, in crayons and artists' ink and carbon paper.

Every culture has a name for it. The ancient Romans called it *fuligo* and ate their meals in the *atrium*, a partially roofed courtyard that got its name from the smoke (*ater* in Latin means "gloomy black") given off by open fires for cooking and heating. (Ironic, no, that the atrium, which we now associate with an open light-filled space, has its roots in the color of soot.) And today, Italians call it *fuliggine,* and Brazilians, after the fires ravage their savannahs every summer, call the dark coating left on the buriti palms *fuligem*. Deep in the Amazon, similar-looking stuff still lies buried in a layer that archeologists and soil chemists call *terra preta*, "dark earth," the legacy of a long-lost, prosperous civilization that enriched its soils with charred vegetation, spawning the legend of El Dorado.

Zanskaris call it *shremok*. In another form, they call it *tutpa chakse*, to describe the version that condenses out with water vapor and deposits as *tutsi*, that oily glistening black coating that coats the poplar ceiling beams, and which the people value because of its waterproofing properties. They used to draw patterns with soot on a baby's forehead to mimic the look of *lhande*, a "demon," to scare away evil forces, illness, and jealousy; they still put a dot of *shremok*—called *naktsik*—on the foreheads of newborn babies to ward off the evil eye. Just as in the woods of my own home state of Vermont, in the years before the Revolutionary War, Ethan Allen and his Green Mountain Boys smeared it on their faces to scare off visiting interlopers from New York, surveying disputed lands.

For all these applications, the primary value lay in its darkness and its durability. Ever try to rub soot off your hands without soap and water? The stuff is tenacious. Just imagine, then, how it clings to the inside of your lungs, to your "most precious *Repositories*."

• • •

As we talked on into the evening in Stobdan's kitchen, sipping tea, sipping *chang*, sketching plans for solar houses, all the while smoke continued to pool in the room's hidden eddies of air. There was no aethalometer—a filter-based measuring device whose name comes from *aethalos*, ancient Greek for "blackened with soot"—to track the concentration of black carbon in the kitchen, its spikes and troughs, throughout our meeting. But based on past studies of other north Indian kitchens, some plausible guesses can be made.

A typical wood fire emits about 400 cigarettes' worth of smoke per hour. Dung fires are even dirtier. Dung fires at high altitude, even dirtier still. As Zangmo sat near the fire, she had a higher exposure than the rest of us. With each breath she drew in millions of fine particles. She may have been exposed to an average concentration as high as 400 micrograms per cubic meter of those tiny particles throughout the day—more than forty times the limit recommended by the World Health Organization. And while she was starting the fire in the stove, this number probably spiked higher than 1,000 micrograms. Zangmo is not a smoker. (Stobdan used to smoke, but thanks to Zangmo's persistent efforts he recently quit.) But simply by cooking and tending this stove each winter day she inhales roughly the same amount of smoke as she would by smoking two packs of cigarettes a day.

The rest of us, seated farther from the stove, had more limited exposure to those fine particles—though still, safe to say, far above the thresholds set by most environmental regulators in the industrialized world. So the men of Kumik talked on, oblivious to the marauders in their midst. Most of the particles that rushed in with each breath encountered our bodies' evolved defenses: they were grabbed by the mucous membranes lining our nose and sinuses. The black carbon and other smaller particles penetrated deeper.

"The interactions of particles in the respiratory system is very complicated," Patrick Kinney, a professor of environmental health sciences at Columbia University, would tell me later. He's been studying the health impacts of various kinds of air pollution for decades. "It depends on particle size." The smallest particles, 0.1 microns and smaller, are more likely to get deep inside and deposit on the lung walls. "And that's probably where the black carbon would be; it's mostly in that small size. Once you inhale them, they certainly get into the lung, because they have no inertia to hit the nose or the upper airways . . . once it lands there, for one thing those are more sensitive places, because they're not as protected by mucus as the upper airways are. So you get damaged lung, but also there's this idea of penetrating through the epithelium and the particles getting into the bloodstream and going someplace else. The other effects we would worry about from particles, like cardiovascular death, probably have to do with particles going someplace else, causing atherosclerosis, in the arteries around the lung, near the heart."

So as we sat there around the stove, some particles traveled into our lungs, where, just as John Evelyn surmised, they anchored themselves in the bronchioles and alveoli, taking up precious surface area in those tiny air-filled sacs where oxygen is delivered to our blood. Some of the really small particles might have managed to reach that special place and moment where each breath culminates—the capillary beds that exchange carbon dioxide and oxygen between the air and our bloodstream. And from there a few particles might even have caught a ride throughout our vascular systems, to be parked permanently in our vital organs, and even our brains.

Most, however, came to rest in the cul-de-sacs of our lungs. There they sit patiently for the rest of our lives, like fifth columns, nanoscale sleeper cells, until they co-conspire with some other environmental input, infection, or immune condition to possibly trigger, inflame, or simply weaken our defenses against bronchitis, cardiovascular disease, pulmonary diseases, emphysema, cancer. (It's worth noting: two-thirds of women with lung cancer in India and China are not smokers.)

These same risks are run—in rough proportion to the duration and concentration of exposure (though, as with black carbon's damage to snow, the response is not always linear)—by people the world over who breathe in black carbon and other fine particles produced by incomplete combustion, whether they stream from a diesel engine, a burning pile of straw, a wood fire, or an old coal boiler. These are the high stakes of exposure to "good, clean soot."

In the hour that passed since Stobdan first filled our teacups, somewhere between 900 and 2,000 metric tons of black carbon were added to the global atmosphere around the world. Most of it escaped into the wild blue yonder. But based on a host of epidemiological studies, we can hazard some plausible (and likely conservative) estimates of the damage done by the fraction that humans inhaled, just sitting and breathing by the hearthside.

In those sixty minutes, those dark particles and their fellow byproducts of incomplete combustion killed at least 400 people around the world, including 57 children via pneumonia (and including at least 119 Chinese, 117 Indians, 53 sub-Saharan Africans, 10 Latin Americans, and 3 Russians).* These people succumbed to illnesses caused by exposure to household air pollution from burning wood, dung, or coal—the same fuels humans were burning, in much the same way, before we invented alphabets or agriculture. Most of them were women and children.

If you look at it under an electron microscope, you'll see that black carbon actually manifests as a chain or clump of spherules, which themselves are layers of graphite that form into a kind of hollow shell. The structure resembles "a bunch of dark grapes," says Tami Bond, a professor at the University of Illinois, Urbana-Champaign and one of the world's foremost experts on black carbon.

The carbon spherules themselves are extremely small, with diameters in the tens of nanometers. (A nanometer is one-billionth of a meter, and is used to measure things at the atomic scale; a water molecule is less than one nanometer wide.) During combustion, hundreds or thousands of these spherules quickly combine to form more durable chain structures.

Looking at these chains under the microscope, you can appreciate the two primary characteristics that make black carbon unique: they

*Household air pollution is not a major risk factor in Europe, the United States, and other "high income" countries—but *ambient* air pollution is. In that same hour, 19 Western Europeans and 12 Americans died due to exposure to carbonaceous and other particles suspended in outdoor air.

are very, very small—less than 2.5 microns in diameter, one-thirtieth the width of a human hair.

And they are very, very dark. Of all the things we pump into the atmosphere, these particles are the most efficient at absorbing sunlight, at all visible wavelengths, and turning it into heat. So efficient that "emitting one-third of an ounce to the atmosphere (about the weight of two nickels) is like adding a home furnace, running continuously, to the Earth system for one week," according to Bond. It also has a high surface area, which means that "one ounce of black carbon dispersed in the atmosphere blocks the amount of sunlight that would fall on a tennis court."

It is far outnumbered and out-massed in the atmosphere by carbon dioxide, methane, and nitrous oxide (after water vapor, the most abundant greenhouse gases in the atmosphere). But pound for pound, black carbon packs much more of a wallop, causing about 700 times the warming as the same mass of carbon dioxide staying in the atmosphere for 100 years. As such, it is utterly unlike anything else floating around in the air, from the Himalayan heights to the space around your backyard barbecue.

Black carbon has a few other remarkable properties: it is refractory, meaning that it keeps its form at very high temperatures; and it is insoluble in water and solvents, meaning it is extremely durable and stable (which is the reason india ink was so prized). The stuff is resilient. And black carbon always travels in packs. Like a Zanskari, you'll almost never find it alone. A rogue's gallery of other gases and particles and chemicals are co-emitted alongside black carbon from flaming combustion; these can piggyback onto black carbon chains to catch a ride deep into our lungs, and beyond.

All of these properties help explain the subtle note of alarm you can detect among the scientists who study it for a living. Black carbon is small and pervasive enough to penetrate our lungs and tissue, common enough to be found in significant amounts in both New York's Times Square and the Brahmaputra Valley of remote northeast India, and dark enough to be the likely single biggest factor driving the accelerated melting of the frozen water towers, from the Alps over a century ago to the roof of the world today.

A few months after my first visit to Kumik, I sat in a lecture hall in Berkeley and listened to public health expert Kirk Smith lay out a case for more research and action on black carbon and other short-lived climate forcers, such as methane. "Black carbon is like carbon dioxide on crack," he said to nervous laughter in the audience. "It's hundreds of times worse" in terms of the amount of energy it absorbs per unit of mass in the atmosphere. But his primary concern as a health researcher was the fact that it was partially responsible for millions of deaths from exposure to smoke from cooking fires alone every year.

Five years after Smith's talk, two exhaustive studies released just a few weeks apart confirmed his dire warning and took it even further.

The Global Burden of Disease Study published in *The Lancet* in December 2012 found that exposure to household air pollution, including black carbon and its co-emitted products of incomplete combustion, was the single biggest environmental health risk in the world, responsible for 4 million premature deaths each year—a toll greater than the numbers who die annually from HIV/AIDS, malaria, and tuberculosis combined. (The burden from exposure to outdoor air pollution was over 3.2 million premature deaths per year.)

And in an exhaustive multiyear study of all of the ways in which black carbon contributes to warming in the atmosphere, published in January 2013, Tami Bond and a couple of hundred colleagues concluded that black carbon is second only to carbon dioxide in its contribution to global atmospheric warming. Interestingly, the findings about the climate impacts made a much bigger news splash in the West than those about the enormous health damage.

The two top drivers of climate change differ in several important ways. First off, black carbon is not a gas: Tami Bond advises us to "think of really small, dark rocks." Carbon dioxide, though so dangerous to human health over the long term—acidifying oceans, spurring heat waves, spreading disease vectors, to name just a few of its many impacts—is harmless when breathed in directly. Humans can safely breathe carbon dioxide up to a concentration of 30,000 parts per million (ppm). We're currently at 400 ppm, a terrifying number from the perspective of climate scientists, but pretty innocuous to a toxicologist.

Not so black carbon. As an ultra-fine component of particulate matter, it is a potent health threat at even low concentrations. According to the World Health Organization, "there is no evidence of a safe level of exposure" to fine particulate matter, "or a threshold below which no adverse health effects occur." The upshot: you really don't want to breathe *any* amount of soot. (Recent research suggests that combustion products are more dangerous than other types of particles, and that black carbon might therefore be a more relevant metric for adverse health effects than either PM10 or PM2.5: a 2011 study found that reducing one microgram per cubic meter of black carbon increased life expectancy by four to nine times as much as reducing the same unit of PM2.5.)

In terms of effects on climate, while carbon dioxide lets light through and then traps and absorbs outgoing infrared radiation, black carbon absorbs incident light directly and transforms it into heat. (It also affects cloud formation; biomass smoke in particular, rich in light-scattering sulfates and organic carbon, has a complex effect on clouds, which also scatter light. More on that later.)

Most important, whereas carbon dioxide is mixed evenly in the global atmosphere, black carbon stays relatively close to home. Whereas carbon dioxide is quite long-lived in the atmosphere, black carbon stays aloft for just one to two weeks, on average. Then it's gone, washed out, blown onto the ground or the snow or the foamy sea.

This short lifespan means black carbon has an outsized impact in its own neighborhood, near where it's emitted. It can alter environmental systems on regional scales, as it did the Alpine glaciers in the Little Ice Age paradox. To take another example, black carbon emitted within the Arctic is five times more effective in warming that region than black carbon that is swept in from lower latitudes (black carbon from temperate regions is transported at a higher altitude, where air is more stable, so there is less deposition). Likewise, black carbon that originates on the Gangetic Plain poses a greater threat to Himalayan snow and ice than do diesel emissions streaming eastward from Egypt.

As a result, there are hotspots around the world, places where black carbon is found in greater concentrations, because incomplete combustion happens there more frequently and with greater intensity: The Indo-Gangetic Plain; eastern China; sub-Saharan Africa; parts of Central America, Brazil, and Peru; and Indonesia. Three billion people live in these hotspots.

Because of these outsized effects in certain regions, because of its short lifespan aloft, and because of its potency in driving feedback mechanisms, the standard metric for understanding the climate-changing power of a given pollutant—its global warming potential, or GWP—doesn't fully capture the range of ways that black carbon contributes to global warming. As such, GWP can't fully communicate the range of risks black carbon poses to Earth and climate systems in certain regions, from the Arctic to the South Asian monsoon, and thereby, the risks it poses to human civilization. It's a wild card—the straw that stirs the drink. "We need a new potential warming metric for black carbon," Kirk Smith told me, "something like 'climate disruption potential!' "

Measured over short time horizons, black carbon has a terrifyingly high global warming potential. Testifying on black carbon and climate change before Congress, Bond stated, "One kilogram of black carbon absorbs about 500–700 times more energy in 100 years than 1 kilogram of CO_2, when only direct interaction with sunlight is considered." (And 2,000 times more energy over a twenty-year time frame). Measured over 1,000 years, though, its GWP seems risibly low compared to carbon dioxide. In his book *The Long Thaw*, geophysical scientist David Archer offers a vivid way to understand the far-reaching impact of a molecule of carbon dioxide. When a gallon of gasoline is burned, he writes:

> It yields about 2500 kilocalories of energy, but this is just the beginning. Its carbon is released as CO_2 to the atmosphere, trapping Earth's radiant energy by absorbing infrared radiation. About three-quarters of the CO_2 will go away in a few centuries, but the rest will remain in the atmosphere for thousands of years. If we add up the total amount of energy trapped by the CO_2 from the gallon of gas over its atmospheric lifetime, we find that our gallon of gasoline ultimately traps one hundred billion

(100,000,000,000) kilocalories of useless and unwanted greenhouse heat. The bad energy from burning that gallon ultimately outweighs the good energy by a factor of about 40 million. The enormous world-altering potential of that gallon of gasoline has taken the reins of Earth's climate away from its natural stabilizing feedback systems, and given them to us. May we use our newfound powers wisely.

So picture the human race, here in the Anthropocene, gripping *two* strong sets of reins. One is carbon dioxide and the other is short-lived pollutants like black carbon. If we shake those reins, the carbon dioxide will transmit that energy in smaller, steadier amounts, which will aggregate into a staggering total over centuries to come. Black carbon will transmit a short but much more intense burst, almost immediately.

Because both have a huge disruptive effect on the climate and other natural systems on which humanity depends for survival, if we want to keep them from spiraling out of control (and we *do*, don't we?), we can't afford to ignore either one.

At some point today, you encountered a flame. You lit a stove to prepare a meal. You turned the key in the ignition of your car and felt it rumble to life. Perhaps you covered your nose as you walked past a generator rumbling on a construction site, or wrinkled it as you drove past a belching smokestack, or got caught in traffic behind an idling diesel truck. If you live in colder climes and went for an afternoon walk, you might have caught a whiff of wood smoke snaking from your neighbor's chimney.

Even if you couldn't directly hear, smell, taste, or see them, countless fires have been part of your day. The light from your bedside lamp, the heat circulating from your furnace, the current charging your phone, the motion of the train, bus, car, or ferry that got you to your destination, somewhere at the origin of each of those energy transformations, a fuel—coal, natural gas, gasoline, diesel, wood, or even agricultural waste—was being burned to make it happen. Of those fires you could sense directly, some (the noxious emissions of the generator) irritated, some (the Proustian tang of wood smoke, perhaps recalling childhood camping trips) enticed, and some (the blue jets of the gas stove) maybe didn't stir you at all. Regardless, the same basic phenomenon was driving the turbine, the piston, the draft of hot air, or the nostalgia: combustion.

As each day dawns these fires are lit anew. From them we derive our modern mastery of time and space, as well as our basic survival. There are now well over a billion hearths, where we do our cooking and heating and coming together. The internal combustion engine has expanded and quickened our reach into the farthest corners of the world. We raid the bowels of the earth to stoke blast furnaces, to transform limestone into cement and iron ore into steel. Our modern forges—where we make our electricity, steel, bricks, cement, ceaselessly assembling the vascular system, flesh, and bones of our energy-intensive civilization—are as common and familiar features in the twenty-first-century landscape as rivers and forests. Now we wield computer models of flame dynamics and craft jet engines to withstand fantastically high temperatures. We can turn wood, coal, and even our own waste into gas that becomes heat and light and motion and the thrumming binary dreams of supercomputers.

We light all these fires for largely the same reasons our ancestors first did: burning stuff gives us heat and light. These are the desirable products of combustion, harnessed to cook food, warm spaces, get people and goods from point A to point B, and dispose of unwanted things—needs and desires as universal as fire is ubiquitous. After half a million years or more of practice, now 7 billion strong, we have built civilizations on a foundation of fire. Now the whole world is burning.

A woman in Langtang, Nepal, loads chunks of split pine into the mouth of a boxlike mud stove to bake fresh chapatis for the morning meal. A hundred miles north, a diesel truck bolted together in a roadside shop rattles its way from Lhasa to Shigatse on the Tibetan Plateau. On the other side of the world, a diesel-powered bus idles near the George Washington Bridge bus station in Washington Heights, and a few miles south, boilers full of sludge-filled No. 6 heating oil shudder to life in the basement of an apartment building at 353 East 53rd Street. A farmer in Bhutan's Punakha Valley sets alight a pile of rice straw in his fields. A woman walks past a pile of burning garbage on the outskirts of Nairobi. A father in Sitapur on the Gangetic Plain in north India puts a match to a wick in a kerosene lamp as evening falls, so his children can study. Two hundred miles away the power goes out in a crowded corner of New Delhi, and the owner of a small clothing shop in Paharganj cranks a generator to keep the lights on. A team of laborers shield their faces against the shimmering heat as they load more coal into a brick kiln outside Kathmandu, while a family in Yunnan District in southwest China sips green tea around an open fire leaping through an iron grate. A ship approaches the Port of Oakland, laden with goods from China, and up above in the Berkeley Hills, someone lights a fire in a woodstove at the start of a dinner party. Far away a ship plying the waters of the North Atlantic carries a load of crude oil between Norway and Japan, one of the first to make the newly ice-free crossing, flames churning invisibly in its metal bowels, burning oil to bring oil for more burning.

Most of us can relate to Captain Ramsay, the ornery, old-school nuclear submarine commander in the movie *Crimson Tide*, played by Gene Hackman. Before his vessel dives down to the depths for several weeks, he takes one last puff on his cigar out in the open air and declares, "I don't trust air I can't see."

The odorlessness and invisibility of carbon dioxide has long confounded those on the front lines of understanding and communicating the immense risks of global climate change. Combine those properties with the fact that this gas's impacts via climate change—from storm surge due to sea level rise, to intensifying droughts to heat waves to increased incidence of malaria—are mostly (though not always) distant in time and space from those responsible for putting most of it into the atmosphere, and you get a perfect recipe for inaction. We humans are hardwired to deal with the fanged threats right in front of us, not terrors that seem to dwell far off and in the future.

Our soot-laden skies are a dashboard that we're still learning how to read. The smoke in the air all around us, like the haze blanketing north India, is in some sense a signal of just how far we have come since Prometheus. Our co-evolution with fire, our dependence on it, is so complete and profound that the phenomenon itself seems as mundane as the acts of breathing and eating, as familiar to us as our own bodies. We like to think that, after all this time we've spent together, we know everything there is to fathom about fire.

Tashi Stobdan's walls and ceiling beams are coated black from centuries of fires lit just to survive. These dark materials give us a rough idea of how well we're burning after all these millennia. Stare at the black carbon for a while, and you start to wonder: Have we really mastered the flames? And have we ever made a full accounting of their cost?

In midcentury America, at the height of the country's industrial might, most of the public concern about particulate pollution centered on visibility. The smoggy haze marred landscapes. Soot caked on tenement windows in Manhattan and soiled linens hung out to dry. It was a nuisance, mostly because it was an *eye*sore. An offensive sight.

But, for the briefest moment during its lifespan, as it streams upward, black carbon is hypnotic, mesmerizing. We can't look away from it. Black carbon is the darkest substance we have, and yet briefly, it throws off the brightest of lights.

"All bright flames contain these solid particles, either during the time they are burning . . . or immediately after being burned," the great nineteenth-century chemist Michael Faraday observed in his famous lectures on the combustion of candles in 1860, in London.

"You would hardly think," he marveled, "that all those substances which fly about London, in the form of soot and blacks, are the very beauty and life of the flame." All this time, as we've been staring mesmerized into the glowing hearth fire, what we've been looking at—but not really *seeing*—is black carbon streaming upward, briefly incandescing.

In a very literal sense, black carbon *is* the flame. The firelight—source of hope, guidance, companion, refuge in the darkness, illuminator of our neighbors' faces around the primeval hearth—is really just black carbon, the ur-pollutant, energized to the point of glowing. It's a Jekyll-and-Hyde relationship—the first and foremost of many paradoxes in the "secret *Cabinets*" where black carbon is found.

"Is it not beautiful to think that such a process is going on, and that such a dirty thing as charcoal can become so incandescent?" said Faraday. And if we can start to see the flame for what it is—perceive both the dangers and opportunities it presents—what else might we see by its light?

Here etymology can help us. The word "soot" contains a clue to one more very important property of black carbon. "Soot" is derived from the Old English word *sittan*, which means "to sit." Soot is what settles. Because black carbon has very low chemical reactivity in the atmosphere, the only way it gets removed from the air is by deposition, washing out with the rain and snow and wind. The next weather system is always around the corner, ready to pull it back to Earth.

This simple physical fact means that if we stopped producing it tomorrow, the skies would be clear of it within a week or two. We would notice the difference right away. And, once tucked safely away in soils or sediments, its lack of volatility means it is effectively sequestered carbon. Its two most dangerous qualities—its fineness, which lets it penetrate lungs, and its blackness, which enables it to soak up light and turn it into heat—are then rendered moot.

"Black carbon offers an opportunity to reduce the projected global warming trends in the short term," Veerabhadran Ramanathan told Congress in 2010. "The life time of BC in the air is of the order of days to several weeks. The BC concentration and its solar warming effect will decrease almost immediately after reduction of its emission."

This is the key to his optimism. Cleaning up black carbon alone wouldn't solve global warming or save the glaciers that remain. Far from it, he is always careful to point out: concurrent, urgent action to reduce emissions of carbon dioxide and other greenhouse gases is essential. In the long term, those longer-lived gases are what will cook us.

But draining soot from the skies would have an immediate, noticeable impact on the air we all breathe. And it would start to immediately slow down the rate of warming from Kumik to Kathmandu to Ramanathan's home state of California.

• • •

A few days after the meeting in Stobdan's kitchen I am sitting, soaked and shivering and stunned, in a dark cave the size of a school bus.

Urgain sits next to me, feeding logs into a fire blazing four feet high. He sends a ladle of hot salt tea and solicitous glances in my direction. It's been four hours of urgent, constant motion since I fell through the ice into the roiling Zanskar River.

Early this morning, our second on the frozen *chadar* trek back to Leh, Urgain had glanced up at a depthless gray sky spitting snow—a sky that augured slow going, perhaps even avalanches—and whistled. "A difficult road today," he muttered. Then he had winked at me and laughed.

The previous day had gone smoothly, but heavy snow had fallen all morning, covering areas of thick and thin ice alike with an insulating blanket. This inscrutable veil made route finding both tedious and nerve-racking.

"There are no ceiling beams or sticks, but the roof is very strong. What is this?" goes one of the old Zanskari *tsot-tsot* riddles. The answer: the *chadar* ice, a magically self-supporting structure, spanning the flowing water.

Well, not always. Urgain had told me about *tarlok*—how the ice is always "turning over"—a frequently dramatic, sometimes deadly demonstration of the immutable truth that change is the only constant. But I guess the lesson didn't really sink in. I was in the lead that morning, not paying close attention to the subtle signs on the surface, daydreaming instead about chicken chow mein in Leh. Unwittingly I had marched straight into a section on the middle of the river where fresh snow masked a weak patch of ice.

Then the floor—or roof, if you will—gave out.

I plunged down and slightly forward, the way you do when you miss a step on the stairs. My bulky pack caught on the edge of the hole I made; it was the only thing keeping me up out of the unseen river's ineluctable sweep toward Pakistan and the Arabian Sea. I flailed about, but each attempt to grab the edge of the ice just broke it off. So I became still, letting the gravity of the situation soak in along with the water filling my boots and pack. My thoughts turned, against my will, toward the Zanskari porter who had disappeared through the ice a few years before, his body discovered weeks later and several kilometers downstream.

I watched as Tsewang Thinles, our third team member, gingerly shuffled toward me. He was about thirty feet behind.

Then he fell through the ice, too.

We stared at each other helplessly for a moment. Then he snapped out of it, struggled for a bit, and magically wriggled out onto the ice on his belly. He inched his way over to me, spread out like a man under enemy fire. When he got within six feet, he reached out his trusty wooden staff. I clenched it like the lifeline it was, and he delicately hauled me out of the water, hand over hand.

Soon we were standing, dripping and apprehensive, looking down at the ragged, slushy hole in the ice—a modest-looking little gate to the Underworld.

Urgain had seen it all from a hundred yards back and started shuffle-running until he caught up with us. (It was the first time I ever saw him ruffled by a turn of events; "Shit! Shit!" he yelled all the way.) He smiled, eyes a bit wide with evident relief. We took a moment to silently acknowledge the literal gravity of what had just happened, and then we started walking with a newfound sense of urgency: moving until we found fuel was the only way to fend off hypothermia.

We hiked all through the late afternoon, until we arrived at this wide-mouthed cave a hundred feet above a broad bend in the river, where Urgain scavenged the banks for scarce driftwood.

Now, here by the cave's mouth, twilight is falling fast. The cold has its knives out; it stabs at our backs while in front we are massaged by the fire's standing wave of heat. The Zanskar River steams slowly out of my socks, boots, pants, and gloves. Blood returns grudgingly to my fingers and toes. Urgain and Thinles laugh in the firelight, telling stories of past *chadar*s, of avalanches and ice dams and epic climb-arounds and old legends. I just sit there and listen as the steam mingles with the smoke, in the half-numb, sheepish manner of one engaged in cosmic recalibration after a close shave.

I silently turn over my mistakes: I had extrapolated disastrously from the sure footing of the day before; trusted too much in the seeming thickness of the ice; strayed too far from my companions. This pile of smoking driftwood—and my friends' timely actions—were all that stood between me and calamity.

"There was the fire, snapping and crackling and promising life with every dancing flame," Jack London wrote a hundred years prior in "To Build a Fire."

London knew well from his years bouncing around the Yukon Territory that a fire is not only the oldest and most effective morale booster there is but also a surrogate companion—in some situations it means life and its absence means death.

The protagonist of his most famous short story breaks through the icy crust of a stream and gets soaked in 50 below temperatures. It's a death sentence unless he can light a fire and thaw out. He struggles with frozen, fumbling fingers to start one. He nurtures it to life, but as the flames flicker higher the rising heat causes snow to slough off a branch and squelch his kindling. He fails again and again to revive the flames. As he slowly freezes to death, he ruefully recalls an old man's warning: never travel alone in such cold in this hard country. "Perhaps the old-timer on Sulphur Creek was right. If he had only had a trail-mate he would have been in no danger now. The trail-mate could have built the fire."

Sitting safe and half-dry in its warm glow, I now vibrate with an animal understanding of both truths. Fire is life. Trail mates are life.

As Urgain and Thinles swap stories, the afternoon's brush with disaster slowly recedes and shrinks into some slapstick set piece whose outcome now seems like it was never really in doubt. Thus entranced, bailed out, forgiven (aided too by the bottle of Old Monk rum that Urgain pulls out of his tattered knapsack and passes around), I scooch closer and closer to the flames.

Until I notice that the cuffs of my synthetic trekking pants are melting. I jerk my feet back, spell broken by the caustic perfume of burning plastic. The breeze shifts. Plugs of smoke rush at our faces. The flames leap and lick at the roof of the cave, illuminating fragments of Tibetan script, interspersed with conical symbols composed of white dots. Urgain watches me as I watch them flicker.

"They are like prayers," he says, suddenly serious again.

The symbols are supplications and mantras penned by generations of travelers, made to protect against the chaotic forces that dwell beyond the edge of their authors' villages, in places like this wild river gorge, and left on a dark canvas made by countless fires. Enigmatic appeals etched in a thick layer of soot. Fragments that speak to a realm beyond the firelight, where unknowable and implacable currents rule.

We watch in silence as they flash in the fire's strobe. Though the markings look recent, just a few decades old, they seem to voice ancient preoccupations—riddles posed with equal parts dread and hope—hovering unspoken and inchoate in the smoke wreathing our heads, since the very first fire was kindled:

What will kill us?

What will save us?

Urgain gently places another log on the blaze. Its sinister smoke invades our lungs even as the heat keeps us from freezing.

The fuel hisses and pops as it sunders. The river below cracks and sighs as it reassembles its wintry lid. The physical world, so often inscrutable, helpfully whispering answers: fire and ice.

3

Water Connection, Fire Connection

I snatched the hidden spring of stolen fire, which is to men a teacher of all arts, their chief resource.

—AESCHYLUS, *PROMETHEUS BOUND*

We believe that man is free. We never see the cord that binds him to wells and fountains, that umbilical cord by which he is tied to the womb of the world.

—ANTOINE DE SAINT-EXUPÉRY, *WIND, SAND AND STARS*

How on earth do these people survive?

—JAMES CROWDEN, ON HIS 1976 WINTER VISIT TO ZANSKAR

Two years have passed, and I am walking alongside Urgain Dorjay once more on the frozen surface of the Zanskar River. Our group of eleven is just half an hour into our four-day southward journey to Zanskar. The sun is shining in a cloudless sky. The ice seems thick when we rap it with our staffs. Our legs are fresh and spirits high.

Urgain comes to a sudden halt. He squints up at the sheer rock face above us, walks a bit ahead, and calls up to a hidden figure. They exchange tense shouts. Urgain starts running back and calls out to the rest of us: "Move back!"

We drop our packs, abandon our sleds, and quickly shuffle in retreat. After fifty yards, we turn around just in time to watch several tons of rock explode outward from the cliff. Boulders tumble down the steep slope above. Dust billows through the canyon, envelops us, coating the ice, coating our faces. "They are blasting," Urgain explains with a laugh, as he wipes the grit from his eyes. "Oh ho, very dangerous!"

The invisible workers above us are building the "*chadar* road," a two-lane highway to be cut like a shelf into the sheer rock walls of the Zanskar River gorge. When it is finished, this road will be kept open year-round, making it possible to travel from Padum to Leh in under six hours, and obviating the current road route, a circuitous two-day, 500-kilometer journey via Kargil over three mountain passes.

When another road, now being built through the Lungnak Valley and over the 5,100-meter Shingo-la pass, is completed, this will become the shortest route linking Leh to the resort town of Manali in Himachal Pradesh, and beyond, to the rest of India. The time and cost of bringing goods to Zanskar will be halved. Padum will become a bustling way station. Land prices along the road will skyrocket. And Marthang, the site of Lower Kumik, will be right next to the new highway, in the thick of it all—a fact that is not lost on younger Kumikpas. Some are already planning to build shops along the road.

But they will have to be patient. The road building is dangerous, grinding work. Indian army engineers oversee teams of laborers from Nepal and poorer states like Bihar, who blast the rock walls and attack the rubble with pickaxes. Several have died in accidents in the past few years, buried by falling rock and landslides or drowned inside excavating machines that have tumbled into the river. In recent years, they've only managed to advance a few hundred meters from each end. There are at least forty more kilometers yet to go, through the most difficult, narrow section of the river gorge. When I asked an army engineer commanding the effort from the Ladakh side in 2009 to estimate how long it would take, he said, "Ten years minimum." Then, after we became friendly over several more cups of tea, he confessed, "More like fifteen years."

When you're walking on the ice, gazing up at the vertical rock face, even this seems a wildly optimistic assessment. (But some Zanskaris point out that if relations with Pakistan deteriorate further, the road will become a much more strategic asset to the Indian Army and the work will be completed quickly. "They eventually want to make Zanskar into one big weapons depot, very strategic," a friend of mine in Leh says, repeating a bazaar rumor he has heard.)

Some older Zanskaris, like my friend Mohammed Amin Zanskari, think the existing road from Kargil to Padum has already weakened Zanskari culture to the point of breaking. Since the road's arrival in 1981, Zanskaris have had a taste of the wider world, its technological wonders and tawdry compromises, its perils and promises. The road has brought gas cylinders and subsidized rice from the Punjab, ceiling beams and concrete from Kashmir, Tata pickup trucks and Toyota 4x4s from Jammu, and many new ideas of what constitutes the good life. Truckloads of government-subsidized rice, sugar, flour, and kerosene arrive throughout the summer and are distributed at steeply discounted prices to the locals. This has removed a traditional source of anxiety—the specter of famine due to a few bad harvests—while also severely undermining their subsistence agriculture economy. The result has been increased food security at the price of a gradual erosion of cultural identity and social cohesion.

It's a tradeoff that few Zanskaris regret. But the older generation is more preoccupied with what has been lost.

"The Zanskari way of life is ending," Amin told me on my very first day in Zanskar, on a sun-drenched July afternoon in 2003. He said the *chadar* road would deal the death blow. "This road will bring many people from Leh, who will buy the land. The traditional ways of using the land, living from the land, will disappear. The seven months of sleeping, drinking, and singing in the winter will be passed differently," he added with a chuckle.

Life will be easier, faster, and, most would venture to say, *better*—but would they still be Zanskaris, he wondered? Amin thinks that their isolation, the self-sufficiency demanded by their difficult environment, has made them who they are.

"Before, grazing the animals between the mountains, we were here," he told me once, pointing emphatically toward the ground. There wasn't so much yearning to be elsewhere. "We were at rest. And if your mind is at rest, your body will be at rest. Now in Zanskar, people's minds are not here, not at rest," he said. "Their minds are in Kargil, Srinagar, Delhi, Leh."

But most young Zanskaris seem ready to let go of those grinding constraints. They are ready to become someone else. They prefer to focus on the far easier access to the medical care, education, goods, and services in Leh that the new road will afford them. One winter I was in Padum as a pregnant woman from an outlying village lay in terrible pain in the unheated hospital ward, experiencing complications in labor. The medical staff had called for an evacuation by army helicopter to Leh. Bad weather delayed the flight. The doctors told me later that the child was lost. With the road, people point out, they could drive straight to Sonam Norboo Memorial Hospital in Leh any time of year in less than half a day.

For now, though, the ephemeral *chadar* ice highway is still the only way in or out of Zanskar in the long winter. So Urgain and I and our companions plod on, picking our way through the jagged boulders that now litter the ice.

A few days later nine of us sit shoulder to shoulder in Urgain Dorjay's kitchen. We sip tea with one hand and *chang* with the other and soak up the heat. The mood is festive. Outside the wind races down the slopes of the mountain called Kapalungpa and rustles loose scraps of plastic on the greenhouse attached to Urgain's home. Inside we slurp *thukpa* stew while Urgain's eight-year-old son Phorboo demonstrates some Shaolin kung fu moves gleaned from his favorite film, *Return to the 36th Chamber*.

We are celebrating our good fortune, having narrowly beaten a raging blizzard that arrived right at the end of our *chadar* walk. Had we reached the terminus of the new road near Kilima just a few hours later, we'd have been stuck there in the road-building camp for several days. Instead, we escaped in a couple of hired jeeps, two of us riding the rear bumpers to weight the vehicle down and keep it from sliding off the switchbacking roads down the slope into the Zanskar River. Rescued by diesel power, we'd gratefully breathed its fumes through the teeth of the storm, all the way to Padum. And now here we are, warm and dry, drinking *chang* and applauding Phorboo's technique.

Until the wind shifts imperceptibly and smoke ricochets back down the chimney into the room, invades our throats, obscures our faces. Within a minute we have to clear out, in a flurry of coughs and tears. We wait under the black dome of the sky, pinpricked with unnumbered stars, for the smoke to clear. Inside, Urgain tries to open a window, jimmies the stovepipe a bit, opens the kitchen door for some draft.

"Oh ho! A very special wind!" he laughs as he comes out, eyes blinking. "Come back inside, it is cold out." ("Special" is the strongest derogatory term Urgain deploys. For example, of corrupt politicians who siphon funds meant for Zanskar's watershed development, Urgain will only say with a hand wave of disgust, "These are very *special* people!")

We all shuffle back inside. A choking miasma still hovers at head height in the kitchen, reminding me of the Angel of Death streaming through doorways in ancient Egypt, in the Cecil B. DeMille classic *The Ten Commandments*.

Once you start paying attention, you start to notice *shremok*—soot—wherever you look in Zanskar.

It climbs up the walls of Stobdan's kitchen and coats the willow sticks around the *tutkhung*—the "smoke hole." It is the dark genie escaping from the barrels of burning garbage on Padum's dusty main street. It spreads like spilt ink across the eastern wall of the Padum mosque, above the opening of the furnace where they heat water for daily ablutions, in a blot that grows bigger each day as the nearby stack of split wood diminishes. It clings to the sides of the great big pots of *chang* that Tsewang Zangmo tends for hours in the small willow grove in the heart of Kumik. It coats the roughly stacked stones that form the walls of the shepherds' huts at the *doksa*, the seasonal encampment where women from each household take turns tending the cattle, an hour's walk above the village.

It covers the leaky kerosene heater that Urgain Dorjay, worried I would freeze in below-zero temperatures, installed in my room a few nights before we left once for *chadar*. It comes away in thick layers, left by centuries of driftwood fires, with a swipe of one's fingers in Tsome Bao and the other caves that line the *chadar* route. It marks the bottom edges of the *lhato*, those stone shrines topped with ibex skulls marking the domain of the *lha*, where juniper boughs have been burned as a cleansing ritual. It gives a poignant tint to certain stones, which discretely signal the boundaries of the *romkhang*, the cremation grounds on the edge of Kumik. It curls up around the fire-lit faces of giddy children out of the big fire made in the heart of Kumik, just down from the *mani* wall, whenever there is a wedding or a celebration or an important meeting. It lines the hearth and the stove, quietly thickening in the same place where the *pha lha*, the god of the hearth, sits on his ever darkening throne.

These days you can see some new flavors of soot, too—ones that speak to the desire for a better life, with more comfort and more opportunity. It wreathes the heads of pedestrians crossing the lone dusty intersection in Padum, where young taxi drivers freshly arrived from Kargil joyfully rev their Tata Sumo four-wheel drives, coughing out dense black plugs of exhaust from their tailpipes. It might as well be paint, so thoroughly does it encase the outer walls of the small stone building perched above the Lungnak River, on the outskirts of Padum, that houses the diesel generators that provide power to the town for a few hours each night.

All this soot is in equal measure survival's residue and a sign of progress, spreading like a rash across the "good white place," as one interpretation of the origin of the name *Zanskar* goes (a Buddhist reference to the supposed purity and piety of its inhabitants). As such, it is perhaps the best way we have to render visible the connections—so often opaque in our modern world—that sustain us, a kind of tracer fluid, illuminating for the brief time it is aloft and in our faces, the twin vascular systems of the global climate and our burning-based modern economy.

Seeing all that *shremok*, all those plumes, we can learn to pay closer attention to other dark clouds on the horizon, looming over California, Greenland, New York, Africa, China. And as any plumber can tell you: finding and understanding the source of the leak is the first step to plugging it.

Spring may be arriving earlier and earlier, but winter at 12,000 feet is still damn cold. And more modern technologies are arriving every summer—even gas and electric space heaters—but the typical Zanskari home is still quite often choked with fumes, thanks to its simple, traditional stoves, presenting its inhabitants with a tough choice: endure the smoke or suffer the cold.

Most people through history, from modern Zanskar to nineteenth-century New England, have chosen the former. The first European ever to visit Zanskar wasn't like most people.

Alexander Csoma de Koros was born in a small, impoverished village in Transylvania, in eastern Hungary. Gripped by nationalist fervor, the scholar left the boarding school where he taught one day in 1819 and ambled off "lightly clad, as if he intended merely taking a walk." He kept walking all the way to Asia, over several years, on a quixotic quest in search of the geographic origins of the Hungarian people. His theory was that they lay somewhere in Central Asia—perhaps the Uighur people were related to Hungarian Magyars, he posited—and he trekked through the Middle East to British India, hoping to eventually reach Kyrgyzstan via the western Himalaya to prove it. Rebuffed by suspicious border agents on his way north to what is now far western China, Csoma met the British government agent William Moorcroft on the trail south of Kargil.

Moorcroft immediately saw an opportunity to take advantage of Csoma's scholarly inclinations, ascetic-minded ways, and prodigious linguistic gifts (Csoma had mastered over a dozen languages). So he persuaded the eccentric young pilgrim to take a detour to Zanskar. There he could study with a local eminent lama, learn the classical Tibetan script, and develop a dictionary that would be of inestimable value to European scholars seeking to unlock the vast trove of Tibetan literature—and, incidentally, of strategic use to the British empire as well, which sought to improve its understanding of the remote region as a potential buffer in its ongoing contest of shadows with czarist Russia.

So in the winter of 1824, Alexander Csoma de Koros departed from the bustling bazaar of Leh and walked for ten grueling days up the valley of the Zanskar River to the village of Zangla, which sits twelve miles north of Kumik. There he met Sangye Phuntsog, the learned abbot of the fortress-monastery that sits high above Zangla on a rocky outcropping.

Csoma spent sixteen months in almost uninterrupted seclusion and study with Phuntsog, poring over the Kangyur and Tengyur, the two great anthologies of the classical Tibetan canon, becoming the first European to thoroughly learn the Tibetan script, a heroic feat that would eventually earn him fame in the West as the "father of Tibetan studies."

But in an even more astonishing achievement, for four straight months through the heart of the Zanskari winter, Csoma did not once leave the nine-square-foot room where he studied with Phuntsog. They paused only, we can surmise, to sleep and to eat in silence: each meal consisted of barley flour mixed with tea, salt, and butter. Master and student sat reading side by side, under blankets. It was so cold that each man, reluctant to extend his arm out, elbowed the other when it was time to turn the page.

Early in his studies with Lama Phuntsog the tiny cell was heated by an open fire in the center of the earthen floor. There was no chimney—the chimney wouldn't arrive in Zanskar for another 150 years—so the pungent smoke filled the room. He soon discovered that the pall stung his eyes and made it difficult to read. So he insisted that the fire not be lit—better to preserve one's vision for study, he reasoned, than to be warm.

Many years later Csoma described his Zangla sojourn to a British doctor in Calcutta. "There he sat," Dr. J. G. Gerard recounted in his papers, "enveloped in a sheep-skin cloak, with his arms folded, and in this situation he read from morning till evening without fire, or light after dusk, the ground to sleep upon, and the bare walls of the building for protection against the rigors of the climate. . . . Some idea may be formed of the climate of Zanskar from the fact that, on the day of the summer solstice, a fall of snow covered the ground; and so early as the 10th September following, when the crops were yet uncut, the soil was again sheeted in snow; such is the horrid aspect of the country and its eternal winter."

Have you ever seen a Zanskari with frostbite?"

It sounds like a rhetorical question or an old Zanskari *tsot-tsot* riddle, though it's a fair one to ask in a place where temperatures used to regularly descend to 30 degrees below zero (and occasionally still do).

Dr. Stenzin Namgyal, one of Zanskar's two doctors (until he was posted to Kargil in 2013) and the son of the old king of Padum (who is a retired schoolteacher), gives me a sober, searching look as he poses it. But I just laugh in response: the very idea seems preposterous and paradoxical, like a koala bear allergic to eucalyptus or a mountaineer afraid of heights. This, of course, is the doctor's point.

"Zanskaris know how to cope with the cold," he says. "But the smoke is hard to manage."

"If I see a hundred patients, maybe 30 to 35 percent or more have respiratory problems," Dr. Stenzin tells me. "From acute to chronic disease, I think respiratory is the number one [health problem] here."

Csoma's patron, the British spy William Moorcroft, traveled widely throughout neighboring Ladakh in the 1820s. A veterinarian by training, he was often approached by villagers for medical help. He was "appalled at the acrid smoke from the yak dung fires which filled every Ladakhi living room in winter and he often insisted that the fire be put out, or lay on the floor to clear his streaming eyes, before he could treat some of his sick patients."

When medical anthropologists Keith Ball and Jonathan Elford visited Zanskar in the early 1980s, they found conditions little changed since the time of Csoma and Moorcroft. "During the winter Zangskaris spend a great deal of time in the small windowless 'winter kitchens' on the ground floor of their houses," they wrote. "These are heated by brushwood and yak-dung stoves which generate much smoke, the chimney usually being a simple hole in the roof. The blackened walls and soot-covered cobwebs in the 'winter kitchens' we visited indicated heavy domestic smoke pollution."

Glass for windows didn't reach Zanskar in large quantities until the road was completed a year after the British researchers' visit. Prior to that, only the very wealthiest families could afford to have such luxuries carried in over the passes (the same was true for once-rare treats like rice), and builders made the most of the limited palate of natural materials available to them—poplar beams, willow sticks, earthen bricks, and stones. So for centuries the Zanskari home retained largely the same features: a few windows to admit light, kept small to minimize heat loss and covered with wooden shutters or wax paper; a partially sunken ground-floor kitchen surrounded by stables for the animals (to take advantage of the creatures' body heat and the insulating effect of straw and hay); and a hearth in the middle of the earthen floor with a hole in the roof overhead to evacuate at least some of the smoke (and with it, alas, much of the heat). Traditional Zanskari houses thus resemble their makers in their sturdy pragmatism. They are studies in careful tradeoffs: light versus heat, cleanliness versus insulation, ample space versus the fuel needed to keep it habitable.

Then the Indian government, worried about maintaining access to the region in the event of yet another war with Pakistan, built the road from Kargil in the early 1980s. Alongside manufactured goods like glass and metal stoves and Portland cement, another key item started to arrive in the Padum bazaar. Today, most kitchens in Zanskar have chimneys, a device that reached most American homes by the mid-nineteenth century. These have improved ventilation, but smoke still often fills kitchens, even without a "very special wind."

Stoves rust and leak, fires smolder for lack of proper draft, the gaps between pots and the openings in which they sit allow fine particles and toxins to escape into the room. Fires being temperamental things, they demand a careful mix of oxygen, dry fuel, and temperature to burn optimally. What's more, installing a chimney doesn't really solve the problem: it just moves it outside, where the combustion products linger in the air around the home, in the spaces where people work, play, and gather. Sometimes the wind carries it away. Sometimes it doesn't.

And those chimneys get a lot of use. In Kumik, women like Tsewang Zangmo get up before dawn to milk the cows and start cooking breakfast around six o'clock. They keep the dung fire burning in the stove for two hours, to boil water for the copious cups of salt-and-butter tea that are consumed (over a dozen per day; Zanskaris love their tea as much as their *chang*), and baking chapatis or *tagi khambir*, a yogurt-infused local bread. Lunch might be a simple recombination of leftover rice and vegetables or *paba* (a boiled brick of barley, wheat, and pea flour taken with butter or yogurt), eaten out in the fields or quickly in the kitchen. Then the stove is fired up again around four or five o'clock in the afternoon, stoked and fed well into the night, for up to eight hours. Kumikpas estimate that an average household burns 10 to 15 kilograms of dried dung in a typical day. In winter, they consume at least twice that amount—maybe one full *tsepo* (a deep woven basket the length of a tall man's torso) load every day for both cooking and heating.

Those ten hours of burning are an act of survival, but like Prometheus' gift they come at a steep price.

Most of us don't spend several hours a day hovering around a wood or biomass fire. Our exposure is limited to summer days grilling in the backyard or sitting around the campfire. Thus we tend to regard the smoke as a mere irritant, or we may even welcome its familiar tang.

Zanskaris, meanwhile, still face Csoma's choice on every winter day. And while he chose to freeze and forbear, they have long made the same choice most of us would: to burn the fuel at hand, dung and brushwood gathered from the hills, to stay warm—hell, to stay *sane*—and simply put up with the *tutpa* and *shremok*, the smoke and the soot.

Three billion people around the world still face this same dilemma.

When Urgain was a young man, he sometimes traveled down the *chadar* to procure two essential items for his family from the bazaars of Leh: tools for carding wool and matches.

This was before the road, when manufactured goods from *chigyal* (outside the kingdom) were scarce. Carding and spinning wool was a form of productive entertainment during those long dark winter nights and it also led to the practical outcome of being clothed. And matches kept the hearth fires going. Back at home, Urgain and his family ate in the dark or by moonlight. They could get by without much light, but a source of heat was essential when temperatures plunged to 30 or 40 below zero.

Before they went to sleep, his mother piled ashes over the coals in the stove to keep them warm. But sometimes they would wake up to find the embers had died out. There was no electricity, and kerosene fuel was a rare commodity. If their fire went out there was just one thing to do.

"At that time many people are not having matches," Urgain says.

"If matches are finished, they must go to neighbors. I went to the neighbor's house with a metal plate. People would keep their dung in *bokhari* [metal heating stoves] always burning."

The neighbor would extract some glowing embers from her own hearth and deposit them on the plate. Urgain would hurry home in the bitter cold to deliver this precious gift to his mother. She took the coals, placed them carefully in the stove, and added kindling. Then she would blow to bring the fire back to life, adding pieces of yak and sheep dung as the flames caught. She would put a pot of tea to boil on the stove, and the family would cluster around it for warmth. Crisis averted.

And should their neighbors' fire die out one cold morning, Urgain's family would share their embers in turn, paying it forever forward, a kind of eternal flame. In the Zanskar of old, Prometheus lived right next door.

The presence and path of water has long dictated the siting and evolution of settlements, from Rome to Manhattan to London. Zanskar is no different. In fact, that generally sensible tendency is distilled here to an axiom: oasis-like farming villages sprout exclusively along watercourses throughout its central valley, forming a bright green-and-yellow patchwork quilt of barley, wheat, peas, mustard, and alfalfa. Due to the scarcity of flowing water, some researchers have estimated that just a quarter of 1 percent of Zanskar's land is cultivated.

While all Ladakhis and Zanskaris are extraordinarily attuned to their water constraints, Kumikpas are even more sensitive than most. Every spring the members of each of Kumik's thirty-nine households look up at the snow-covered mountainside, reckon how much water it holds, and decide how many of their fields to sow. On May 25 or thereabouts, depending on the Tibetan calendar, they harness their *dzo*s and plow and plant their fields in teams, one household helping the others in each labor-sharing group, for several days until everyone's fields are finished. In lean years they plant only the fields closest to the water channels; since some water is lost from seepage through the soil, shorter distances are more efficient. Each household has a brief window—perhaps just a couple of hours—available to use its share of the water until every *nang*, or "room," of the field is slaked.

This practice demands an attentive eye and care commensurate with the narrow tolerances at work. It also requires a cooperative framework animated by the same principle behind the tradition of carrying embers from hearth to hearth. There's a finite amount of water available for *chu tangches* (literally, "giving water"), and if someone takes more than, or is deprived of, their fair share, the whole system collapses. If you want to irrigate your fields, you need to be on the same page with the households upstream and downstream of you.

So each household takes its turn drawing water for its fields roughly once a week. There are eighteen *khangchen* (big houses) and twenty-three *khangchung* (little houses). The big houses are the ancestral homes, passed down through the generations to the oldest son. The little houses are offshoots, formed to accommodate parents who move out of the big house when their son has his firstborn child or when a younger brother decides he wants a family and home of his own (every few generations or so).

In the Indus Valley villages of Ladakh, a man is elected by the villagers to be the *churpon*, or "lord of the water." He polices water withdrawals, adjudicating irrigation disputes on the spot, assessing penalties on those who take more than their due. In Zanskar there is no *churpon*, as the villages tend to be somewhat smaller and more self-regulating.

"Everyone knows the rules," says Tashi Stobdan.

In Kumik, there are three irrigation groups of six big houses each, and two groups of ten little houses each. On every fifth day one of these five groups gets water. The members work out the timing of the division of the flow internally. Those who get the short end of the stick must sometimes draw their water in the middle of the night if the water is very low and the *zing* are empty. The *zing* are elliptical ponds, twenty by sixty feet by fifteen feet deep, that store water during the night for the next day's use; the village can't afford to let it flow unused into the Zanskar River below.

There is occasional fudging among neighbors. "People are always digging [their] channel deeper," Stobdan says with a laugh. "Some people go in the night and take water from the *zing*" when they think everyone is asleep.

In a village as small as Kumik, it's hard to keep such minor transgressions secret, and they mostly fuel jokes rather than fights. Disputes sometimes flare, but they are rare, and the village leaders and elder men try to resolve them swiftly. Festering conflicts are a source of dread to Zanskaris and Ladakhis. With a few key exceptions, most arguments I have witnessed over the course of ten years of visiting Zanskar have been put out with greater urgency than fires. Those that aren't stamped out quickly are regarded as a serious threat to everyone in the village—for in any lifeboat (next to a lack of water) discord is the greatest danger of all. Water is life, as Tsering Motup had pointed out. And life must be lived together.

This reciprocity, this clear-eyed recognition of how their fates interlock, is perhaps the core answer to the question the geographer James Crowden, one of the first Europeans to spend a winter in Zanskar since Csoma, asked as he traveled about in 1976, marveling at the juxtaposition of all this apparent scarcity and Zanskaris' vigor, cheer, and sense of abundance: "How on earth do these people survive?"

With fire, with water, and with each other.

This ethos even manifests in Zanskaris' small talk. Spend a bit of time in Zanskar, and you will start to hear the same questions, over and over, in any season or setting:

Karu cha-chey? Karu songpin? "Where are you going? Where have you come from?"

These two questions function as a greeting of sorts, in lieu of "How are you," and are invariably exchanged between acquaintances in the Mani Ringmo bazaar of Padum or neighbors passing each other on the well-worn footpaths between the fields. (This might be because Zanskar is—or until recently was—a kingdom of homebodies. As Mohammed Amin once told me, "There are some older people in Padum who have never been as far as Karsha," less than five miles distant across the shimmering plain. "They move only in connection with work.") These two questions sound perfunctory, but they are to be answered literally—"I'm going to the market. I've come from my aunt's house"—as you continue on your way.

Trangmo mi-rak-a? "Are you not cold?"

This question speaks for itself.

Suna-mi-rak-a? "You're not bored or lonely?"

This question speaks to Zanskaris' perennial assumption that guests from afar—or even from neighboring villages—will be homesick, missing friends and family and familiar environs, because, well, that's how they would feel. Zanskaris are a tough lot, game for all manner of hardship. Just not alone. (Corollary: Most Zanskaris have little need for, or understanding of, the related concepts of "me time" and privacy.) So among them there is no shame in answering this question truthfully: "Yes, I'm a bit lonely."

Yato meta?

It may take a few minutes into the conversation for this one to surface, but it is bound to be encountered by anyone—Zanskari, Ladakhi, Punjabi, Czech, American, Japanese, German, Inuit—who travels alone. "You don't have help?" is the literal translation, but colloquially it means something like "Where are your friends?" and carries with it an implicit scolding: "C'mon, what's the matter with you? You should know better." *Yato*, "help," not only makes survival possible; it also happens to make life worth living. Traveling alone is not only dangerous, inviting unnecessary risks; it's not any fun, either.

Which is why a reply of *Met, nga chikpo*—"Nope, I'm alone"—is typically met with a mix of pity, polite dismay, and disarming bewilderment at the speaker's foolhardiness. A tsk-tsk tongue-clucking sound may sometimes accompany it.*

Over the years, I have come to regard *Yato meta?* as expressing something essential about the place and its people. For Zanskaris, the

---

*Or even stronger voicing of disapproval. One time, upon departing from an extended visit to Kumik, I went to say good-bye to Ishay Paldan and his family. Over tea, his wife, who's hard of hearing, asked if I was traveling out of Zanskar. Yes, I shouted. "*Yato*?" she shouted. "*Met*," I replied. Alone. She leaned back with a look of frank concern. She turned to her daughter and said, at a slightly lower volume, "He'll get killed." Then she drew her finger across her throat, in case I didn't catch it.

absolute worst thing, the only true disaster, is to find yourself utterly, irrevocably alone.

The flip side of the same truth is enshrined, somewhat darkly, in another old Ladakhi saying: *Chu len me len chaden*. This means "The water connection and the fire connection are cut."

This was a kind of nuclear option, an extreme measure only resorted to in the event of severe transgressions of communal norms. If one individual or family failed to do their fair share to keep the lifeboat afloat and moving forward, or upset the delicate equilibrium of communal use of finite resources, it threatened the entire community's survival. If someone habitually started fights and feuds, or failed to contribute materially to the maintenance of the village infrastructure (canal repair, temple upkeep), or refused to take his or her turn performing the duties expected of every household (like the *lorapa* who keeps cattle out of the crops), the village would threaten them with *Chu len me len chaden*.

Effectively, this amounted to a social boycott. Should the offending household's fire die out, no one would carry embers to relight the family's hearth. Water would not be permitted to flow to that family's fields. No one would visit their hearth, or even speak with them.

The severing of these two connections amounted to a kind of death-in-life. As one anthropologist observed, "It is the ultimate sanction that can be applied since in a village it would be impossible to continue to live under such conditions." For who can live without fire and water? And how many of us would want to live without that other kind of warmth and sustenance, to be found in a neighbor's greeting?

So, though it was rarely deployed, *Chu len me len chaden* functioned as a powerful deterrent. But the advent of modernity has largely drained this threat of its power. A Ladakhi friend recounted to me a telling episode from a few years back. A local politician's actions broke up his political party into feuding factions. His former allies told him, *Chu len me len chaden*. His fire and water were cut off.

To this he replied with nonchalance, "What do you mean, *Chu len me len chaden*? For 50 paise [half a rupee] I can get a matchbox in the market. The Public Works Department brings water to my door. So what will you cut?"

The politician thus succinctly summed up the last three decades of dramatic change in Ladakhi and Zanskari society: their dependence on each other has declined, in more or less direct proportion to the extent to which their dependence on the state and on the global economy has dramatically increased.

The water and fire connections have been frayed not through any single conscious act but by the availability of new goods and services, by access to modern fuels and bureaucratic safety nets, and by human nature's gradual accustomization to these comforts. Since the road's advent in 1981, pent-up technology has burst into Zanskar like water breaching a dam. People have more purchasing power, thanks to income from government and tourism-related jobs. The government provides kerosene, propane gas, and diesel for Ladakhis and Zanskaris at subsidized rates. The bazaars and roadside shops are bursting with packaged foods and goods—even bottled water—trucked in on garishly painted diesel Tata trucks from all over India. So who needs neighbors anymore? Tell me: what will you cut?

But let's say a landslide cuts the road into Zanskar (a not infrequent occurrence), and the army can't get its bulldozers out there to clear it for a few weeks. Suppose, then, that matches run out in the market and gas cylinders aren't restocked in the depot in Padum and the PWD water tanker doesn't make its rounds because there's no more diesel fuel.

Then the politician, the man who thinks he is beyond those ancient constraints, is in a bind. There's a very good reason water and fire are the first two topics on any survivalist course's syllabus: none of us would last long without them, whether in New York (as Superstorm Sandy's aftermath so grimly demonstrated) or in Kumik. As "modern" people, we may consider ourselves to be independent of the Prometheus next door. But we still need fire and water to survive.

Stobdan today teaches ten young students in a bright new white-and-green concrete building. On one classroom door he has written: "Where there is a will there is a way."

Kumik's old primary school houses young Nepali laborers during the summers, working to reconcrete the canal. It's a small, modest, tin-roofed building. Some kids have scrawled "Hip hop party tonight" in black marker across its south wall.

So it's hard to believe this old school set off a year-long feud almost two decades ago. The village of Kumik was divided into two warring factions over the question of whether to build a new primary school. Some villagers didn't like the fact that it seemed like a boondoggle: a handful of people would directly benefit from the government contract to build it. This would fray the social fabric. Others supported the school simply so their kids wouldn't have to walk to a neighboring village every day.

The result was an internal social boycott: the two sides didn't speak to each other, or help each other in the fields, for many months, until some elder Zanskari leaders from other villages came and brokered an end to the dispute, and the school was built (as the law required for any village of Kumik's size).

Then, the following year, the villagers came together to send their young men up to move a mountain.

Stobdan keeps a tattered photo in his wallet of himself with five companions halfway up their climb of Sultan Largo in 1997. Then in his midtwenties, he strikes a tough guy pose, unsmiling, chin jutting out—the serious bearing of a young warrior out to battle a monster imperiling his people.

"The *lha* possesses a *nilim*," Rigzin Tantar, Stobdan's father, once explained to me. The *lha*, the spirit that dwells on the mountain's summit, closely guards this *nilim*, a precious sapphire that fell from the heavens. "When people approach, it always rains or snows because the *lha* is jealous and thinks they will take it."

Whatever the cause, storms can indeed gather quickly at those heights, and occasionally rage on the summit for days, mysteriously, even as the rest of Zanskar is sun-drenched all around. And there are other hazards. From Kumik to the top of Sultan Largo is not far as the eagle flies, just a few kilometers. But it's a climb of over 6,000 feet, up into the zone where nothing grows, and where temperatures can drop into frostbite range as soon as the sun goes down, even in summer. On the slopes below, car-sized rocks lubricated by snowmelt tumble down the scree fields every so often. If you climb too quickly, the thin air will bite you.

Back in 1997, Stobdan and Tsewang Norboo and several other young men climbed, not in search of the *nilim* but of something far more precious. Armed only with picks and shovels, the young men spent a day and a half hacking a shallow canal at the base of the summit glaciers, trying to direct more meltwater toward their village.

The photo was taken on the way up. Stobdan was not feeling so defiantly confident on the way down. After spending a day and a night working up on the heights, around 18,300 feet, the men descended the steep snowfields. Stobdan came down with a bad case of altitude sickness and had to be half carried down the mountain by his friends. He remembers the descent like it was yesterday.

"At that time I come like this, one person this side, one person this side, vomiting, vomiting. Then down, down, down, then I am feeling better." He shakes his head and flaps one hand in a way that suggests he can still feel the pain in his nerve endings. "Oh! This kind of memory in my life!"

Reshaping the mountain didn't seem so outlandish because it had been attempted before. Ishay Paldan, the eldest Kumikpa, remembers the first time just as clearly. When he was a young man, in the 1960s, he joined a party to Sultan Largo to make canals. He pulls a calloused hand out of the folds of his woolen *goncha* to mime the swollen fingers and describe the splitting headaches of those who lingered for three long days on the mountaintop to gather stones and dig trenches in the spongy soil. They were spurred on by the memory of a disastrous Zanskar-wide drought in 1948 that still echoed in Kumik (according to the abbot of the Karsha monastery, some people had even died from the food shortage) and by the slow tightening of their water supply that began, he says, around that time.

"The water didn't come, because the snow didn't come," he tells Stobdan and me one evening, talking about that famine year in his youth. His brow creases as he journeys back through the decades. "The glacier was much bigger then, but there was no water."

"We went up to Bu Namchok," Meme Paldan says, the Donkey's

Ears, a mountain ridge stretching up above Kumik where some of the summer pastures are located, "and we gathered wild vegetables to boil and eat." Other villagers were forced to beg for food from the monastery grain stores and from relatives in other less hard-hit villages.

This brief famine in the late 1940s had alarmed them, but the first stirrings of chronic drought in the 1960s galvanized the villagers. So they mustered a man from every household, and together they climbed up the slopes of the *kangri*, the ice mountain.

"And one *meme* had no shoes," says Ishay Paldan.

"He went like this, barefoot, on the top," says Stobdan. "Poor, I think. Meme Sonam Norboo. Bared feet on the glacier he walked."

The group's ambitious goal was the same as Stobdan's would be almost forty years later: to dig a shallow canal to channel more meltwater toward Kumik. For three days the men labored up at the glacier's foot. Meme Ishay Paldan shakes his head and chuckles, and murmurs something low that I can't quite make out. Stobdan cocks his head as he listens, then translates. "He says, three days they work. Too cold, no digesting food, and some had frozen face. Special problems."

"The warmth didn't come," Meme Paldan concludes simply.

The late Meme Gelag, Paldan's contemporary, recalling this same trip in 2001, told his interviewers that the men nearly froze and had to retreat several times. He would rather go down the *chadar* in a thin shirt, he said, than go back up there again: "At least on the *chadar* there is plentiful driftwood to make a huge fire at night for warmth, while there is not a speck of wood or dung to be found near the . . . snowfields."

Using willow sticks they had brought from the village, Paldan and Gelag and their companions made a small fire at the foot of the glacier to stave off exposure. They had none of the tools and weather-buffering gear of today's mountain climbers, no tents or down sleeping bags. "At that time there is no kerosene, no stove," says Stobdan. "And my father said, till the marriage they have no pants."

Just a *goncha*, in winter, maybe two layered *gonchas*. Stobdan smiles. "After marriage, then the mother gives some pants." Pants in those days were an indulgence, an expensive prestige item. (Urgain Dorjay characterizes this long, pantsless period of Zanskar's history thus: "Before: just open!")

"We didn't see the water going from glacier to Shila side," Meme Paldan continues. This observation, he notes, seemed to contradict the old proverb: *Kha Kumik, chu Shila.*

"Actually the glacier is in middle," Stobdan clarifies. "The water absorbs and makes springs and comes down two sides: Pipcha and Shila. Yeah, the saying is true. But this side [to Kumik] it doesn't come, because this side, you see, is totally rock. If you dig a tunnel, then water will come here."

"So maybe every house should go and blast a tunnel."

Stobdan frowns, considers my lame joke for a moment, humors me. "Maybe five or six *crore* we need." Sixty million rupees—more than $1 million.

Maybe not. Although tunnels weren't practicable, the Kumikpas considered other options. They considered burying a pipe down to the village from the snowfields. The labor, while daunting, was conceivable, but who has the money for several kilometers of wide-gauge metal pipe? (Especially after corrupt government and aid organization officials have taken their usual cut of designated funds and materials.) At various times the concrete channel that constitutes the bottom third of the stream has been repaired and relined to prevent seepage, but raw Zanskari winters heave, fracture, and erode it, so upkeep is expensive and quixotic.

"So this is an old tradition?" I ask. "Every house helped making the *yura* [canal] in Marthang. And in Meme's day someone from every house went up to the glacier."

"Same, same," Stobdan waggles his head.

"*Nyampo?*"

"Yeah," Stobdan says. "*Nyampo.*" Together.

While the village water budget kept tightening since Meme Paldan's youth, with the drought especially accelerating in the past two decades, the old techniques of communal problem solving were still deployed. In lieu of conventional wealth—$1 million, or, you know, pants and shoes—the people of Kumik continued to leverage some more reliable, inexhaustible resources: each other and copious elbow grease.

Even so, in the end, those early mountain-shaping efforts failed. "It was just too difficult," Ishay Paldan says with a stooped shrug.

When Stobdan and his group went up in 1997 and again in 1998, for their own ultimately futile efforts—when they had finished, the wind mockingly blew the small trickle pouring off the edge of the mountain face back in their faces as a fine spray—they saw the evidence of Meme Paldan's labor more than four decades earlier.

"We see the old people have worked this side, on the east. Their channel is like a small path. So when we go there at that time it is totally finished. Nothing there, only the channel shape." And another thing had changed since Meme Paldan's visit, Stobdan says.

"Totally snow is finished, glacier is finished." The edge of the ice had retreated dramatically from where Meme Paldan had labored four decades prior.

We all fall silent for a moment and contemplate our cups of *chang*.

"But in Marthang you will have plenty of water," I point out, trying to lighten the mood, "so everyone can have a greenhouse and grow vegetables."

"Yeah!" Stobdan's face brightens briefly. "Then is good life."

But his thoughts quickly pivot back to their present misfortune—the "bad luck" of living in old Kumik, at least until they gather the money and stones and time needed to build new homes below. Stobdan points out how they don't have enough water to grow vegetables. Kumikpas have to buy them at great expense, in the Padum market seven kilometers away. "And if you like to grow trees, then impossible!" He adds wistfully, "In Shila you see they have *lot* of vegetable! Oh ho!"

Having failed to move the mountain, or the unyielding *lha*, the Kumikpas found themselves between the increasingly naked rock of Sultan Largo and the increasingly dry, hard place of Marthang.

Once they decided to relocate, the Kumikpas soon confronted a new problem. The Lungnak River drained all the glaciers and snowfields of eastern Zanskar. It held plenty of water, if you could tap it. But how to get that water to the "red place"?

At their meeting in 2000, the villagers agreed to start working immediately on the seven-kilometer-long channel to bring river water to the site of their new village. But first they needed a committee to guide the effort, which Stobdan was elected to lead. He chose two other young neighbors, Tsering Motup and Tsewang Norboo, to join him in organizing all the villagers into rotating teams. That first autumn, Kumikpas from every household worked for twenty straight days, using picks and shovels to hack a small trench across a plain next to the river near the bridge to Pipiting. It was backbreaking work.

But once they reached a forty-foot-high cliff looming in their way, right next to the river, tedium turned to danger: there was no way around it, so they would have to go through it by carving a shelf into the cliff. Several times the cliff wall collapsed, raining soil and rock down on top of them. Two men were injured, one with a fractured rib, and taken to the hospital. Another time, two women were injured in a similar collapse. Because of these risks, and their simple tools, it was slow going.

"We only used *genti* [pickaxes] and spades," Stobdan recalls, digging at most "ten meters per day, with maybe forty people, sometimes more." We work after *khuyus* [threshing] from 2001, every year."

"Did you have to ask permission to dig the *yura*?"

"Nobody is saying nothing about *yura*." Stobdan grins.

They just went for it. Everyone in Zanskar could see what they were up to, though. The local government didn't voice any objection to their chosen route through the wasteland. So the Kumikpas plunged onward, without a blueprint. Stobdan brought in a relative who was an experienced surveyor to help them level the canal properly over long distances, but they didn't bother to calculate the exact depth and width and flow rate and volume that fifty-seven new households would require. Instead they eyeballed it, picking a size that seemed reasonable, and started digging, figuring they could always come back later and make it bigger. The important thing was getting started.

In some places, spades weren't enough. A few hundred meters after they began, they ran into a huge slab of rock. Stobdan went to the public works department and secured an official letter as well as some dynamite. A few days later, he, Tashi Dawa, and Tsewang Norboo jammed the explosives into a few holes and lit the fuse. It brought down the rock, but it also brought the police from Padum, who accused them of blasting without permission. The three men were carted off to jail for half the day before the PWD cleared it all up and they were released.

And so it went. For three years, straight through the month of October, after the fall harvest and threshing, every house in Kumik sent someone to dig, blast, and hack at the earth, to fashion a lifeline into the desert for their future selves, their children, their children's children, in an act of collective, obstinate imagination. In the place where he stashes all his most important records, Stobdan still keeps the logbook in which he recorded the name of every person who worked on the *yura*, the dates, the money they spent on tea and fuel and tools—and white ceremonial scarves for when the *tehsildar* visited.

"We are digging, the wind comes, everyone is whited with the soil," Stobdan recalls, in the rhythmic cadence he deploys when sharing old memories. "Our eyes are dirty. Very, very hard work we have done. We are thinking, 'More people are coming. We have the opportunity, the possibility.' Every one of fifty-seven houses [gave labor];* those absent pay a fine, something like 200 rupees. Everybody made a good effort, that one day they will have success in getting the water."

During this period Stobdan and others repeatedly beseeched government officials for aid. The first winter, he and Tsewang Norboo and Tashi Dawa walked on *chadar* to Leh, and waited for days to see two influential politicians. The three men from Kumik offered them the rich Zanskari butter they had carried, and asked for their support—but because the men were officials in Leh District, there was little they could do. The next year, Stobdan traveled to Kargil, where he waited one month to see the deputy commissioner. He returned with a promise that the district government would provide 60 quintals of rice to feed the villagers while they labored, but no other concrete support.

After the first few years of working by hand, during the narrow window of time between threshing and the onset of winter, it became apparent that the canal would take too long to finish at this rate. With just pickaxes and spades, the Kumikpas would be digging for another thirty years. Stobdan would be approaching Ishay Paldan's age before

*Old Kumik's thirty-nine households will become at least fifty-seven in the new village, mostly because of younger brothers starting new households.

his vision could begin to bear fruit. And by then the water in Kumik would be long gone.

So after the initial stage, two senior monks from Kumik, along with Stobdan's father, Rigzin Tantar, had replaced the young men on the *yura* committee. They got a loan of 300,000 rupees (about $6,000 at the time; a loan the government would later partially reimburse) from the Stongde Monastery and went off to the distant city of Leh in search of a way to hot-rod this thing. There the three men found a yellow monster called a JCB (the Caterpillar of India). The following summer, in 2006, a truck hauled it into Zanskar, over the long and dubious road.

It was the first diesel-powered heavy excavating machine ever to reach Zanskar. These days they're all over the place, but back then, people marveled as they watched the JCB and its hired operator dig almost the rest of the canal in one busy summer.*

Were it not for this rock-plucking, soot-belching beast, Kumikpas would today still be chipping at the hard sediment, inching slowly toward the promised land, with more than two-thirds of the way to go. Now, thanks to the fire in the JCB's steel belly, the new water connection is complete and silty water from the Lungnak River flows to the virgin fields of Marthang.

Legend says that Kumik was first settled by refugees from a collapsing civilization, wandering through the mountains looking for a new place to call home.

Many say this story gave the village its name: *gu* and *mi*—"man from Guge." The founders were escaping from the ancient kingdom of Guge, centered on the city of Tsaparang in what is now western Tibet, which was racked with war and intrigue in the tenth and eleventh centuries.

---

*Starting in 2002, the Zanskar Block Office began compensating the villagers for some of their labor (at the paltry rate of 500 rupees, or about $10 per household for an entire season). After 2003, the local public works department paid for the remainder of the canal work and maintenance, engaging local contractors directly—a process over which Kumikpas had no control.

Kumikpas are thus descendants of pioneers forced to start over when their familiar world came to a sudden end.

There are other interpretations of Kumik's name. Some say it is derived from *kun* and *min*—"everything grows"—describing a place where once upon a time all kinds of grains and vegetables (and snowmelt to water them) were found in abundance. This explanation is Tashi Stobdan's favorite. It hearkens back to a happier time when Kumik was known far and wide for its prosperity rather than for its privation. A reminder that they had it pretty good for a long while and a sad comment on just how much things have changed.

Another school of thought says it's a combination of *kun* and *mik*—"everything" and "eye"—a reference to Kumik's elevated, easily defensible perch on a steep hillside, from which its early residents could survey most of central Zanskar and see imminent danger with great clarity coming from far off.

The Buddhist missionary and mystic Padmasambhava—known to locals as Guru Rinpoche, "Precious Master"—visited Zanskar in the ninth century, staying just long enough to subdue the "bad features of the area" (code for human and supernatural enemies of the Buddhist teachings). He stayed to meditate in a cave high above what is now the village of Sani, and before moving on to wage metaphysical battle with recalcitrant spirits throughout Tibet and Bhutan and Nepal, he prophesied that Zanskar would be "like the happy cemetery of Sukhavati in India." Indian mystics of the day sought out burial grounds for their contemplative practice, places well stocked with reminders of impermanence, decay, the folly of holding on. "The trees of the cemetery are the bushes at Sani and the birds there are the vultures thereof. So said Padmasambhava."

Local legend has it that Guru Rinpoche took a liking to Zanskar and stuck around for a few years because it was conducive to meditation, to the perception of those foundational truths: the interdependence of all things, the mutability and impermanence of all things. Everything's connected. And nothing lasts forever.

He may have been the first visitor to prize Zanskar for this rarefied clarity, but he was far from the last. Newcomers often remark on how close things seem, like you can reach out and touch them. From Kumik, it does almost seem like you could toss a paper airplane to the monastery of Karsha, several miles distant on the other side of the main valley.

This is true of the whole "rain-shadowed" roof of the world: the clarity at these heights, an utterly singular vernacular of light and shadow, makes visible both promise and peril. The physical constraints are clearly discernible in the thin, dry, crystalline air. Being a kind of closed system, in Zanskar it's easier to see the origins of needful things, like water and fire. The contingency of human existence—the short, fragile chain of human dependence on the sun, the clouds, the seasons, the soil, the sparse bushes, the cycle of freezing, melting, flowing, plowing and planting and harvest—is not unique to Zanskar. It pertains everywhere, though often hidden behind the veil of our modern infrastructure and institutions, which mediates, if not obscures, our relationship to the very elements that sustain us. In Zanskar, however, this dependence is transparent to a degree as in few other places on Earth. It's also easier to see the counterfactuals of survival, the consequences of severing the water and fire connections.

With its relative lack of safety nets, its position at the bitter end of global supply chains, and its singularly demanding environment, the pace of change and the arc of development is made uncannily visible in Zanskar, too. It's as if the entire Industrial Revolution has been compressed into the space of a generation or two.

There is now less imperative for *bes*, the informal system of collaborative labor sharing that has long defined Ladakhi and Zanskari agriculture. Groups of households have long been bound in these reciprocal relationships, pooling resources, helping each other during tough times, sharing good fortune if they have surplus labor. These ties, however, are rapidly deteriorating.

"Before, everyone doing for each other," Urgain says. If I have a powerful *dzo* that I use for plowing my fields, but you don't, no problem. Together we will use my *dzo* to plow your fields in the spring, and you will help me harvest mine in the fall. Families used to help each other harvest their fields, for however long it took. "When they finish the field work, they are waiting for each other. Now, oh! Everyone wants to finish quick. Now, not helping. Now they don't have time. Time is money.

They need to finish as soon as [possible]. Then they start again another work, another business."

These days fewer Zanskaris depend exclusively on agriculture for their livelihoods. The government rations of flour and rice, and vegetables purchased from the market, now constitute a significant portion of their diets. But more than that, the logic of *bes* has been undermined, as the politician, with his matches and PWD-delivered water, astutely observed, by improved access to modern goods and services. No one has had to carry embers from a neighbor's hearth these past couple of decades.

This fraying of the ties that bind is as true in Kumik as it is in other Zanskari villages. Some villagers with government jobs now hire migrant laborers to harvest their grass and barley. They are less and less interested in traditional rotating jobs like the *lorapa*, who ensures hungry livestock are kept out of everyone's fields and levies fines against careless yak owners, and the *rarzepa*, who spends the day tending each household's sheep and goats on the slopes around the village. The way of the world has arrived, and so the *chu len me len* has been almost severed even as Kumik's connection to the rest of the world—with its packaged Maggi Noodle soups displacing barley *thukpa*, its Old Monk Rum flowing in lieu of *chang*—has grown stronger.

The economic logic of contributing to villagewide projects like the canal also strikes some households as less than persuasive. Some are more enthusiastic about such schemes than others. Sometimes people fail to show, and resentments over money and inequality and who isn't pulling their weight flare into open arguments in Kumik's placid fields. And as in any village the world over, some have more time and money and resources to share, and some have less. The Gonpapa household, for example, has labor to spare: ten children and several daughters-in-law. Tsewang Rigzin's house, on the other hand, has no young strong backs to offer, only his grandmother, his mother, and two fathers (from a time when polyandry was common in the region, to keep populations in check and family farms from being fragmented). Yet equal sacrifice would be expected of each house.

The result is a paradox: the Kumikpas are both more free from ancient resource constraints and more vulnerable than ever. Individually perhaps more resilient but collectively less so. There's a poignant irony of which they are well aware: even as the village faces its gravest challenge to date—the move to Marthang will require them to come together as one unit more than ever before—the social fabric is tearing under the influence of these global and national economic forces.

It's tough timing. The cord of the water and fire connection is down to a few strands and may not be strong enough to do all the heavy lifting still ahead of them in the "red place."

Watching this manifest in a thousand different subtle ways, the more time I spent in Kumik, the more I came to wonder: how would Kumikpas come together to build a new village in the Marthang desert, their most daunting undertaking since the "men from Guge" arrived in the empty valley? The clock was ticking with each year of failed snows.

Could they rebraid the water and fire connection in time to save themselves, and each other?

Each time I returned to the United States from a long stay in Zanskar, I began to ask the same questions at a larger scale. For most of the rest of us, though, our fire and water connections are much more opaque. So, too, are the contours of our own looming Marthangs.

"We're looking at a scenario where there's no more agriculture in California."

That was former Secretary of Energy and Nobel laureate Steven Chu in a 2009 interview, discussing the likely future climate change would visit upon his native state—which has the eighth largest economy in the world.

Just as Zanskar was once known as the "food palace" of the arid western Trans-Himalaya, just as Kumik was the village where "everything grows," California is today the "salad bowl" of the United States. The state produces 21 percent of the country's milk and half of all fruits, nuts, and vegetables grown in the United States. Most of this produce is grown in the 400-mile-long Central Valley. And, as in Zanskar, this bounty is made possible by melting snow.

Kumik's water delivery system is a simple affair. Start with melting snowfields and a glacier and then let gravity (with the help of some small sections of pipe and concrete) take care of the rest. The distance between the snowfields of Sultan Largo and the fields of Kumik is just a few kilometers. The whole thing, from top to bottom, source to spring, is visible from any rooftop in town.

In contrast, the California State Water Project is the world's largest piece of public water delivery infrastructure: a complex network of 701 miles of pipelines and canals and tunnels, 20 pumping plants, 34 reservoirs and lakes and storage aquifers, and 21 dams, including Oroville Dam, the tallest in the United States (at 770 feet, it's about as high as the elevation difference between old Kumik and Marthang).

The SWP supplies water to 25 million people, about two-thirds of the state's population, and irrigates 750,000 acres of farmland. It moves 2.4 million acre-feet of water a year from where it's most abundant (mostly in the north) to the most populous, thirsty parts of the state (mostly in the south). About two-thirds of the water is destined for residential and industrial use in California's bigger cities; the rest goes to farms in the Central Valley. The Central Valley Project, which predates it, stores 13 million acre-feet of water in 20 reservoirs; this is then channeled to the San Joaquin Valley and other important agricultural areas.

California's prodigious water-moving efforts began a century ago. When its first section was completed in 1913, the 233-mile-long Los Angeles Aqueduct was one of the largest infrastructure projects of its kind in the world. Its series of canals and pipes and pumps bring water from the Owens Valley (which lies in a sort of "rain shadow" on the east side of the Sierras) all the way to water Angelenos' lawns and fill their mugs of herbal tea. Much of that water heading south is melted snow.

The snows of the Sierra Nevada provide more than a third of the state's total water supply. The snowpack serves a critical storage function, like a vast frozen reservoir, releasing meltwater in the spring and summer when it's most needed by thirsty farms. The downstream reservoirs store water when it's flowing in high volume (just like the small *zing* ponds in Kumik), so that it can be parceled out later in the hot, dry summer.

And, much like Kumik, the most populous U.S. state faces a gloomy long-term prognosis for its snow-fed agriculture. This fact was the reason for Chu's downbeat, provocative assessment. California's agricultural sector is particularly vulnerable to warming temperatures. While climate change won't necessarily change the total amount of water in circulation in the region, it will dramatically reduce the amount of snowfall. More precipitation will fall as rain, depriving California of its frozen water stores, and adding runoff that will overwhelm the current capacity of its reservoirs. Analysts at the state's Department of Water Resources are bracing for a 25 to 40 percent reduction in the Sierra snowpack by 2050. "And data suggest it could be even worse than that," John Andrews, the department's lead climate change analyst, told me. Model projections for the end of this century suggest the end-of-season snowpack in those mountains, and in other western mountain ranges, will likely decrease by 40 to 90 percent.

Some scientists and analysts, including Andrews, politely disagreed with Chu's forecast, noting Californians' genius for adaptation and self-reinvention. "Even under the most severe projections for climate change," Andrews told me, "we're going to have the same level of precipitation; it's just not going to be in the form that we want. It's going to be more rain. The timing is going to suck compared to what we have now, but we've got eighty-eight years to adapt to that. So far, it could be better, but we've adapted to a climate that most of the Europeans who arrived here were not used to. The rest of the country doesn't do this, going eight months of the year without rain."

All the same, Chu's comments pulled back the veil briefly on some linkages and hard truths that often remain hidden, or at least unacknowledged, to most Americans. If current rates of greenhouse gas emissions continue, average temperatures are projected to rise between 4 and 10 degrees Fahrenheit in the American Southwest. (So will Californians' internal thermostats: By 2090, a California resident will be up to seven times as likely to die in an extreme heat spell.) Uncontroversial climate models suggest that by 2100, California's water supply, and its enormous agricultural industry—and by extension, the U.S. food system—will likely be severely strained by the shift from snow to winter rain and the subsequent loss of the snowpack's storage function.

Despite their differences in design and scale, the function of Kumik's and California's systems is identical: to carry melted snow and ice through an arid land, to where people who need it can use it. And both places are now in the grips of severe drought.

In September 2013, federal agriculture officials declared that almost all of California—every county except San Francisco—was a federal "drought disaster area." (Over 1,000 counties across the United States were also declared drought disaster areas, the vast majority of them in the western United States.) The year 2013 was the state's driest in almost 120 years of record keeping. In January 2014, Governor Jerry Brown declared a drought emergency, and officials at the state's Department of Water Resources took the unprecedented step of cutting off all water to local water agencies, which serve 25 million Californians. Their goal was to preserve water supplies in upstream reservoirs so there would be more available for everyone, including the state's farmers, during the hot months of summer.

The snowpack was then at 12 percent of its normal levels. Close to zero precipitation had fallen in the usually wet month of January. Shasta Lake was 100 feet below its average level. Plans were made to deliver drinking water in tanker trucks to dozens of rural communities that would run out of water before summer. Ranchers were selling off their animals because they couldn't harvest enough hay to feed them and purchasing feed was getting too expensive. Farmers planned to let more and more of their fields of lettuce, tomatoes, cotton, and wheat lie fallow.

Meanwhile, without rain to scavenge particles out of the air, pollution levels spiked in the Los Angeles Basin and the San Joaquin Valley. In early January—well after the normal end of the wildfire season—fires consumed 2,000 acres in the San Gabriel Mountains above Los Angeles. "We've been living off the snowpack from a couple of seasons ago. It has become a very serious issue and probably exacerbated some of the other issues we are seeing, such as the [wildland] fires," an agriculture official told the Stockton *Record*.

As temperatures ratchet upward, longer and more severe droughts are expected in the future. Warming leads to reduced snowpacks. Less snow means less moisture in the soil, which, together with higher temperatures, leads to drier summer conditions—which means more wildfires. More wildfires also mean more soot released into the atmosphere, including light-absorbing black carbon and brown carbon, the lighter-colored form of organic carbon produced in large quantities by burning biomass. And this all leads to more warming.*

More wildfires in the American West also mean more soot will plume upward into the higher part of the atmosphere, where it will be blown across the continent and possibly into the Arctic, speeding up melting there. Speeded-up melting hurries us toward the day when some invisible tipping threshold is crossed, and the ice sheet's outflow feeds rising sea levels, which threaten coastal cities like San Francisco, but also islands in the San Joaquin Delta, which are up to twenty feet below sea level and protected by levees.

"Some parts of the state are going to have to deal with sea level rise," John Andrews of the DWR told me. "Most of our urban population is along the coast, so that's going to be their problem. But if you go to the interior of the state, they're going to be less interested in sea level rise and maybe more interested in snowpack."

In addition to reducing the amount of snow that falls in the Sierra—and increasing the amount of winter rain and flood events—rising temperatures will shift melting to earlier in the year. It could begin as early as February in the future, leaving less snowmelt runoff available during the dry summer months, when farmers need water most. Meanwhile, increasing winter rain will likely overwhelm the storage capacity of the state's reservoirs.

It's as if those ancient Vedic gods, Vritra and Indra and Agni, have brought their never-ending struggle to the skies over the American West.

In 2009, researchers from the Himalayan region and California gathered for a workshop in La Jolla to discuss all these parallels between the observed, ongoing environmental changes in their respective regions

---

*Ramanathan and colleagues, whose lab at the Scripps Institution of Oceanography in San Diego has done as much work sampling and measuring air pollution over California as he has over India, flew drones over southern California during the extensive wildfires of July 2008 to sample what was in their dense plumes of smoke. They found that black carbon and other aerosols from those fires contributed significant warming, at least 0.6 degrees Celsius per day, in the lowest 10,000 feet of the atmosphere. Black carbon accounted for 80 percent of the heating.

and to sound a warning about the impact of air pollution and climate change on future water availability in the western United States and high Asia. The researchers cited a troubling litany of changes reported from Asia and the American West, making Chu sound pretty reasonable: "The monsoon, which provides water to over 1 billion people, has been weakening and becoming more erratic in recent years. . . . In 2006 Assam, East India, one of the wettest places on Earth, experienced a drought. . . . The Sierra Nevada snowpack, a vital summer water source already in decline, will be reduced by 40–90% by the end of this century. . . . The flow in many of the major rivers in China is decreasing . . . the Yellow River sometimes fails to reach the sea as does the Colorado River. . . . Melting glaciers in Asia are forming larger mountain lakes and are likely to cause potential catastrophic floods." On top of it all, "Water demand is projected to rise at a faster pace than that of the world's population growth, the latter of which is expected to increase by 50% by 2050."

This compounded threat of ever-tightening water supply, overlaid on increasing demand, hovers over the entire American West. The western states get 75 percent of their water—for power generation, industrial and domestic use, and irrigation—from snowmelt stored in reservoirs. The snow-fed Colorado River system provides water for more than 30 million people in Los Angeles, Phoenix, Las Vegas, and Denver. "I don't actually see how they can keep their cities going," Chu continued in his unusually candid interview. "This is true not only of California, this is true for all the Western states."

That future already seems to be arriving. By the end of 2012, roughly 55 percent of the United States was in drought. A below-average snowpack in the western United States the winter before had been compounded by record-high temperatures to create arid conditions, which led to a record-setting wildfire season. Another subpar snowpack in the winter of 2013 raised fears that another summer of drought and wildfire was in the offing. Reservoirs were low in places beyond California—one beet and alfalfa farmer in northern Colorado told the *New York Times* that, due to the low snowfall, he might reduce his planting by a third in 2013, fallowing 1,000 acres, and that he "was praying the spring snow and rains would come to save him."

Some government forecasters, the *Times* noted, "now end phone calls by saying, 'Pray for snow.' " Sounding like a Kumikpa speculating on the mysterious motives of the *lha*, Senator Mark Udall of Colorado told the *Times*, "Mother Nature is testing us."

What connects droughts in Kumik and California, waning ice on the slopes of Everest and Greenland, hazy skies over Beijing and smog over Los Angeles?

Scientists are already starting to uncover some hidden linkages. Carbon dioxide remains a prime suspect in the unusual severity and duration of California's droughts and wildfires, as it adds energy to the global atmospheric system that influences almost all weather patterns, even if it can't be said to "cause" any particular storm or event.

But black carbon seems to play a critical role, too, with its strong, disruptive influence on regional climate, but also because of its surprisingly long, intercontinental reach in some cases—like some dark, twisted version of the "It's a small world, after all" ride at Disneyland. As in the Himalaya, black carbon seems to upend the timing of water flow wherever it's emitted, disrupting the agricultural systems that depend on reliable irrigation patterns.

In 2010, Odelle Hadley, then a researcher at Lawrence Berkeley National Laboratory, led a team that analyzed snow samples taken from different monitoring stations throughout the Sierra Nevada. They found concentrations of black carbon that were high enough to significantly reduce the albedo and increase melting of the snowpack in early spring. The majority of the black carbon was local in origin, mostly from diesel exhaust. But Hadley also made a surprising finding: up to one-third of the black carbon they found at high elevation sites in the northern and central Sierras originated in Asia. Hadley and her coauthors noted that "between 1988 and 2001, the annual average atmospheric BC concentration in the San Francisco Bay Area"—the major upwind source of black carbon reaching the mountains—"has decreased from 2 µg/m3 [micrograms per cubic meter] to less than 1 µg/m3, while at the same time, BC emissions from Asia have risen dramatically."

There were more dark fingerprints if you looked further. The extreme drought plaguing California in 2013 and early 2014 was blamed by meteorologists on a high-pressure ridge parked over the West Coast that had blocked the storm tracks that typically brought precipitation in those months, deflecting them farther north. California had seen these kinds of systems before, suggesting natural variability was at work. But this ridge was unusually stable and long-lasting. The way it unfolded looked eerily like a scenario outlined by Lisa Sloan and Jacob Sewall at the University of California, Santa Cruz, a decade prior, in research they had done examining how a warming Arctic might alter the jet stream that drives much of the familiar weather patterns in North America. Their many model runs yielded a robust forecast: as the Arctic warmed, in part due to the shrinking sea ice and associated feedback mechanisms, the western United States would get noticeably drier.

In interviews, Sloan pointed out to reporters all the other factors that could be contributing to California's record-busting drought: a strong El Niño system, for instance. Or global warming might be increasing the thickness of the atmosphere, which can increase the persistence of high-pressure ridges. Still, she couldn't help noting, as many other experts did, that the resemblance of the 2014 drought to her and Sewall's scenario was somewhat uncanny. And if they're right that the changes in the Arctic are at least one strong contributing factor to the weird dry weather in the western United States, then the question remains: why is the Arctic warming so fast—faster than any other part of the planet, and far faster than scientists had predicted?

Oh, lots of things: Arctic climate offers an equally (which is to say, inordinately) complex picture as California's. Rising air temperatures due to greenhouse gas emissions are surely a major factor. The feedback mechanism of increasing areas of exposed dark seas heating up nearby light-colored snow and ice is another. But increasingly, scientists' attention is trained on a certain potent bad actor up in those high latitudes: our old friend black carbon.

Charles Zender, a professor at the University of California, Irvine, and an expert on aerosols' impact on the Arctic, told Congress in 2007 that "black carbon appears to warm the Arctic more than any other agent except CO_2." Black carbon poses a "triple threat" during its stay in the higher latitudes. "Because the Arctic is so very bright . . . the sunlight that it can absorb has two chances to be absorbed by it: on its way down, and on its way back up being reflected from the ice sheets." Then, when black carbon gets deposited, it has yet another crack at warming, adding energy directly to the snow and ice. For these reasons, Zender said, "It is the most potent warming agent we know of in the Arctic."

"Because of its short life time and strong effects," he concluded, "reducing Arctic black carbon concentrations sooner rather than later is the most efficient way that we know of to retard Arctic warming."

I've been keeping a nervous eye on the calendar. The *chadar* ice way melts in late February, blocking all exit from the "wolf trap" of Zanskar until summer returns. (*Chadar* used to last until April, but, you know, winters are warmer these days, so . . .) I head to Kumik for one more visit before I walk out on the ice. There is much to plan.

I soon find myself sitting in Tashi Stobdan's *yokhang* again, close to the long, box-shaped metal stove. Nearly three years have passed since we first met in 2008. Some things have changed. There are a couple more house foundations down in Marthang. I learn that Stenzin Namgyal, the enterprising mason who had built his own Trombe wall solar system two doors up from Stobdan's house, had died. (From drinking too much *chang* and rum, the villagers say.) Abi Tsering Dolma had died, and the day after I arrived I watched from a distance as she was cremated out on the edge of the village, and a column of bluish smoke merged into the clear blue sky.

Now I watch Stobdan pad around in search of cups, then place the teakettle on the stovetop, grimace, and convulse briefly in a hacking cough. He exchanges some low words with his father, Rigzin Tantar, a man of few words and an irascible, winking sense of humor, who is seated with Tsewang Rigzin's father and Ishay Paldan along the far wall. Wrapped in two layers each of homespun woolen robes, the three elders talk to each other in that easy, unhurried way of customers in a small-town barbershop, the way of men who have worked, suffered, and celebrated shoulder to shoulder all their lives.

Ishay Paldan motions toward my cup of *chang*, urging me to drink.

The old men chat idly about the weather. It hasn't snowed much this winter, again, they say, and leave it at that. No need to spell out what it means for the coming summer. Above their graying heads the window frames the mountain and its snowfields, and the stout mud brick houses that pile up the hillside against an electric blue sky. Sheep bleat and children shout outside in the alley. Zangmo, ever bustling, flits in and out, pours out boiling salt tea, and spoons lumps of butter from a battered tin can into our cups. Stobdan tosses more dung into the firebox of the stove. He pokes at the embers until they sputter back to life.

Yellow flames lunge and light up his face. He looks tired, but there is a gleam in his eye that I've seen before, the same patient yet pregnant glance he wore the first time we met, when he slid the design for his new home across the table at me. The same look I saw on that first tour of Marthang. Now he sits cross-legged on the floor in the center of the kitchen.

He clears his throat and recaps the story that everyone already knows because they have lived it. He talks about the long drought and the lost grass harvests of recent years; about the declining snowfalls, the shrinking glacier, the warmer springs; about the ever-shifting time line for leaving their village; about the dozen or so families that have built new homes down on the dry and dusty plain a thousand feet below, but have yet to make the leap of living there year-round; about the canal that they have labored together to dig. A short tale of action born of necessity. The men listen closely to their own story and punctuate it with solemn nods.

Stobdan pauses in his discourse, looks at his hands, rubs one wearily across his brow. He rearranges the glowing dung coals in the stove's firebox with a willow stick, sending a series of small self-contained gray puffs up through the gap between the stovetop and the edges of the blackened kettle. Now he comes to the reason for this meeting. A small sum of money has been donated from *chigyal* for construction in the new settlement in Marthang. "And we are asking everyone, how can we use it to benefit not just one or two households but the whole village?"

With his mild-mannered bearing and his soft, almost lisping voice and youthful features, Stobdan's unmistakable air of authority can seem incongruous. Though he is more than three decades their junior, the men listen respectfully and murmur their assent. Ishay Paldan leans forward, adjusts his soot-stained yellow woolen hat, cocks his head to one side to hear better.

"We have an idea for this," Stobdan continues.

He recounts yesterday's scene, when he shooed his dozen sheep into the ground-floor stables of the house, and people from every household in the village crowded into his dung-strewn courtyard, laughing and jostling. Stobdan had stood in front of them and announced that, after a decade of shared labor on a new canal, it was time to take another step together. "So what should we do?" he had asked.

There was a raucous minute that sounded like debate but was really just a collective voicing of spontaneous consensus: Kumik's grandmothers, teenagers, government workers, masons, mothers, soldiers, sisters, and sons decided to build a new *lhakhang*. If the hearth served as the physical and spiritual focal point of every household, then the *lhakhang*—prayer hall, community center, gathering place—would be the hearth for the whole of the new village they were slowly raising on the plain below.

"A solar *lhakhang*," Stobdan tells the old men, as the smoke races up the stovepipe. A gathering place heated by the sun. A new fire connection for life in Marthang. "And everyone in Kumik will help build it."

The elders quickly warm to this idea, and the conversation turns to details of execution. Better to use mud bricks or stone? Would every family be expected to contribute the same amount of labor regardless of size or wealth? Where will the high lamas sleep when they visit to give blessings and teachings?

"And it should have a space for the women to weave together in winter," Zangmo suggests firmly.

After about an hour of discussion, the fire has died down. The tea and *chang* cups are still full, because in Zanskar a good host never lets a cup sit empty, and Zangmo is a very good host. And the elders of Kumik have endorsed a course of action: construction on the solar *lhakhang* will begin in the summer. Each household will contribute at least ten days of labor. The structure will be made of earth, reinforced to withstand earthquakes, and designed to soak up the sun's rays and turn them into heat, in this valley over 12,000 feet high, where the sun shines 315 days out of 365.

Satisfied, the three men get up slowly, rheumatically, to leave. Ishay Paldan stoops in my direction and shakes my hand, still wearing that look, like a kindly uncle winking with his entire face, the one that says, *One day, kid, I'll let you in on the joke.* The men shuffle out the door. It's almost dark out there, a windless night, and in the waning light the smoke from thirty-nine hearth fires hovers over the village, still visible, like a brooding afterthought.

Back inside, Stobdan passes me the tray of biscuits. He radiates a slow-burning optimism that throws off at least as much warmth as the stove. In its glow, the hidden punch line of his joke by the metal sign is partly illuminated: his was the laugh of a man who has a plan, who has embraced the change forced on him, because in it, he sees great opportunity. It was also the laugh of a man who knew he was not alone. He had *yato*—help.

He flashes his boyish grin. "It's a good plan, isn't it?"

PART TWO

The World, Burning

Bhikkhus, all is burning.

—BUDDHA

As combustion is ubiquitous, carbon soot is found virtually everywhere.

—ROBERT BERGSTROM ET AL., 2000

4

The Road to Shangri-La

Human history becomes more and more a race between education and catastrophe.

—H. G. WELLS, *THE OUTLINES OF HUMAN HISTORY*

"You are a strong committee!"

It's summertime 2011—five months after the meeting with the elders, three years after I first arrived in Kumik, over a decade since the villagers' decision to build a new home in Marthang—and late at night in the "waterless village." Four men review the state of things down in Marthang. The first need—water for irrigation—has been met: the canal is finished. Work on the solar *lhakhang* will begin soon, and the villagers have chosen Tashi Stobdan and Tsewang Rigzin to lead the building committee. Their neighbor, Tashi Dawa, is thrilled.

"STRONG COMMITTEE!" he repeats, at higher volume.

Tashi Dawa is a hardworking, no-bull kind of guy. A mason by trade, and like all his neighbors a farmer, too, he now works for the Indian government's road-building outfit. He's been *goba* (mayor) and held every other leadership position in the village in the course of his fifty-some years, soberly executing his various responsibilities. He is normally a man of few words. But, like many Zanskaris, after several cups of *chang* he gets voluble and exudes warm feeling. His gray mustache begins to work up and down with praise.

Stobdan's feeling good, too. He switches to English. "We make a model not only for Kumik," he declares, "but for whole the villagers of Zanskar!"

Stobdan leans forward to emphasize the word "whole" such that I think he might fall over onto the *choktse*, the small, ornately carved table that serves as the only piece of furniture in Zanskari kitchens. But he keeps it together and reaches for his cup of *chang* instead.

"Kumik people will move after one year, two years, three years," he says, "but it must be model for Zanskar. So most people—intelligent people—they will come to see how they want to make themselves."

Tsewang Rigzin, who's kept his foot less heavily on the *chang* pedal these past three hours, nods with a cryptic smile. He and his oldest friend (who is listed as "Stoby" in his mobile phone's address book, and who in turn has entered Rigzin's number under his nickname *Garshung*, "Little Gara") have been chosen because they are the two most educated sons of Kumik, because they are forward thinking, initiative-takers.

"You have the support of the whole village!" Tashi Dawa says, pounding his fist on the *choktse* in front of him. "We are ready to come at night and build, if you give the order. You decide and we will follow you into battle!"

"On war footing!" shouts Rigzin, laughing.

Tashi Dawa continues along this line, in slightly less coherent but still emphatic terms of encouragement. "*Song, song,*" Rigzin keeps intoning with patient nods. "Agreed, agreed." Their neighbor Gara, whose permanent hangdog expression belies a quicksilver sense of humor, chimes in with some familiar context.

"The government has not helped us," he says. It's a perennial grievance: how little the authorities have done for the village. "We have asked the member of parliament, the MLA [member of the state assembly]—nothing." Heads around the room waggle in accord. "*Song, song . . .*" No matter, they will do it themselves, as usual. They are used to this kind of bootstrapping.

Stobdan is staring straight ahead over his cup into the middle distance, where perhaps he can see the buildings of the new village sprouting like green shoots of young barley from the plain. "It is good for people in Zanskar. I want to show people the modern house. If my new *yokhang* is 4 or 5 degrees Celsius [in winter] then I am very, very happy, really!" Vigorous nods all around.

Everyone is in agreement: the *lhakhang* should offer a model for villagers building their new homes below in Marthang—and for other Zanskaris—of how to heat their living spaces using less dung, with fewer fires and cleaner air.

"As compared to Leh, in Zanskar is very, very different," Stobdan reminds us all, a bit needlessly. "Because when we are in Leh we wash in the early morning. But if in Zanskar you stay [washing] for long time—is too cold!"

"*Song, song . . .*"

In the meantime I've lost Tashi Dawa's thread, which has continued apace. But Stobdan, also deep in his cups, seems to grasp this new language, which only permits enthusiastic phrasing, so he translates for the rest of us:

"He says one day Kumik will become the model village! He says we are very lucky persons . . ."

Before we all stagger home, Stobdan and Tashi Dawa spend fifteen minutes hunched over, heads close together, trying to liberate a beetle that has gotten its legs tangled up in a piece of thread.

Stobdan and I are a bit sluggish the next morning. Over breakfast he looks up at me sharply, as though he just remembered something important.

"Ishay Paldan has gone to the *lhakhang*," he says, "but no one is there."

It's our first full day of work. Kumik's most senior citizen is down there, alone in the hot sun, waiting for us *chang*-guzzling slackers to start working on a building he may never use.

We wolf down our bread and head outside, hop on Stobdan's old Yamaha motorcycle, parts of which, I notice for the first time, are held together with twine. We coast halfway down the rough switchbacking road, dodging basketball-sized rocks, and soon come upon Meme Paldan leaning into the slope. Hands clasped behind his stooped back, he is already making his way slowly back up the nearly thousand-foot climb to Kumik.

"We're coming now to work!" Stobdan shouts.

Meme Paldan stops and raises his head. He nods and gives a gentle laugh, turns right around, and walks back down the hill, head down and hands still clasped like a monk on his pre-matins stroll. We roll off down the hill ahead of him, accelerating into Marthang.

"Why don't I get off and you give him a ride?" I shout to Stobdan.

"Oh he's fine! Meme is strong!"

A half hour later, down below, on a small plateau in the heart of Marthang, Meme and I are swinging pickaxes and shovels, hacking at the earth's baked crust. (The rest of the "strong committee" has demurred: Stobdan has to teach at the school, and Rigzin has hurt his hand somehow, so he cooks lunch instead.) Our goal is to carve a shallow 200-meter-long canal that will bring river water from the new main canal to the site of the *lhakhang*.

Soon that water would fuel the growth of new fields full of barley, wheat, peas, and generations of Kumikpas yet to be born. But first the new village must be built. "It all depends on water," Stobdan says. "If the river is low, water is stopped, everything stops." And by everything he means not just agriculture but construction: water is needed for making mud bricks and mixing concrete and mortar and plaster.

So Meme Paldan and I attack the hard ground. The morning wears on and heats up, and we leapfrog around each other, trading off the shovel and the pickaxe as we make glacial progress toward the little promontory where the *lhakhang*'s dimensions have been marked out with stones and twine. Meme hunches over as he works, breathing audibly and coughing loudly. But the promise of a smile never quite leaves his face, framed as always by his dirty yellow wool hat and sparse white whiskers.

Our humble channel, two shovel blades wide and six to eight inches deep, advances slowly but steadily. At intervals Meme makes a simple announcement—*Ngalte rak*, "I feel tired"—and sits down on the sand to watch me work.

He asks after his relative Lobzhang Tashi, with whom I first visited Kumik three years prior. I remind him of what we discussed at that first meeting—the disappearing waters, the snow of the old days, the past and future, and how things are now *hamo,* difficult, and what is there to do but keep working. His smile lines deepen at the recollection. We talk about the previous winter's scant snows and about his son, who drives the school bus in the neighboring village of Stongde. Meme Paldan has two daughters who live in villages across the river and two more in Kumik. One of them is unmarried and lives with him and his wife in their unassuming little mud brick house on the upper edge of the village. This day spent sweating in Marthang has turned his thoughts to them. He doesn't expect to be around to see the new village in full swing. But he is invested: the flow of the water is inseparable from the fate of his children, and of their children, and of their neighbors, and everything he knows and prizes in this world. So I imagine as we work that with each shovelful of sand he is chipping away at uncertainty, at the mountain of his concern for the life that awaits his family here, in much the same way he labored on the shoulder of Sultan Largo fifty years ago—exerting some last act of gravity to bend the flow of events. He leans on his shovel and looks up at me.

"Do you have children?"

"No," I say, and because an explanation is always required in this part of the world, where family and household constitute much of one's identity, I add reflexively, "haven't found a wife."

"*Topchen*," he says. "You will find her."

He says this one word with such simple sincerity and confidence and level gaze—like some gentle soothsayer—that I lean on my pickaxe and regard him anew. He picks up the shovel again and points out that I've gotten off course. He's right: my line is all crooked. He scrapes out a straighter one for me to take toward the *lhakhang*. I heft the pick again, and we work through the balance of the afternoon.

When Stobdan comes to check on our progress, the sun has started to slant into the Stod valley and our shadows lengthen toward Kumik. We call it a day. As we gather our tools and thermoses, Ishay Paldan invites me to come to his little house for tea, tonight or any time. Then he departs, and I watch him shuffle slowly off across the darkening plain, head down, hands clasped.

Stobdan and I linger for a while to discuss work plans for the week.

We appraise the fruit of the day's labor. Without a JCB, our little canal is two-thirds complete.

"You were right," I say, "Meme is strong."

If ever a Himalayan kingdom deserves the title Shangri-la the remote valley of Zanskar would surely be the prime candidate."

That's a sentence from the opening passage of the book that Tsewang Rigzin started writing about his homeland several years ago. (It's still in progress.) The comparison to the elusive paradise of James Hilton's 1933 novel *Lost Horizon* is, at least on the surface, apt: if Zanskar has penetrated the wider consciousness at all, it is indeed as a place apart, inaccessible, unspoiled, a remote zone that is home to a pious and self-sufficient people. (One of the first of the few books written about Zanskar was aptly subtitled *The Hidden Kingdom.*)

And one doesn't have to work too hard to find uncanny similarities between the real place known as Zanskar, located on planet Earth roughly between latitudes 33 and 34 degrees north and longitudes 76 and 77 degrees east, and Hilton's fictional refuge of unspecified coordinates.

Early in *Lost Horizon*, a hijacked plane, bearing a motley crew of disaffected westerners trying to escape a local revolution, crash-lands high in the western Himalaya. The survivors are rescued (or gently kidnapped) by villagers who materialize out of a raging blizzard and take them over steep mountain paths to a breathtaking valley ringed by high peaks, where the local people live in pastoral harmony: Shangri-La. While strife and storms rage on in the outside world, in this valley the weather—political, emotional, and atmospheric—is sunny and calm.

"You see, we are sheltered by mountains on every side," the mysterious Chang, a senior member of the lamasery of Shangri-La, tells his visitors upon their arrival, in the 1937 film version of *Lost Horizon*. "A strange phenomenon for which we are grateful."

"How do you deal with incorrigibles? Criminals?" asks Conway, the thoughtful man-of-action protagonist, played by the suave Ronald Colman.

"Why, we have no crime here!" says Chang. "What makes a criminal? Lack, usually. Avariciousness, envy . . . there can be no crime where there is a sufficiency of everything."

Compare this exchange to an anecdote relayed in the pages of *Himalayan Buddhist Villages* by John Crook, who traveled extensively to Zanskar before the road to Kargil was finished:

> In the summer of 1977, the first of our parties to visit Zangskar [sic] was approaching the Umasi-la on foot by way of the Bhutna valley and stayed overnight in the small Kashmiri town of Atholi. The police chief, "Thakur Sahib," treated us royally and, in the course of a celebratory supper, remarked that once we were over the pass we would find no police. "There is no crime in Zangskar," he remarked. "So why should we police go there?" Pressed to explain this extraordinary fact, he was slightly embarrassed to admit it was something to do with the peaceful character of Zangskari farming life and Buddhist religion. There was a remarkable lack of violent disputation, theft was almost unknown and any altercations that did occur were about land, irrigation or marital infidelities which the villagers sorted out themselves. And, indeed, on reaching Zangskar we found a land of cheerful honesty, a sharp contrast to the valley of Kashmir.

Corroboration comes from James Crowden, who recounted how, on a separate visit in 1976, the *tehsildar* (the government revenue officer, hailing from another part of India) in Padum told him, "These people are perfectly civilized. They have no need of government. It is we who should be learning from them." And yet more from Michel Peissel, a French explorer and author who visited Zanskar in 1976 and had to listen to the Sikh police officer posted in Padum complain about how bored he was, due to the lack of crime: "At best we get a few fist fights when the people have drunk too much at festivals."

Zanskar has some other eyebrow-raising parallels with Shangri-la, beyond the mountain walls that block the storm clouds and quicksilver obsessions pulsing through the world beyond. Improbably cliff-hugging monasteries that once ruled over the villages below—check. Sophisticated traditions of communitarian, zero-waste pastoralism and agriculture—check. Indomitably cheerful residents who give you the distinct impression they know something important about life that you don't—check.

Even the name—"food palace"—suggests a past abundance similar to the Arcadian self-sufficiency Conway and his fellow refugees from the "civilized" world discover when they reach Shangri-La.* (Yet another reading of Zanskar's name—"land of white copper"—alludes to its abundant deposits of the valuable metal; in the book Shangri-La's mountainsides are veined with gold.) Then there's the credo by which Shangri-la's inhabitants live—"everything in moderation, including moderation"—which could function pretty well as Zanskaris' motto, too. (The Zanskari analogue might be *sheep sheep,* most often invoked when one is being served *chang* at celebrations, meaning, "Okay, I will drink slowly—but continuously.")

I could go on for a while in this vein. But there are some equally obvious problems with Rigzin's formulation.

For one, we all tend to think of Shangri-la as a lush place. As Hilton describes it, the mountains that keep out both storms and unwanted interlopers somehow manage to wring enough moisture out of the sky to keep the place looking like Tuscany. There are forests and abundant fields. There are no waterless villages. Shangri-la would seem to have mild Californian winters, if it has winters at all.

Contrast this with ice-encrusted and sand-blasted Zanskar, where people are obsessed with the color green—with *lchangma* (trees), *hariyali* (green spaces), *spang* (grassy areas)—because it is as scarce and precious in their world as gold. Ask any Kumikpa what the top priority is for work in Marthang, and she will tell you straight away, as another Zangmo, the formidable and forthright wife of Tsewang Norboo and a village leader herself, is always reminding me and other members of the "strong committee": "Trees! Trees! We need to make it green."

---

*Michel Peissel, in his book about his 1976 sojourn, wrote, "There can be few more closed economies than that of Zanskar, whose 12,000-odd inhabitants not only make their own clothes and shoes and grow all their food, but mine their own gold, silver and copper. They also gather upon their apparently barren mountains the herbs that produce their medicines. The result is a strong, sturdy and independent race of men and women who would make the average Asian peasant look sickly by comparison."

Then there's the remoteness. Not a big deal if you live in Shangri-La, because there's something about the water and the climate and the local herbs and the lifestyle and the level of medical insight that enables people to live fantastically long, disease-free lives, of 200 to 300 years. (If they leave the valley, though, they age rapidly, as all those years catch up with them all at once. Meanwhile, in Zanskar the inverse happens: the elements weather people's faces and hands prematurely, and women of thirty or forty years of age can often look decades older.) But the isolation that the residents of Shangri-la so prize is, from the point of view of most Zanskaris, a serious liability.

Perhaps the biggest problem with the comparison is that Shangri-La is in equilibrium. It doesn't change. That's the whole point: it's hermetically sealed off from the contagion, the greed and hatred and ignorance, that plagues the fallen world of shortsighted empires in which Conway happens to be a major diplomatic player. That's why this secret mini-civilization will survive, stewarding the glowing embers of humanity's best thought and creations for future generations. It is precisely because it is so remote, so hidden, that Shangri-La is heroically stable.

It is easy to forget that Shangri-La is a *modern* fiction, a response to a very particular point in time and space: industrialized Europe between the two great wars. Exploiting a nascent fascination with the forbidden land of Tibet, Hilton conjured up a new Valhalla to appeal to a generation made cynical by an age of mechanized mayhem and still reeling from the nihilistic devastation of World War I. *Lost Horizon* was thus a tale meant to both stoke and allay the fears quickened by technology run amok—by the consequences of Prometheus's gift. The tale of Shangri-La, writes one modern commentator, had a powerful appeal for those living through "the increasingly pessimistic 1930s, when Western civilisation seemed bent on a path to self-destruction—and when, as Carl Jung put it, 'the smell of burning was in the air.' "

To be sure, that burning smell can be a sign of disaster, of things becoming unhinged, of accelerating entropy. But it can also be a comfort, with sweetly acrid wood smoke conjuring the security and comity of the hearth. So if, as Rigzin claims, Zanskar merits the comparison to paradise, it seems fair to ask: is there smoke in Shangri-La?

There is in the film version, at least. In one scene, Conway ambles around the cozy village on the valley floor. On his rounds, he pauses to observe a blacksmith at work. Smoke curls out of the village man's forge—a welcome signal that all is right with the world, that tools are being made, progress and plenty are being secured. The sight is redolent of human ingenuity, togetherness, and Promethean leaps, but at a human scale. It conjures all the "good fires": the cozy campfire, the hearth, the sacred incense, the winter focal point, the clearing of land for planting. Neither man coughs or waves the fumes away. They smile warmly and nod at each other. If we noticed it at all, humanity has long worn its smoke almost as a badge of pride, even in paradise.

Prior to 1982, Zanskar was indeed almost as stable as Shangri-La. The rhythms of life hadn't changed much for centuries. New rulers came and went, demanding taxes and tribute—the Tibetans several hundred years ago, Hindu Dogra generals in the early nineteenth century, the distant government of the new nation of India in the mid-twentieth—but through it all Zanskaris' days and seasons remained cyclically limned by *zhing mos*, the spring plowing, and *khuyus*, the fall threshing, and *kha tangches*, the snowfall. As James Crowden observed, in 1977 Zanskar probably looked then much as it did in the 1820s: "Little I suspect had changed: that is, until the road came."

With the arrival of the road, the seal was punctured. The veil of the hidden kingdom was thrown aside.

Just a few hundred meters from where the new village of Lower Kumik is rising, on the main road there is a yellow sign erected by the Indian government's road-building outfit. It proclaims: "BRO brings people of remote to the mainstream." However one cares to read it—a promise, a threat—it's a succinct statement of both India's Border Road Organization's mission and its effect on the rural regions where it operates. The logic of settlement, as well as Zanskari's economic decisions, used to be dictated by the presence and volume of flowing water. Now there's another source of gravity: flowing people and goods.

Thanks to the rutted jeep track connecting them to the global economy, Urgain, Stobdan, and their contemporaries have experienced 150 years' worth of technological change—the arrival of roads, vehicles, electricity, radio, computers, lights, modern building materials, cooking gas, and cell phones—in the space of a generation.

Urgain's new house of stone and cement, with tiled bathrooms and solar-powered lights, is just a stone's throw from his old one made of mud brick, where as children he and his siblings ate in the dark or by moonlight. Back then, if they had a little kerosene, or some homegrown mustard oil, they would put it in a bottle and fill it with water to push up the fuel, because they could only afford a short wick. They would play word games over meals of boiled barley flour and butter tea. Now Urgain reads his email on his smartphone as he cooks rice from Punjab and vegetables from Kashmir in an aluminum pressure cooker. Stobdan rides his motorcycle, instead of a horse or donkey, to Padum to get online to register for his university exams.

It's a familiar story, one you can find in communities across the developing world, from Kenya to China to Brazil. The key difference is that, unlike many of those communities, Zanskar really *was,* quite literally, walled off from the wider world, just like Shangri-la. And unlike many other rural parts of the world, where villagers have had to migrate to big cities to taste these changes, it has come to them. Like the JCB that Kumikpas brought from Leh, the mainstream has been brought to Zanskar.

And what are the most striking changes in the wake of its arrival?

"The ration system," says Ishay Paldan. "Since then people don't worry about going hungry anymore if a harvest is bad."

In 2006, for example, a plague of locusts descended on Zanskar. The bugs stayed for two years and ate just about everything in sight. No one had seen anything like it. Harvests were lost. The price of barley in the market shot up fourfold. Had it happened fifty years ago, hunger would have been widespread. Today the government's ration depot provides a critical buffer for such vagaries. One can still go hungry in Zanskar, but it's a lot harder than it used to be.

"Machines," says Sonam Chondol, Urgain's wife. "Work is easier now."

In their youth there were no tractors, no vehicles, no generators. Everything—plowing, threshing, washing clothes, churning butter, carding wool, and sewing clothes—was done by hand, using people power.

Now Zanskar, like most of the rest of the world, harnesses the awesome power stored in the molecular C-H-O bonds of hydrocarbons such as diesel and kerosene and liquefied petroleum gas. The result has been an exponential increase in the amount of useful energy available to the average Zanskari. Tasks that once required human and animal muscle are now performed by flipping a switch or revving an engine. The wooden paddle that Chondol and Urgain used to use to churn butter now hangs mostly dormant on the wood post in the kitchen; a few years back they started using an electric butter churn.

And what does Urgain regard as the greatest change in Zanskar?

"Time." He smiles at a secret memory.

"Sometime I'm thinking when I was young, eighteen, twenty years old, after field work, we have many time, many long time. We went to mountain, playing cricket, time for sleeping in field. Sometimes I am thinking, 'Where is that time going?' " He chuckles. "Now people are saying, 'Time is changing, time is very short!' Everyone says. When we wake up and start doing, quick, we want to go somewhere. We feel days are very short."

He clarifies that it's not time itself, but rather people's perception of time that has been transformed. "Many people say, 'Oh, people are changing, lifestyle are changing.' But actually time is same, like before." He recalls a prophecy made by Guru Rinpoche thirteen centuries ago. "Is there in the book: we say, *tui mingyur mi namgyur*. *Tui* means lifestyle, seasons—time! *Mi* means human. The time is never changing, the seasons also same. But human beings' mind is changing, human beings' activities, everything, is changing. This is happening, almost every year, everywhere."

On balance, almost all Zanskaris you ask regard these changes as improvements. No one would like to go back to the days of huddling in the dark around a fire pit in the kitchen—least of all older folks who have passed the majority of their winter nights that way, like Ishay Paldan. If, as some outside observers have written, Ladakh and Zanskar before the intrusion of roads and tourism and the modern cash economy was some kind of Himalayan Eden before the fall, then why do local people clamor for more access, more connectedness, more goods and services? For shoes?

Because, as barefoot Meme Sonam might have pointed out, shoes are useful. Pants are useful. Kerosene and gas stoves are useful. JCBs are *very* useful. Urgain no longer has to trek for weeks to get some matches. That is an unambiguously good thing. "This time, people have a choice to eat, they have very good choice for wearing clothes, for everything," Urgain notes approvingly.

All these changes—food security, labor-saving machines, even the perceived compression of time itself—have one thing in common: they were brought by, or on, the road. But something else has arrived on the road, too. And people have begun to notice it.

"This is effect from development," he says. "Many cars are here, so we have pollution. Every year in Zanskar, fifteen or twenty new Sumo and buses coming here. Zanskari people can buy only secondhand, thirdhand vehicle. So it makes more smoke, more pollution. And right now, this time, [there are] also small factories and generators for lighting. Many smoke is produced.

"People need development. One side, people need light, they need very quick services from vehicle, everything, no? So the other side, they can lose their . . . they get problem, with the glacier . . ." Urgain pauses a few times, then falls silent.

He smiles and wags his head, a slight, eloquent gesture that asks: what to do?

One option is to just wait. History teaches that all that soot will settle after a while.

There's a famous story that biology teachers like to tell to illustrate the principle of natural selection. It's also a succinct parable of the arc that black carbon has traced through modern history, and of its potent influence on the natural world.

Peppered moths in England used to have light-colored wings that blended in nicely with the light-colored trees and lichens on which they alighted. Pollution during the Industrial Revolution, mostly from burning coal, killed off the lichens and darkened the tree bark with soot. Light-colored moths became easy to spot, and thus easy meals for birds. Rare, darker-colored mutants thrived in the sooty era, and before long almost all peppered moths were dark. The first reported sighting of a dark moth in Manchester was in 1848; by 1895, 98 percent of the moths there were dark—children of the soot (an adaptation to manmade pollution called "industrial melanism"). In areas less affected by industrial pollution, researchers found more of the light-colored moths. By the 1970s, after Britain enacted pollution controls and air quality improved, the light-colored moths were making a comeback.

This little evolutionary textbook episode neatly encapsulates the modern arc of burning. Black carbon emissions have historically followed a parabolic trajectory over the decades, not unlike the path that the particles trace through the sky on much shorter time scales, rising, traveling a ways, and then settling back to earth. Early on the development path of economic development, there's a lot of dark smoke in everyone's face: the price of survival and the price of getting beyond survival. People burn solid fuels with crude technology in order to heat rooms, cook their food, get around, and dispose of waste. Think of Tashi Stobdan in his house circa midwinter 1980, seated with his six sisters around the open hearth as smoke rose toward a hole in the ceiling. Or picture early-nineteenth-century Americans squinting as they tucked chunks of wood under blackened pots in their large, room-sized hearths.

Eventually, as incomes rise and people can afford to get fed up with nagging lung infections (like the chronic bronchitis I seem to get every time I go to Leh or New Delhi) and outraged by blackened curtains, cleaner technology and better fuels and regulations are deployed and noxious soot emissions decrease. Picture Stobdan in that same winter kitchen with Zangmo and their three children circa 2010, still burning dung for heating, but more efficiently in a metal stove, watching most (if not all) of that smoke exit the kitchen through a chimney pipe, and burning propane gas for some of their cooking needs. Or my great-grandfather installing a new gas pipe, connected to the municipal distribution system in turn-of-the-century Brooklyn, in what was once an old coal furnace.

This is the arc of our burning. We indulge in many decades of dirty, poverty-alleviating development. Then, after increased productivity has made us relatively wealthy, to the point where we can afford to care, our dismay peaks along with soot concentrations in the ambient air, people get as angry as John Evelyn was back in 1661, and governments respond, passing laws to control the smoke. More of the soot settles out, and the skies brighten somewhat, and people start to feel like the problem's been solved.

Tami Bond likes to tell this story through perhaps its best case study: the United States. Bond, an associate professor of civil and environmental engineering at the University of Illinois Urbana-Champaign, and one of the world's foremost authorities on black carbon, has made an illustrious career out of burning stuff and trying to understand what happens afterward.

Bond testified in a hearing at the U.S. House of Representatives in 2010, as bipartisan policy interest in black carbon was starting to ratchet up in the wake of several new studies demonstrating the troubling extent of its damage to human health and the environment. She presented a striking graph to the committee, showing how U.S. emissions of black carbon rose steadily from about 250,000 tons per year in 1850 to 1.2 million tons by 1925. Then they dropped off steeply, in part due to the Great Depression but in part due to transitions to cleaner fuels and better technology, especially for burning coal. Residential coal combustion went from being the largest source to barely a sliver starting around 1970, and after that it shrunk even more. Residential biofuel, namely wood, was the source of practically the entire pie's worth of emissions in 1850; by midcentury it was only a tenth, and by 2000 an even smaller fraction. Industry emissions peaked in 1920, then steadily declined throughout the century, as regulations tightened up over time.

"The history of the United States illustrates how black carbon emitted from energy use changes with development," Bond explained. And vice versa: the history of black carbon emissions illustrates how the United States has developed. It's the shadow cast by the country's meteoric economic ascent. "In the late 1800s," Bond said, "U.S. black carbon emissions were dominated by residential solid fuel, especially coal. Industry was on the increase, too. Making the coke needed to feed the steel furnaces of Pittsburgh created a lot of black carbon."

Just how much black carbon? Anthony Trollope, the Victorian-era British novelist, once visited Pittsburgh. As a native Londoner, he was no stranger to unhealthy urban environments. (Charles Dickens evoked his native ambient environment in *A Christmas Carol*: "candles were flaring in the windows . . . like ruddy smears upon the palpable brown air.") But his visit to Pittsburgh prompted him to declare it "without exception . . . the blackest place which I ever saw."

And he saw the city well before its soot levels peaked. During the 1940s and 1950s, it was one of the most polluted places in the United States; its noxious air had earned it the sobriquets "the Smoky City" and "Hell with the lid off." Pittsburgh's denizens burned locally mined bituminous coal in their furnaces, and the local steel industry and busy train depots burned even more of the stuff. Photos of a busy street in Pittsburgh, on a winter day in 1945, show neon lights blazing under a dark sky; it looks like the middle of the night, but it's morning. Cars had to keep their headlights on all day.

Eventually the citizens of Pittsburgh finally got so tired with their city being compared to—and feeling like—hell that they teamed up with local industry to pass a smoke control ordinance in the 1950s. Natural gas was piped in to replace much of the coal used in the city's homes, and locomotives switched from coal to diesel-electric.

But as incomes rose, a new source of black carbon was on the rise. In Tami Bond's graph, starting in the early 1890s—right around when Rudolf Diesel, inventor of the engine that would go on to conquer the world, was dodging blown pistons in a laboratory in Augsburg, Germany—transportation appeared on the scene as a significant source of black carbon and grew steadily through the early decades of the twentieth century. Emissions from transport—mostly diesel trucks, passenger vehicles, ships, and trains—declined a bit after the war, and then, starting in the 1970s (when interstate highway building in the United States was peaking), became by far the largest source of black carbon pollution in the United States at the end of the century.

Rudolf Diesel had developed an engine that relied on a simple physical law: when a gas is compressed, its temperature rises. Instead of applying a spark or flame to a mixture of fuel and air, combustion in a diesel engine is driven by compression. The fuel (in the early days Diesel experimented with promising biofuels such as peanut oil, before settling on the stable and widely available petroleum distillate by-product that would eventually bear his name) is mixed with air in the combustion chamber; when that fuel-air mixture is compressed, it heats up and the fuel ignites.

Rudolf Diesel correctly anticipated that, thanks to its higher compression ratios, his invention could achieve higher efficiency than other "prime movers" that reigned in the late nineteenth century. He was confident that his engine would become the dominant force of global industry and transport, and pointed to rising public disgust with the coal-powered steam engine's "foul-smelling, smogging clouds of dark smoke and unending rain of unburned particles of coal dust." Diesel, one of his biographers has written, "dwelled perhaps prematurely on the perils of air pollution. As he had once explained to his children when they seemed distressed by the occasional puffs of exhaust smoke from the central chimney of the Diesel engine pavilion at the Munich exposition, any working engine emits some amount of exhaust waste, but the Diesel engine was the cleanest running of all." He probably did not anticipate that his name would one day become synonymous with dark, choking clouds of pollution.

The same high-compression process that makes the diesel engine more fuel efficient also explains why, for a long time, no one wanted to get stuck behind a diesel in traffic. Diesels became notorious as soot producers, in part because the fuel and air in a diesel engine aren't mixed as evenly as in a gasoline engine. This leads to the formation of some pockets of air in the combustion chamber that are fuel-rich, especially deeper into the "power stroke" of the engine (which is why you often see plugs of soot emanating from older trucks climbing hills or accelerating). During combustion, some of the fuel in these pockets doesn't react with oxygen, remaining unburned, producing black carbon and other superfine particles (which are also supertoxic: diesel exhaust was collectively classified in 2012 by the WHO as a Group 1 carcinogen).

Despite this inconvenient fact—and because diesel is such a prime mover of growth—India's is a "dieselized" economy: diesel accounts for over 43 percent of the petroleum products consumed, one of the highest shares of any major economy. (In the United States, diesel accounts for about 25 percent of all petroleum consumption.) The Indian government heavily subsidizes diesel, ostensibly to keep prices down for low-income citizens. But most of that fuel gets diverted to commercial use and to wealthier individuals who don't need the discount. This subsidy drives diesel consumption ever higher—India burns 522 million barrels of diesel fuel each year—and represents a $14.8 billion expenditure for the government (not much less than what the cash-strapped central government spends on education).

Diesel is a magically potent energy source. Take the forest green Eicher 5660 tractor that rolled up the newly paved link road just below Urgain's house near Padum one day in late July. It can take four attachments: a thresher, a plow, a flatbed cart, and a grain grinder. The tractor—new, still shiny, with bright orange hubcaps and a clean canvas-colored canopy—was owned by a man from Padum; the operator was a migrant worker from Bihar, India's poorest state. Its diesel engine turned a blue crankshaft that spun a wheel that spun a belt that turned a grinding wheel housed in a blue metal casement. A candy red hopper on top, sporting the label "Hindustan Agro Industrial Corporation," mimicked the woven basket that slowly drips barley grains into the hole at the center of the millstone in the *ranthaks*, the traditional water-powered mill of Ladakh and Zanskar.

One hour's use of the Eicher grinds all of the household's barley, and costs 720 rupees (about $12). The old way demanded one full day and night, with someone watching over and filling the hopper in the *ranthaks*, to grind a household's entire barley harvest. By using a tractor for *zhing mos*, the spring plowing, a family can finish work in half a day that used to take several days with a *dzo*.

"Now everything is machine. So easy, yeah?" Urgain observed.

The overwhelming impression of watching this tractor in action is of *speed*. It alters the fabric of time in a place like Zanskar, so bound by its seasonal rhythms and the caloric limits of its primary movers. (Humans' maximum work output is about 100 watts; draft animals, about 500 to 800 watts. The power output of the Eicher 5660 is 18,000 watts.)

In a sense, diesel is building the new village in Marthang. A diesel-powered JCB helped dig the canal bringing water to Marthang. It digs the trench for the pipe bringing drinking water from the mountain. Diesel-powered tractors bring loads of stone and dump piles of sand

and aggregate and soil shoveled from the banks of the silt-choked Lungnak River. Once those piles become walls (assembled largely by Nepali laborers, who have been conveyed from their desperate home villages by a series of diesel buses and trucks) diesel-powered Tata Mobile pickup trucks will deliver poplar beams, willow poles, steel rebar, cement, wood, and glass.

Everything for sale in the Padum market has been brought across the Himalaya on a diesel truck. Mango Fruity drink boxes and bags of Amul Milkmaid to put in tea, they are all soaked in diesel, lifted over the world's tallest mountains by diesel. I have walked to Zanskar three times. Six times, I rode: diesel brought me there. In the eighth century Guru Rinpoche had his pregnant flying tigress ferrying him around the Himalaya; today the traveler to Zanskar has the Tata diesel engine, arguably even more powerful.

As the oil company Chevron crows in one of its technical briefs, diesel "helps keep the world on the move." Global diesel consumption has been rising steadily, from roughly 8 million barrels per day in 1985 to almost 25 million per day in 2008. All told, the world's diesel engines—mostly for moving goods and people but also for powering generators for electricity—burn about 383 billion gallons of crude oil distillate each year. Fully 10 percent of that is consumed in the United States alone (which is home to less than 5 percent of the world's population). Thanks to the efforts of Tami Bond and her colleagues, we know that burning diesel for transport contributes about 20 percent of the (at least) 7.5 million tons of black carbon we spew into the air every year.

As its inventor predicted, the diesel engine has indeed conquered the world.

Bond has led comprehensive efforts to develop global emissions inventories for black carbon—quantities, sources, emission factors. In the process she has measured everything from coal boilers in East German factories during her graduate studies to coal heating stoves that she built based on photos sent to her from China. As anyone who has ever tended a bonfire or a woodstove knows, this is a notoriously difficult task: smoke emissions vary widely, and wildly, depending on fuel type, moisture content, wind strength, combustion method, even the season (and the operator's mood, as well). Data on some sources—such as burning of agricultural waste in Asia—are frustratingly scarce. All her efforts culminated years later in the production of the definitive global inventory in 2004, which led to an updated, even more definitive, 232-page report, published in January 2013, "Bounding the Role of Black Carbon in the Climate System."

In it, Bond and her coauthors confirmed the outlines of the arc of burning, which traces our Promethean surge since 1750, when there were 1.4 million tons of black carbon sent aloft (mostly from naturally occurring wildfires) to 2000, when we pumped out at least 7.5 million tons of the stuff (mostly from our economic activity), and possibly as much as 17 million tons, according to recent research out of MIT. Where does it all come from?

Diesel engines.

Small coal-fired industries, especially brick kilns and boilers.

People burning wood, dung, waste, and coal in simple stoves for cooking and heating.

Open burning of biomass (including both intentional burning of agricultural residue such as straw and wildfires).

That's it. Those four categories of burning stuff account for over 90 percent of the black carbon we send aloft around the world every year.*

Bond's inventories offer a few key lessons. First, as societies develop, black carbon emissions will decrease: "Black carbon emission sources are changing rapidly due to greater energy consumption, which increases emissions, and cleaner technology and fuels, which decreases them."

The United States' story is typically offered up as the hopeful ex-

---

*Globally, the emissions breakdown is roughly 20 percent from diesel engines (including on-road cars and trucks, agriculture construction heavy equipment), 9 percent from industry, 25 percent from household combustion (of wood, agricultural waste, dung, and coal) for cooking and heating, and 40 percent from open burning (wildfires and burning of agricultural waste). The rest comes mostly from small diesel generators used for residential power generation, shipping, on-road gasoline engines, flaring of natural gas from oil wells, kerosene for lighting (in the developing world), and non-coal industry (mostly using biofuels). Biofuel burning contributes about 65 percent, and fossil fuel 35 percent, of black carbon emissions.

emplar: given enough time, the sheer momentum of technological development, plus the feedback loop of environmental and health awareness it makes possible via rising standards of living, a society will take care of its own black carbon pollution, like some kind of civilization-scale self-cleaning oven. To wit, black carbon emissions in the developed world are declining. The United States, once the primary source of the world's black carbon, today ranks seventh, and produces 6 to 8 percent of the global total. (This glosses over the fact that, though Americans' emissions of black carbon peaked around 1925 and have been declining ever since, largely thanks to regulations like the Clean Air Act and California's stringent tailpipe emissions standards, the levels in parts of certain urban areas, like Manhattan and central Californian cities like Bakersfield, remain recalcitrantly high. California's Air and Resources Board recently determined that 9,000 people still die every year from exposure to fine particles like black carbon in California's air.)

Second, people in Asia and Africa seem to be just a bit further back on that development-burning arc. Whereas the developed world's black carbon is mostly produced by transport (diesel engines constitute about 70 percent of emissions in North America, Europe, and Latin America), in Asia and Africa, on the other hand, 60 to 80 percent of emissions are from residential combustion of coal and biomass. (Over half of all people cooking with biomass in the world live in India and China.) The wealthy countries produce soot mostly from the luxury of mobility; the world's poorer inhabitants produce soot mostly from the simple acts of cooking and staying warm.

But studies suggest that emissions in Asia have increased by 30 percent between 2000 and 2005. The rapid increase in energy consumption in countries like India, Indonesia, and China, due to their expanding middle classes, coupled with the demographic momentum of those populous nations, place them halfway up the very steeply rising part of the curve in Bond's graph. China and India together account for about three-quarters of Asia's black carbon, and between 25 to 35 percent of global black carbon emissions. Emissions in China alone doubled between 2000 and 2006.

Zanskar, like most of India and China and sub-Saharan Africa, actually has a foot in both worlds of burning. In those places the arc has doubled up on itself. India contains both megalopolises like New Delhi, home to over 20 million striving souls, and tiny villages like Kumik, with its 250-some residents. In both places you will find people burning solid fuels—wood and dung and even trash—to survive, and you will find people burning "modern" fuels like diesel and kerosene to keep the lights on, get around, and generally improve their lives. Just as Zanskaris have overwhelmingly welcomed the rough road that has brought them modern medicine and access to useful things like cylinder gas, the world should rightly hail the Chinese government's investments in infrastructure and access to modern fuels that have enabled the modern growth that has lifted 500 million people out of poverty in China over the past thirty years. But that increased economic activity means more burning, especially in diesel engines and in brick kilns to feed the buildings that sprout as fast as bamboo from the fringes of fast-growing cities like Chengdu in China and Bangalore in India.

We think we've seen all this before, but in a sense we are in uncharted waters. Instead of, say, 76 million people burning in 1900 America, there are 1.2 billion people burning in India alone, and another 1.3 billion burning in China. Incomes in Asia and Africa, to a lesser extent, are rising. More people are demanding more bricks and electricity and open land, more goods, more mobility, more comfort, more security and opportunity. And, increasingly, they can pay for it. Half of the projected increase in global energy use between 2010 and 2040 will come from growth in just two countries: India and China.

The third lesson from Bond's inventory is this: soot is as illuminating at the macroscale as it is in the microscale of the flame. Black carbon is a useful proxy for understanding the speed and direction and consequences of our economic development ever since we started to transition from burning wood to burning fossil fuels in the 1700s. Looking at it helps us understand the choices we have made, how far we've come since the first hearth fire, and how far we have yet to go. Apple sold over 500 million iPhones between 2007 and 2014. But we still have over 3 billion people cooking much the same way humans did a few hundred thousand years ago. (Some of those demographics overlap; I had my first iPhone sighting in Zanskar in 2012.) It links up questions of environmental and human health with an urgency unmatched by civilization's other by-products. Trace black carbon's rise and fall through the centuries, in different nations and cities and villages, and you have a reasonable indicator of collective progress on the roads to our imagined Shangri-las.

Black carbon's story is thus *our* story, the story of humankind's march out of the cold, dark steppes of our hunter-gatherer days all the way to the smoggy gathering places of Piccadilly Circus and Times Square. And like our story, much is yet to be written.

But in the meantime, at the rate things are going, Urgain and his fellow Zanskaris may be waiting quite a while for all that soot to settle.

One evening, taking a break from the dusty *lhakhang* site, I walk the five kilometers from Marthang to Padum to meet with Tsewang Rigzin.

Just past the petrol pump, where government-subsidized diesel fuel is selling for 55 rupees per liter (roughly $4 per gallon), and just across the river from the hill of Pipiting where Guru Rinpoche pinned down the demoness that once ruled Zanskar by building a temple over her heart, I am overtaken by a bright yellow school bus. Meme Paldan's son is at the wheel, bringing children home from the Marpaling Lamdon School in Stongde. I exchange waves with him and a knot of smiling students, their noses pressed against the window. As the bus trundles past, it belches a cartoonish cloud of soot around my head.

A half hour later I reach the Mani Ringmo market at the same time as the battered Dorje Coach bus, arriving at the end of its epic two-day journey from Leh. It disgorges battered boxes and red gas cylinders and a few dozen weary-looking passengers and leaves behind another dark cloud that envelops us all. Farther on, I pass the Jammu and Kashmir Bank, where a rope of smoke twists up from a metal barrel full of smoldering garbage. A young guy in a *Playboy* T-shirt and aviator sunglasses loiters nearby. "Next year Padum may have an ATM," he tells me, "then you can use your credit card."

I have time to kill before meeting Rigzin, so I walk on, toward the stately, crumbling houses of Old Padum. On the other side of the glacial moraine upon which Padum Castle sits, perched just above the Lungnak River, is a rectangular stone building. From afar it looks like someone painted half of it black, then left and forgot to finish the job. Up close I can see two exhaust pipes poking through the roof: the exterior walls are encrusted not with paint but with chalky black soot, the residue of two enormous diesel generators that provide electricity to Padum and a few surrounding villages each night for a few hours.

I circle back toward the main market. A Tata Mobile, driven by a laughing Zanskari youth, speeds past me issuing staccato puffs of soot. Near the gate to the Mont Blanc Guest House, I pass a shop where Sonam "Bura" Tantar, a Kumikpa now retired from the Indian army, repairs kerosene and gas stoves. (There's a delightful, famous song about Tantar that everyone in Padum used to sing in the 1970s, when he was a young, tall, handsome soldier on leave, rounding in the market: "Swinging his arms wide, wearing his army parka, there goes lucky Bura Tantar!" Now lucky Bura Tantar and his wife have a modest little house in Marthang, near the solar *lhakhang*.)

"If everything were turned to smoke," Heraclitus observed, "the nose would be the seat of judgment." On days like this, it almost seems as though everything really *is* being turned to smoke. What Zanskaris burn and how they burn it is thus an apt proxy measure of the broader changes that have taken place in the past thirty years.

Not long ago, Padum's soot came almost exclusively from the burning of dung and scarce wood in household hearths. Today the town has become a veritable buffet of black carbon: a wide variety of flavors are represented, including a few conspicuous new sources. The dark smoke has become more and more a sign of Zanskaris' wanting what most of the world wants: mobility, convenience, comfort, a decent standard of living. Pants.

One new source is garbage, mostly plastic packaging, arriving daily with the truckloads of goods brought in from Kashmir and beyond. Thirty years ago the very concept of waste was alien to Zanskar; everything was recycled into the soil: clothes, shoes, building materials, since everything was biodegradable and a vehicle for precious, scarce nutrients. Today, this manufactured waste is set on fire in discarded army fifty-five-gallon drums on the street and then dumped into the atmosphere (much as residents of Los Angeles and New York did well into the 1960s, before clean-air ordinances banned trash incinerators within their city limits). And all this new stuff has to be brought up to 12,000 feet somehow, and right now that somehow is a steady stream of brightly colored trucks made by the ubiquitous Indian conglomerate Tata, burning diesel as they strain over the Himalayan passes.

Bura Tantar's stove repair shop, the speeding truck, the generators all represent the technological advances made possible by the road. Kerosene-burning lights to replace open flames, and then electricity to replace the kerosene lights. More efficient cookstoves and metal chimneys to replace open hearths. Diesel trucks and buses and passenger vehicles to replace donkeys, horses, one's own two feet. The Eicher tractor has replaced the *dzo*—the yak-cow hybrid—for moving heavy things, pulling a plow across the fields, even grinding and threshing the grain. The Tata Mobile, with LOAD CARRIER printed across the windshield, has likewise replaced the yak, the horse, and human labor for moving goods long distances.

Here, Zanskar poses another paradox: the road has brought some technologies and techniques that have undoubtedly *reduced* soot and other pollution. The wide acceptance of LPG gas stoves, for example, has helped clean up the air in kitchens. At the same time, the road has just as clearly *increased* soot and other pollution, as it has brought these exceedingly useful vehicles and electricity-generating machines, which also happen to spout black carbon into the streets and skies above their glaciers.

So Kumik—and Zanskar as a whole, and India and China, for that matter—is caught on the sharp yak horns of a dilemma. Zanskaris are getting penalized for trying to survive—as they continue to breathe in vast quantities of black carbon and other particulates in their kitchens—and for trying to get *beyond* survival—as they embrace technologies that raise their standard of living and lower the time and cost of moving about, building, lighting, and manufacturing.

It's like battling two raging fires at the same time. The hearth smoke of Kumik is now reinforced by the dark plumes from the steel exhaust pipes of those Eicher tractors in Marthang, delivering stones day after day. Stovepipes and tailpipes have become the twin symbols of the twenty-first-century Shangri-la.

In 1980, there were precisely zero scientific journal articles with "black carbon" in the title. By 2001, there were almost 800. Black carbon has become a growth industry.

The man who discovered black carbon is somewhat bemused by this explosion in interest. There was a time, he says, back in the 1970s, when he was accused of blasphemy just for pointing out that there was still soot in the air.

Tica Novakov grew up in a small city in far northern Serbia, the son of a veterinarian. As a high school student he built his own radios and x-ray tubes. He completed a PhD in physics at the University of Belgrade, and soon thereafter emigrated from Yugoslavia to the United States. ("Why did I leave? Communism! And I was a scientist. I wanted to do science!") In 1963, he started a research job at Lawrence Berkeley National Laboratory. Soon he was leading a group studying the formation of aerosols (fine solid particles or liquid droplets suspended in a gas) in the atmosphere.

At the time, concern over air pollution had reached a fever pitch among the public and policy makers, especially in cities like Los Angeles. "California was very much obsessed by visibility degradation," Novakov says.

The conventional wisdom on smog (coined by combining "smoke" and "fog") was that it was caused by ozone, and further, that particulate matter—like the kinds of toxic stews of black carbon and other fine particles that troubled coal-powered and -heated cities like London and Pittsburgh in earlier decades—was a thing of the distant past. We had evolved too far from the "sooty bowers" of Chaucer's time, or even the soot-laden streets of Dickensian London, to worry about it anymore. Researchers assumed that most of the stuff clogging nasal passages, stinging throats, and turning the sky that hazy brown color was sulfate, nitrate, and organic carbon particles. But even into the 1970s, despite a variety of scientific breakthroughs into the origins of ground-level ozone (which is formed when nitrogen oxides react with unburned hydrocarbons in the presence of sunlight) and after implementation of strict emissions controls, brown, hazy pollution remained a persistent problem in Los Angeles.

Novakov and his team members made waves with a 1974 paper in *Science* that purported to answer a question few had thought to ask: why was LA's air pollution dark? They described results they had obtained using an innovative technique developed in their own lab to measure how much light was absorbed and how much reflected or refracted, by the materials gathered on a filter. Their study found that 50 percent of all particulate matter in urban areas such as LA was carbon, and that up to 80 percent of that carbon was soot (what we now call black carbon—but Novakov hadn't coined the term yet, so soot was used interchangeably by scientists, and still is, in some cases).

"So the blasphemy is," he explains, "the air today is not different from that in London" at its midcentury, coal-burning worst.* "It's only the matter of amounts."

Their results shocked others in the field—the numbers were way beyond what authorities assumed was present in the air. Not surprisingly, the finding floated like a lead balloon. "Basically, 'soot is nonsense, it's there but it's a minor fraction of aerosols,' " is how Novakov sums up the typical response among researchers at the time.

At a certain point in the early 1980s, as his lab accumulated more evidence and interest in his theories was mounting, Novakov had to figure out what to call the elusive but apparently ubiquitous stuff his group had discovered, hiding in plain sight. Other phrases that floated

---

*In December 1952 a temperature inversion and a cold fog parked itself over London, leading people to burn more coal in their furnaces to stay warm, producing more black carbon and other particulates, which were trapped at ground level by the fog. Over the course of four days, subsequent analyses found, 11,000 people died from exposure to the soot and other pollution before the deadly cloud lifted. Many were elderly and children; many succumbed to pneumonia and bronchitis. Royal London Hospital reported twice as many deaths from chronic obstructive pulmonary disease as normal. Later examination of lung tissue from those who died in the event found soot and other ultrafine carbonaceous particles. The episode sparked the first air quality legislation in Britain, the Clean Air Act of 1956, which imposed controls on smoke emissions in urban areas and pushed for a shift to cleaner fuels like electricity and gas.

about referred to fundamentally different properties: "graphitic carbon" implied pure carbon linked in a crystalline structure; "elemental carbon" meant just that, pure carbon and nothing else mixed in.

But the key thing about what Novakov and his colleagues were detecting, from their point of view, was that "it burns, and it's black." So he chose to emphasize its light-absorbing behavior over the fact that it was a form of elemental carbon. He consciously inverted the widely known "carbon black" (aka "lampblack") used in printing and paint making and other commercial applications for centuries.

"We called it black carbon," he says with a shrug. He debuted the term at a conference in Austria in 1983, less than a year after the first diesel truck reached Zanskar, and, like soot on a filter, it stuck.

It would be another fifteen years of painstaking sampling, measurements and development of totally new techniques before Novakov's black carbon hypothesis went mainstream. Novakov worked with two other physicists in his group, Tony Hansen and Hal Rosen, to develop a device they called an aethalometer, from the Greek word *aethalos*, meaning "blackened with soot." A beam of light was shined on particles collected on a filter from ambient air, and the device measured the amount of light absorbed or attenuated by the particles. Invented in the late 1970s, it remains the standard tool for measuring black carbon around the world.

Tom Kirchstetter, Novakov's former student, now a renowned expert on black carbon and aerosols in his own right, has been conducting a study using the device to measure concentrations of black carbon in downtown Oakland, near the port (the fifth busiest in the country). Certain Oakland neighborhoods were black carbon hotspots, where concentrations remained recalcitrantly high, despite decades of EPA and CARB air quality regulations and the gradual improvement of engine efficiency. Many of the trucks exiting and entering the port were older, dirtier models. (California is still home to the country's most particulate-choked cities, mostly located in the San Joaquin Valley and the Los Angeles Basin. The EPA projects that by 2020, only seven counties in the United States will not meet its new annual soot standard: all seven of them will be in California.)

In response to the stubbornly bad air wafting into West Oakland—mostly due to port-related traffic—the CARB instituted new regulations for trucks carrying freight to and from the port in 2010. Truck owners were given a choice: purchase a 2007 model or newer truck, which has diesel particulate filters (DPF) built in, or retrofit their existing vehicles with DPFs. A grant program helped defray the $15,000 to $20,000 cost of installing a DPF, which can reduce particulate emissions by 90 percent.

Old diesel trucks can keep going and going, meaning turnover to cleaner-burning models can take years to decades. (Nowhere is this more apparent than in rural India, where ingenious "barefoot mechanics" are adept at extending the lifespan of ancient Tata trucks retired from urban use. They ply the rutted roads with their loads of mangos and rebar and petrol, belching out truly prodigious, tongue-coating quantities of soot.) The CARB program was designed to accelerate that transition, through a careful mix of incentives and penalties, monitoring and enforcement.

Kirchstetter and his colleagues took independent measurements in 2009, parking a specially equipped van on an overpass above West 7th Street, which connects the nation's fifth busiest port to Interstate 880 and other highways, and did so again for a period after the new rules took effect. They found that the program had more or less worked: black carbon emissions declined 50 percent from November 2009 to June 2010.

Kirchstetter now heads the LBNL aerosol lab that his mentor, Novakov, started. As I sit listening to Novakov reminisce about the cloak-and-dagger days of his early black carbon research, Kirchstetter pulls out the latest evolution of the measurement device that was born just up the road at Berkeley Lab. It's a smartphone-sized device called the microAeth made by a small firm in San Francisco, which some people and institutions use for personal-exposure monitoring. He pulls out a tiny filter strip and holds it up for us to see.

"And thirty years later!"

The filter has a tiny dark circle on it.

"So there's still soot in the air," Kirchstetter adds wryly.

• • •

"Crazy bad."

That was the description of Beijing's air that subscribers to the normally staid, automated @BeijingAir Twitter feed awoke to on November 19, 2010.

The U.S. embassy maintains an air monitor on its rooftop in Beijing that takes hourly measurements of PM2.5 and ozone and tweets the numbers out to a list of 35,000 followers. The particulate matter concentrations the night before and early that morning were literally off the charts. Some wag in the embassy staff had programmed the equipment to trigger the "crazy bad" designation in the event that the air quality index topped out at 500, perhaps expecting it to never reach those levels (or perhaps certain that it would, and hoping to generate some buzz and a few laughs). The scale goes from "good" to "unhealthy" to "hazardous" to "extremely hazardous." "Hazardous" means people shouldn't go outside, and corresponds to air quality index levels over 300. On the "crazy bad" day, the AQI maxed out at 500. Concentrations of PM2.5 reached 569 micrograms per cubic meter, more than sixteen times the U.S. Environmental Protection Agency's twenty-four-hour standard.

The public got a good, though anxious, chuckle at the frank, irreverent assessment. The Chinese government, which used a less stringent measurement index for public dissemination, was not amused. The embassy quickly deleted the tweets, but the kerfuffle drew attention to the alarmingly poor quality of Beijing's air, which was already starting to resemble Pittsburgh on its most hellish days in the 1930s and 1940s, when cars had to keep their headlights on at noon.

Over two years later, in January 2013, another soot-laced cloud squatted over Beijing for several days. The ghostly outlines of skyscrapers could barely be made out in the thick haze. Flights were canceled because the pilots couldn't see beyond the runway. The number of hospital admissions for heart attacks doubled at Peking University People's Hospital. The state-run Xinhua news agency reported that one pediatric hospital treated a "record 9,000 children this month, mostly flu, pneumonia tracheitis, bronchitis and asthma patients." The concentration of PM2.5 peaked at a whopping 993 micrograms per cubic meter. This time the embassy's twitter feed merely offered the anodyne assessment: "beyond index."

With the problem so ridiculously bad and public outrage suddenly impossible to ignore—after all, what weight is there in the threat of imprisonment or fines if you can't breathe?—the government was forced to respond.* It shut down over 100 factories temporarily and took one-third of the government fleet of vehicles off of Beijing's streets.

The causes of Beijing's January 2013 pollution crisis were familiar to veteran air quality observers. There was little wind to blow the pollution away. Many of the city's 20 million residents burned coal to heat their homes in the cold weather. There were also more cars than ever, including diesel vehicles, plying the new highways ringing the city—4 million new vehicles have been added to Beijing's streets in the past fifteen years. The factories and power plants driving China's continuing meteoric economic growth kept burning more and more coal: 87 percent of the total worldwide rise in coal consumption in 2011 came from China, which now burns almost half of all coal consumed globally. The pollution haze has gotten so thick, so dark, that it's interfering with the government's Skynet program of surveillance; some security cameras in public areas have been rendered useless by the pollution.

In September 2013, the government announced that it would make some of the emergency measures permanent, banning heavy-emitting vehicles and reducing coal consumption, with the goal of reducing Beijing's PM2.5 concentrations by 25 percent by 2017. The city even plans to crack down on outdoor barbecues. Skeptics think this goal is impossible to achieve, given China's overwhelming dependence on coal.

*Data on levels of emissions in China, with its tight control of information and different measurement standards, are hard to come by. A recent study of a region in north China where the government provided free coal for winter heating between 1950 and 1980 found that levels of particulate matter were 55 percent higher than levels reported south of the Huai River. There was an associated increase in cardiorespiratory deaths; on average, human lives in this part of northern China were a staggering 5.5 years shorter than those of residents of southern China.

A few weeks later, on October 21, 2013, in the northern city of Harbin—which is famous for its winter snow and ice sculpture festival—coal-fired heating systems were fired up in buildings across the city for the start of the cold season. "I couldn't see anything outside the window of my apartment, and I thought it was snowing," Wu Kai, thirty-three, told the Associated Press. "Then I realized it wasn't snow. I have not seen the sun for a long time." PM2.5 readings at several monitors in Harbin were well above 600 micrograms. Some readings were exactly 1,000, the first known readings of 1,000 since China began releasing figures on PM2.5 in January 2012. Schools were closed, flights were canceled, buses stopped running—just like in a snowstorm. But it was actually a soot storm.

There are signs that the Chinese government, realizing what it's up against—the 8-percent-a-year juggernaut of economic growth, a twin-headed monster of its own making, alleviating poverty and churning out pollution at the same time—might be backpedaling on its nascent tough-on-soot stance.

At the end of 2013, in a surreal segment, a state media outlet highlighted the "benefits" of the off-the-charts air pollution. In the event of a conflict, smog would make enemy missiles less likely to lock onto and hit their targets. It makes everyone equal, since rich and poor alike have to breathe the same nasty air. It's even made people funnier, the broadcaster claimed, and "that sense of humor is the source of strength for defeating the smog." The sooty skies had also educated the citizenry about English words like "smog" and "haze" and historical events like London's Great Smog of 1952.

And last, China Central Television pointed out, the terrible pollution educated the Chinese people about the costs of their country's rapid economic development.

On January 3, 1947, as the cold war was just ramping up, the *New York Times* published a letter to the editor asking if the Grey Lady could "possibly divert a portion of your valuable space from comments about the 'iron curtain' alleged to hang over Europe to the very real curtain that strangles, besmirches and mars the people and habitations of New York City?" Another reader from the Bronx wrote to complain that her brand-new curtains were blackened after just a week, and her lungs never enjoyed "good, clean air."

"The soot is a menace to all of us," she concluded.

Steven Chillrud, an environmental geochemist at Columbia's Lamont-Doherty Earth Observatory, studies particle pollutants and human exposure to them. With his colleague Patrick Louchouarn, he has studied pollution trapped in layers of sediment from lakes in Manhattan's Central Park and Brooklyn's Prospect Park. These columns of mud tell the familiar arc-of-burning story: black carbon concentrations were indeed peaking across New York in the mid-twentieth century.

Burning garbage had made it one of the sootiest places in America. In early twentieth-century New York, hotels, restaurants, apartment buildings, even Macy's department store, all had their own in-house incinerators. By 1937, more than 2 million metric tons of garbage were burned each year in New York.

In those sediment cores, the two scientists were looking for traces of lead from when it was added to gasoline in the 1960s. But instead the cores told them a lot about the city's history of airborne pollution. "We saw that, indeed, New York City being one of the megacities of the world, had incredibly high fluxes of black carbon into the city over time," Chillrud says, "and that the incineration was part of the story. I wasn't living in the city back then, the 1930s to the 1960s, but some of the other professors that we collaborated with on our paper did, and they said it used to just be raining down ash on them all the time.

"New York City, due to a couple of crises in solid waste, bought in big and early to the incineration business, and by coming in early they did some things you wouldn't have wanted them to do," Chillrud explains. "They put the incinerators right in the residential areas and spread throughout the city, to try to be efficient so you didn't have to haul the trash very far. From an engineering point of view you could see why they would do that. And then, even worse probably, they said buildings should burn their own trash. And it's one thing to have a centralized incinerator that you have engineers controlling and trying to [burn] in proper ways. It's another thing to have your super just shoving it in any old way."

By 1952, residential incinerators accounted for 30 percent of the city's particulate air pollution. By the mid-1960s—when a series of extreme air pollution events, like the ones that strike China's cities today, killed hundreds of New Yorkers—17,000 apartment building incinerators and 11 municipal facilities were churning out 35 percent of the particulate pollution produced citywide.

New laws finally phased out many of those residential incinerators in the 1970s, until the City Council made them illegal in the early 1990s. It went a long way toward clearing the air. But other soot sources had already begun to take their place.

Cleaning up the city's dirty fleet of diesel-powered buses took longer. West Harlem Environmental Action (WE ACT), a nonprofit that engages citizens in clean air and clean water advocacy, mobilized community members in neighborhoods with particularly high exposures to diesel exhaust—in high-traffic corridors—from the Bronx to Brooklyn and took out provocative bus shelter advertisements around the city, showing someone waiting for a bus with a gas mask on.*

"It was a campaign to raise the awareness of residents and push back at the MTA, keeping it in their face," says Peggy Shepard, WE ACT's executive director. The Metropolitan Transportation Authority, which runs the city's buses, eventually agreed to clean up its fleet, the largest in the country. "They never had a record of dealing with communities before. Getting them to make that investment and pushing them along, we made it happen a lot sooner."

This grassroots pressure, backed up by data collected by Columbia researchers and others, yielded results: the city's fleet of 4,500 buses is

*New York City's Health Department estimates that 3,200 New Yorkers die prematurely each year—more than the casualties of September 11, 2001—due to exposure to PM2.5 (not to mention the 1,200 hospitalizations for respiratory conditions, 900 cardiovascular hospitalizations, and 2,400 child and 3,600 adult emergency department visits for asthma that it causes). In terms of deaths of New Yorkers, that's like 1.23 September 11s taking place every year. One study puts the number of annual early deaths from breathing PM2.5 in the broader New York metropolitan area as high as 12,040. A recent MIT study found that the ultrafine particles emitted by road vehicles kills 53,000 Americans every year, much more than the 43,000 killed by traffic accidents—meaning that, statistically, simply breathing on your morning commute can be more dangerous than tailgating a tractor trailer with bald tires.

now made up entirely of either hybrid-electric or four-stroke diesels retrofitted with DPFs and burning ultra-low sulfur fuel.

The city can't prevent private diesel trucks and buses from entering without running afoul of interstate rules, says Shepard. This means soot falls through the cracks in EPA's broad regulations. Private diesel truck and bus fleets still ply the city streets without DPFs, and stream in and out of depots located in north Manhattan, like the one by the George Washington Bridge at 178th Street. There are laws against idling, but they are poorly enforced. And because 2 million New Yorkers live within 500 feet of major roadways (over 35 percent of health facilities and playgrounds are in that zone, too), many remain at risk from breathing soot produced by these private diesel vehicles.

In September 2013, around when Beijing was announcing its extraordinary, "war-footing" measures against its own soot storms, Mayor Michael Bloomberg announced that New York City's air was the cleanest it had been in more than fifty years. The New York Community Air Survey found that soot concentrations had declined by 23 percent since 2008. Bloomberg was trumpeting the achievements of his PlaNYC, an ambitious agenda launched in 2007 to reduce the city's climate-warming pollution while improving the energy efficiency of buildings, conserving water, and improving air quality, covering everything from expanding bike lanes to increasing the size of parks.

Viewed from afar, New York's regulators and planners indeed seem to have done their job: there are no more killing fogs like the ones in the 1960s. But residents might be surprised to learn that the Big Apple remains a black carbon hotspot. There are some parts of the city that still have dangerously high concentrations of black carbon in the air. It's just that the soot is now a menace to some more than others.

New York has twice the national average hospitalization rate for childhood asthma. The prevalence of childhood asthma in the city varies widely, and lower-income neighborhoods (which are predominantly home to minority residents) tend to have a much higher incidence. Steve Chillrud co-led a study in the Bronx, placing monitors in high school students' backpacks, that found a significant correlation in asthma inflammation and proximity to buildings that burn No. 4 or No. 6 heating oil (as well as diesel truck routes). The team also measured concentrations of black carbon *inside* homes, which had penetrated from the ambient air.

"We could show that the black carbon that we were measuring inside the homes was associated with both the density of the traffic, the number of roads within some buffer," he told me, "but *also* equally associated with the density of the residual fuel oil within a five-hundred-meter buffer." He and his coauthors noted that "strikingly, most of the homes near residual oil burners are in neighborhoods with a higher asthma prevalence."

Older urban residents are at risk, too. A 2011 study by Melinda Power, of the Harvard School of Public Health, measured cognition in 680 older men in Boston and estimated their exposure to airborne black carbon based on where they lived. Those who were exposed to high levels of black carbon demonstrated reduced cognitive performance equivalent to two years' worth of aging. "Traffic-related air pollution appears to cause inflammation and oxidative stress in the brain," Power told the *Telegraph* newspaper in Britain. "There is also evidence that ultra-fine particulates can get into the brain and cause dysfunction." Research on the brains of dogs in Mexico City by Lilian Calderón-Garcidueñas, a doctor and neuropathologist at the University of Montana and the National Institute of Pediatrics in Mexico City, has showed increased levels of inflammation, including amyloid plaques that are associated with Alzheimer's disease—suggesting a possible link to particulate air pollution exposure. This research is still young and fast-evolving, but there are hints that black carbon and other ultra-fine particulates may be invading our most "precious repository" of all, impeding our ability to think, reason, and remember.

One February afternoon in 2012 (I knew it was Ash Wednesday, from all the sooty marks on the foreheads of those quietly scurrying about the Reading Room) I exited the New York Public Library on Fifth Avenue and noticed a dark plume rising out of the smokestack of 445 Fifth Avenue, a thirty-three-floor condominium. It was the same density and Cimmerian shade as the smoke I've seen pouring out of brick kilns on the Gangetic Plain between Delhi and Lucknow. I wondered what it was burning.

Fortunately, the Environmental Defense Fund keeps track. If you're a New Yorker, you can search its online interactive map to see what your building's owners burn to keep you warm.

Several thousand buildings across the city still burn dirty residual fuel—No. 6 or No. 4 fuel oil—in their boilers, coughing dense curls of dark smoke out into the air that 8 million people share. Over 3,200 of these are in Manhattan, all over the island, from Harlem to the East Village to the tony Upper East Side. These create, in the language of the EDF report, which reads like some modern-day *Fumifugium*, "a rain of toxic soot that aggravates asthma, increases the risk of cancer, exacerbates respiratory illnesses and can cause premature death."

Until quite recently, if you scanned the skyline on a winter afternoon, you would have seen dark smoke guttering up from the roofs of some of Manhattan's priciest real estate, including The Dakota, where John Lennon lived. As of 2012, according to data compiled by the EDF from city government sources, 445 Fifth Avenue was annually burning 106,000 gallons of No. 6 heating oil (or "sludge," as it's called in the industry, since No. 6 often contains solids—thus even in the best furnaces, and especially in the older ones, it produces epic quantities of soot). The same data showed that 175 Fifth Avenue, the iconic Flatiron Building—home to the publisher of this book—was burning 54,900 gallons of No. 6 oil each year to heat its twenty-one floors of offices.

New York became so dependent on burning residual fuel oil to heat its buildings largely because of geography and path dependency. "New York City is a big port town, and it also had a good history of its own refineries," Chillrud explained. "So the ships have to be able to burn anything, because they don't know what fuel they're going to get in the next port. And they're designed to burn the cruddiest stuff. Because there was refining here and there was a demand for that—that being the cheapest stuff—a lot of the buildings, when they were made, took advantage of that cheap fuel and put in boilers that could deal with a really viscous, cruddy residual fuel."

Since the city's Department of Environmental Protection issued new rules in 2011 requiring all existing buildings to phase out the use of No. 6 heating oil by by 2015 and No. 4 by 2030, and banning the installation of boilers using these dirty fuels in new buildings, many of these landmarks have begun to convert to cleaner-burning options like natural gas, ultra low-sulfur No. 2 oil, and biodiesel. The NYC Clean Heat program was a key component of Mayor Bloomberg's PlaNYC air quality efforts, providing information and financial incentives to encourage building owners to convert their heating systems more quickly than the law requires. Some owners have pushed back at the new requirements, or dragged their feet, because of the considerable expense of conversion: switching to natural gas can cost several hundred thousand dollars.

The new rules have accelerated New York's transition to cleaner air. As of late 2013, the city estimated that 2,700 of the 10,000 buildings that had been burning the dirtiest fuels had made the switch to cleaner fuels, and 2,500 more were working on conversions. The Dakota has applied for permits to switch to a dual boiler system that can run on natural gas or No. 2 oil. According to the EDF's latest data, both the Flatiron and 445 Fifth Avenue are in the process of converting to much cleaner, ultra-low sulfur No. 2 oil.

The city claims that the resulting improvements in air quality from these cumulative conversions have prevented 800 premature deaths and 2,000 hospital visits each year, relative to 2008. (A 2009 study of 8,000 cardiac arrests in New York City found that breathing in soot substantially increased the risk of heart attacks for Gotham's residents.) But those hotspots remain, especially in Harlem and the South Bronx, but even in more well off parts of Manhattan.

"Even though we know there is a disproportionate impact in low-income communities," Peggy Shepard of WE ACT says, "we know that Manhattan in general is not in 'attainment.' Park Avenue buildings are in hotspots. Because no one is thinking about hotspots, they are broadly thinking about air quality. They are not thinking, 'Here's a hotspot, here's our strategy.' " WE ACT has been working to educate building owners and tenants about the new rules and options.

But because New York's black carbon comes from so many different point sources, as opposed to one big centralized power plant or other industrial facility, these holdout hotspots will be challenging to deal with.

"Out in California the diesel initiative had a huge impact and they're seeing [black carbon levels] come right down," Chillrud said. "In New

York, it's holding pretty steady. So that has two potential interpretations. One is that it is the older fleet: although the MTA has got all these clean buses out here, the rest of the transportation system might still be much older and might be a larger fraction of the 11 million dirty diesels that are still running around. The other interpretation is that it's the residential fuel oil."

Chillrud thinks both sources are to blame, and that air-monitoring efforts are "never going to see a big impact of the diesel initiatives until you get rid of fuel oil, because it's being driven by both."

Even in pockets of one of the wealthiest cities in the world, staying warm remains a dirty, dangerous business. In that respect, New York is sadly unexceptional: a 2014 World Health Organization study found that almost 90 percent of people living in urban areas all around the world breathe air with particulate levels that exceed the WHO's safety thresholds.

Back in Zanskar, at the end of my walk from Marthang, I meet Rigzin outside of a tea stall. We wedge ourselves into a booth and he lights a cigarette. Rigzin is short, a bit stocky, and his intelligent features often sport a bemused and skeptical expression.

On paper, Rigzin trumpets his home as paradise. So I'm a bit surprised when he tells me that he can't wait to leave. When he looks around at Zanskar today from the perch of his subdivisional agriculture office on the edge of Padum, he doesn't see Shangri-la.

He wants to move to the fast-growing city of Leh, where he lived during his early days of training for his government post, with all its chaos and hassle and pollution and frustrations and opportunities and excitement and comforts.

He even talked to his boss in Kargil about a possible transfer to Leh, where his wife now lives with his two young sons, so they can get a better education than they would at the neglected government schools in Zanskar. But he says he would have to get some minister to give a recommendation. "My boss told me, 'You are Zanskari, you should stay and work in Zanskar. If you don't, who will?' "

"He has a point."

"Yes, it's true." Rigzin smiles in resignation. He savors a good joke, even when it's on him. He finds it hard to get anything done in Zanskar. "After four or five years of my service here, I think it's so hard. The work culture in Leh is 90 percent better than in Zanskar." People in government postings don't work hard because there's no accountability—once they get the coveted position, they may not even show up for weeks at a time. They know it's almost impossible to get fired.

"Everyone wants to become rich here overnight, since the last twenty years," he continues. "Even Stobdan, when he doesn't open the school some days, and the children are coming, I think, should I say something? But he is my neighbor, he is my friend. We have to live together."

I point out the irony of some Zanskaris sending their children to private schools built by Nepali laborers, where they are taught by Tibetan refugees and their tuition is sponsored by European tourists. He nods wryly.

"They are not really poor," he says of those Zanskaris. "It's their way of living that makes them look poor, do you know what I mean? Zanskaris are strange. A man will open a restaurant, but if he sees he can make more money with trekking [guiding or handling horses], he will close the restaurant for weeks and go trekking instead. People aren't having professions."

But isn't that rational, I ask, to pursue the opportunity that will yield the highest wage?

He counters by pointing out the squandered investment, the lack of long-term strategic thinking. In the past two decades, people have lost the patience that characterized their work in the fields and with the animals. Everyone's in a hurry to make money.

Rigzin sees the good traditions—*bes* and other collaborative structures, for example—fading fast, and the useless ones—growing the labor- and water-intensive crop of barley just to make buckets and buckets of *chang*, for example—holding fast. He laments the difficulty of getting farmers to think rationally, economically, which is what his job is all about. Barley barely fetches 10 rupees per kilogram in the market, whereas peas get more than 20 rupees per kilo. But if he suggests planting more vegetables and pulses like lentils and peas, people scoff. "Even my mother, when I tell her to weed the vegetable patch, she won't do it. But she'll spend all day pulling weeds out of the barley field!"

He goes on to describe all the challenges of development here: substandard schools, a lack of connectivity and communication infrastructure, no local media, sporadic electricity, bad roads, deeply entrenched corruption and nepotism. These are problems that plague rural India at large, of course, but somehow in a small society like Zanskar, where everyone seems to know—or be related to—each other, progress seems exponentially harder and the politics and inertia even more daunting. In Zanskar, he says, everyone is too close; social pressures can be overwhelming. Innovation is not rewarded—to the contrary, those who stand out are quietly chastised. He is skeptical that a leader will emerge to change the course that Zanskar seems to be on, as everyone races to cash in.

"He would need to be someone like Sonam Wangchuk." Rigzin once met Wangchuk, the solar guru and education reformer, when he was a skinny, eager college student tutoring dropouts and Rigzin was in high school in Leh.

"Is there a Zanskari who can lead? Someone like Wangchuk?"

He considers for a moment, then shrugs and frowns. No one comes to mind. "Zanskar is too hard to change," he concludes.

By this I assume he means the trajectory on which the society is traveling—the same path that the world's industrialized countries have trod—and how it seems to be a physical fact of the universe, like water flowing downhill. Rigzin sees little chance of the Hidden Kingdom hanging onto the best of its traditions, discarding only those that no longer serve its denizens' core interests, charting its own course on the vast river of commerce it has recently merged into. Zanskar is changing, but fighting the ineluctable current of progress—that seems too hard.

I know what he means. That sense of inevitability, infused by the overwhelming momentum of the status quo, can quickly set up like concrete and calcify the imagination. As we speak, I still have a velvety taste in my mouth from my walk over from Marthang—all that smoke from the Tata Mobiles and diesel generators had the tang of progress. And everyone knows you can't fight progress, even if you wanted to.

And why would you want to? The new road to Zanskar will mean that a pregnant woman from Kumik will have access to far, far better care than she does today. The road also means she'll have another source of soot in her lungs to contend with (and studies increasingly suggest that diesel exhaust exposure leads to impaired cognitive development in children). A catch-22 . . . but what else is new? The snow falls above Kumik, but the water goes to Shila.

But before that next section of the global arc of burning is traced out, from rural India to rural China, it's worth looking to history for one more lesson: the black carbon trajectory, much like the path of development itself, is not a physical law like the nuclear curve of binding energy. It can be bent into a different shape and pursued at different paces. Just as Zanskar has compressed the Industrial Revolution experience into a generation or two, there's hope that, for the rest of the world, the peppered moth–lightening phase can be sped up. There's no law of nature—or human nature—that says China has to spend another fifty years in Pittsburgh-like darkness at noon before it can both breathe easy and keep the lights on. Development can be uncoupled from the wasteful production of black carbon (and carbon dioxide, too).

"History indicates that black carbon emissions can and will be reduced as development occurs," Tami Bond told members of Congress in 2010, concluding her tale of humanity's burning habit. But development is not something that just happens, like the tides. It's driven by human choices.

"However," she added pointedly, "this transition can be accelerated."

A couple of mornings after my walk to Padum I arrive at the site of the *lhakhang* to find the day's volunteer work crew, about a dozen people from different households in Kumik, standing around looking stymied. The cliff above the canal has collapsed; a small mountain of soil is blocking the water. This happens a few times a year, someone says with a shrug.

The *yura* committee has sprung into action. They send a message to Padum: bring a JCB to come dig it out. Until that happens, without water we can't mix mud for bricks. As the villagers head off to return home, where there is always plenty of work to be done, they tell me they won't be coming tomorrow. Sadbhavana, the "goodwill"-generating, outreach-focused branch of the Indian Army, is hosting the Kisan Mela, the Farmer's Festival. Officers will be presenting gifts to Kumik and to other villages, and they have demanded that at least forty Kumikpas turn out for the event, in a show of gratitude, to receive the largesse in person.

That night, Stobdan tells me it's hard to get people to come work on the *lhakhang*: "'We are too busy, too many work,' they say." As is common with any collective undertaking, like digging the canal and even prayer meetings, those who are absent pay a small fine. The money is used for construction materials. "I request them to give money, they say it's too much!"

He wonders out loud if the recent councilor elections between Congress and the Union Territory Front are to blame. It was the first time that Kumikpas divided along party lines, and politicians from both sides had come to campaign.

"This election was very bad for Kumik, for Zanskar," Stobdan says. "It has split the whole village." Stobdan supports Congress—are the UTF supporters staying away from the *lhakhang* because he helps lead the project?

A few days later, Rigzin will echo this concern to me. "This bloody politics has made hell for us," he mutters disgustedly, following his own train of thought as we walked down the deserted evening street from Mani Ringmo. "In March when we met and discussed, everyone was so happy." Then the elections happened in May. This was the first year that villagers themselves have stood for elections, and it's had some ripple effects on all of their collective undertakings. Rigzin thinks politics, the timeless scourge of jealousy and the universal practice of jockeying for prestige, is to blame for why more people haven't contributed labor to the *lhakhang*.

Early the next morning, villagers from around central Zanskar stream into the army camp across the river from Padum. A red carpet has been rolled out on the sand. A dais has been set up under a large orange-and-blue-striped tent, with a sleek black podium and four red velvet upholstered chairs flanking an incongruous divan. Under a sign that says FOREVER IN OPERATIONS, a bunch of tall Rajasthani and Sikh soldiers stand around, clutching whistles, looking stern, and waiting for orders—or for trouble.

We all sit in plastic chairs and wait for the commanding general in charge of Sadbhavana to arrive. Old men and women in their finest *goncha*s gossip and then doze as the day heats up. Young children wander off to play in the midday sun. After three hours of waiting, with no explanation for the delay, we hear the helicopter before we see it, and suddenly it's there, touching down a few hundred meters away on the plain, kicking up a prodigious cloud of dust. The distinguished-looking 14th Corps commander from Leh emerges, his salt-and-pepper hair tousled by the rotor wash. He walks down the red carpet and arrives to a blizzard of ceremonial white scarves. He enters the tent, shaking outstretched hands, a huge smile plastered on his face.

"Very good, very good," he says, smiling and nodding. He has the brisk, upbeat manner of a politician campaigning and already late for his next rally (after this he says he's going to Stongde Monastery). His task is to win and maintain the allegiance, if not the affection, of the peoples (often ethnic and religious minorities) living in the remote, sometimes restive areas along India's long border.

The general ascends the dais, and after a few flowery introductions, flanked by local Zanskari politicians, he gives a short, bullet-pointed speech. With my limited grasp of Hindi, I catch the highlights, a disjointed series of promises and bromides:

> Education and good teachers are important. Ex-army men will soon be able to draw their pension money from the J and K Bank in Padum. Zanskar is beautiful. You in Kumik are family. Switzerland is advanced. It will take a while to become like Switzerland. The *chadar* road is central to achieving that objective. The road will take time. But we'll get drills and finish it as fast as we can. . . .

This last pledge generates the most fervent, extended applause of the morning.

Now it's time to give out the gifts. Grandmothers from each village are called up to the stage, where they are handed gift-wrapped boxes containing kitchen sets for the *ama tsogspa* and other groups. One of them goes back to her seat weeping loudly, apparently with joy, and I speculate on the lifetime of hard, unremitting domestic labor she has performed, or privation she has braved, to occasion such relief at handling these goods. Other village representatives are called up and given framed photos of all sorts of useful, manufactured detritus borne hither by the mainstream: plastic chairs, tarpaulins, computers. The village of Ruk Ruk gets a HY Duty Gas Stove.

The Kumik representative is handed a giant photo of the diesel generator the army is donating to the village. He hoists it awkwardly as he steps down onto the red carpet, smiling, like a sweepstakes winner with an oversized check. And though the general looks nothing like Bob Barker, I can't help thinking of the 1980s game show I grew up watching, *The Price Is Right*: I can see the shiny blue generator, the real thing, off to the side, waiting to be loaded into a Tata Mobile to be brought to its new home in the Kumik community center. The machine practically shouts: *You win! Modernity! Power! Progress!* After all, grid electricity hasn't yet reached Kumik, so it's a big deal.

An announcement is made: it's time for refreshments. And all at once a couple of hundred villagers surge out of the tent toward the tables in back, where Sadbhavana staff ladle samosas and tea out of huge pots. There is much pushing and shoving—nothing angry or aggressive, just what you might expect to happen when 300 people all make for the same table of freebies at the same time. Elderly women stumble unceremoniously in a jostling sea of their neighbors. I hang back, apprehensive of a stampede, and notice Gara is there at my elbow. Watching his clamoring neighbors, his usual hangdog expression is tinged with disdain and sadness. "Just like animals," he says, shaking his head.

Later that day, back in Marthang, I find Tashi Stobdan sitting on the concrete veranda of his new half-finished solar house with a sour look. He had left the event early. "Today I am so upset," he says. "That's why I came back."

He takes a bitter drag on his cigarette, a lapse I decide not to communicate to Zangmo. We watch as a JCB grabs a chunk of soil and pushes it up against the crumbling concrete wall of his new water storage tank, the one that will provide irrigation for his saplings. The concrete hadn't cured yet, so it collapsed a few days earlier, when his son forgot to close off a small canal. Stobdan had taken out a loan to build it. It's a blow.

Stobdan had also been slated to receive the generator gift on behalf of the village, an honor and a rare chance to be more than a schoolteacher, to be *mi chenmo*—big man. Stobdan does not claim to be immune to the seductive allure of prestige. A few days prior, he showed me the new pants he had bought in Padum just for the occasion, a rare splurge. But his name was struck from the official list of recipients—was it possibly his old buddy, the current *goba* Tsewang Norboo?

"Some people ask the Executive Councilor to strike my name, for political reasons. I say to the captain, 'You have insulted me today.' "

For the next few days, Stobdan remains mired in a funk, wearing a distracted look. After one long day's work on the *lhakhang*, making mud bricks and building up the stone foundation, we sit on the verandah patio watching vehicles go by on their way to Padum, making desultory conversation.

Then a mental bolt strikes. He leaps up, grabs a black marker from my tool bag, and walks to the front door. On it he scrawls: "Success is the ability to go from failure to failure without losing your enthusiasm."

Then he flashes a huge grin—much like that cryptic one by the sign back in 2008, on my first visit—and tosses the marker back to me. He jumps on his rickety motorcycle, crosses the road, and accelerates off up the dusty track to his other, more ancient home.

Two days later there is a meeting in the village to discuss what the priorities for building in Marthang should be.

The villagers split into clusters: the male leaders in one corner, the *azhangs*—crusty late-middle-aged graybeards—in another. Young men like Stenzin Konchok and his friends crouch together with a group of older women. Three other groups of *amas* caucus, as does one group of grandmothers, patiently thumbing their prayer beads. This last one is the only group not engaged in animated discussion. In my role as scribe for the session, I walk over to them.

"What would you like to see in the new village?" I ask.

They look back at me, puzzled.

"But we'll be dead within two or three years," one woman says.

"Well, what would you like to see for the children?"

With that, they arch their eyebrows and nod to each other, and the floodgates are opened. They confer for a bit but soon announce their consensus: "A warm *lhakhang*, a school, a medical clinic . . ."

For the most part, everyone in the hall is on the same page. A school. A medical clinic. A vegetable greenhouse, for shared income. A self-help group and weaving center for the women. A flower garden and open space. ("Make it green! Trees! Trees!" shouts the other Zangmo.) Food and water storage. Only the young men deviate a bit, ranking a cricket ground as their third most important priority in the new village.

I glance around the room. People seem pleased and relieved at this revelation, that they share a vision for the future, maybe in light of all the recent political divisions. It could just be because of the huge pots of *chang* that are now being brought out, but I prefer to chalk up the bright mood in the room, the animated voices bouncing off its walls, to the villagers' rising hope that Zanskar's "waterless village" could one day become its "model village."

5

The Burdens of Fire

To summarize bluntly, any increase in fine particles in the atmosphere kills someone. The victims remain nameless, but they have been deprived of life all the same.

—PETER MONTAGUE, ENVIRONMENTAL RESEARCH FOUNDATION

Where God cooks, there is no smoke.

—ZAMBIAN PROVERB

A late summer evening in downtown Kumik: sheep-and-goat rush hour. I wend through the woolly traffic, past the boy who's always pushing his little toy car fashioned from cast-off wire, always deep in concentration as he vaults the dry concrete channel of the main stream. I push through the clattering gate of Stobdan's old house and climb the steep concrete steps.

I enter the main corridor; it takes a moment for my eyes to adjust to the darkness. Smoke is streaming out from the storeroom, in quantities that back home would have the fire department on its way. I poke my head into the room, a cavernous space next to the summer kitchen, where the family stores grain and old plow and other tools and huge old mysterious trunks with unnamed contents.

"Jullay?" I call out at the threshold, in the customary Ladakhi-Zanskari greeting.

Zangmo's voice calls to me from somewhere inside the cloud. I duck through (small door thresholds save scarce wood but cause frequent bruises on *chigyalpa* heads) and find Zangmo sitting next to a large, smooth flat stone called *yampa*—brought, she says, from the village of Ubarak, on the boulder-studded slopes of the Great Himalaya's northern flank, where just last week I heard from Urgain, Zangmo's cousin, that a bear had gone on the rampage and broken down a man's door.

Zangmo has filled a basket with dozens of thick round discs of *tagi khambir*. She wrinkles her nose, leans forward toward the open fire that roars up from the packed-earth floor, and places on it a thick piece of *rigpa*, a dried cake of mixed sheep and goat dung. She slaps a lump of dough back and forth until a perfect circle emerges from her sooty hands, then lays it vertically near the fire on top of the hot stone.

I watch her work with efficient grace for several minutes until my throat starts to burn and my eyes water. I search around the room for an outlet for all the smoke and spy an overmatched eighteen-inch-wide hole in the ceiling. The fumes quickly become too thick for me. I stand up and ask, coughing, "Don't you find it hard to breathe?"

Zangmo's smile spreads quickly across her angular features. Zangmo smiles often, but it's always simultaneous with a small, knotty frown that is permanently etched in the center of her forehead, as though even in the midst of any comic epiphany or pleasure some small part of her remains alert to the next chore, the next burden to be shouldered.

"Yes, my eyes feel it," she says and places another circle of dough near the flames. "I feel the smoke." (The Zanskari word for "smell" is the same as the word meaning "to feel" or "to sense.") She sizes me up with a concerned glance. "Go have some tea. The smoke isn't good for you."

"How much more bread do you have to make?"

She tilts her head away from the fire, squints at the flames, and the thought occurs to me that her perpetual small frown has been etched not by anxiety but by a lifetime of such squinting through the dung smoke at her Sisyphean task.

"Oh, many more pieces," Zangmo says. She laughs, with no trace of bitterness. "There's a lot of work to do in the field tomorrow. Everyone will be hungry!"

Doctor Stenzin Namgyal, in his early forties, compact, thoughtful, holds his head slightly cocked in a way that suggests he doesn't want to miss any word you say. We sit in a large, freshly painted yellow room, empty but for two small decorated *choktse* tables and a new carpet, in the new house that he and his family have just moved into, a stone's throw from Tsewang Rigzin's agriculture office on the outskirts of Padum.

Stenzin served in the Padum hospital from 1997 to 2001, and then in the hospital in Kargil for three years. After another two years doing a postgraduate diploma in pathology in the city of Jammu, he came back to Zanskar in 2007. He is one of the two, sometimes three, doctors serving more than 15,000 villagers spread across an area the size of Rhode Island.*

In winter he sees a spike in respiratory complaints, especially cases of chronic obstructive pulmonary disease in older adults. "In winter people are almost always indoors," he explains. "In minus 25 [degrees], the children go to fetch water, that's the only time when they're out of the house. When there's a sunny day, maybe the older people will go to warm in the sun for a few hours. Otherwise mostly they are inside the houses."

All that sitting around the hearth has consequences. "In Zanskar the chronic diseases, the chronic pulmonary diseases, they are because of smoke," he explains. "I see more women with obstructive problems, more women than men with these problems. And for most of the time women are in the kitchen and men roam around for other work, so they are less prone to have the pulmonary disease. Men do have, but they have less chance, because they are not in direct contact with the cooking." Young children spend most of their time with their mothers, and older children help with chores like gathering fuel and water, feeding the animals, and cooking. So Zanskari children have high exposure rates to indoor pollution as well.

"The children catch acute respiratory infection. They catch very early because the homes are not too hygienic, the rooms are poorly ventilated, and the outside environment is very dusty. So we have many children with pneumonia, frequently. We get cases every other week of bronchial pneumonia, infants getting pneumonia, younger children—that's an acute problem."

---

*In 2013, Dr. Stenzin was posted by the government back to Kargil.

"I think pneumonia is the number one killer here," he says. "Because of pneumonia I have seen deaths in children. But less as compared to many years back. Then it was very common."

This fact makes Zanskar sadly representative of both India and the world at large. According to a recent *Lancet* study, in 2005 pneumonia caused 27.6 percent of the 1.34 million deaths in India among children under five years old. It was the leading cause of child death there, just as it remains the leading cause worldwide. Each year half a million children under the age of five around the world are felled by pneumonia. Public health experts estimate that about half of these deaths are caused by exposure to the airborne stew of pollution from cooking and heating fires. So much for the safety and security of the hearth.

In these Northern Colonies the Inhabitants keep Fires to sit by, generally *Seven Months* in the Year; that is, from the Beginning of October to the End of April; and in some Winters near *Eight Months*, by taking in part of September and May."

That's Benjamin Franklin describing life in northern colonial America, circa 1744, in a pamphlet he wrote about the problem of "smoky chimneys." (Franklin was obsessed with smoky chimneys and tinkered with modified stoves all his life.) But as Dr. Stenzin suggests, it is an apt description of Zanskar circa 2013.

Zanskaris spend the short summers doing everything necessary to make it through the long, looming winter. They grow and store grain and vegetables, gather fuel and fodder. Once these essentials are safely stowed on rooftops or in cool grain storerooms, winters have traditionally brought a Spartan kind of leisure. With no farm work to be done (except for feeding livestock, milking cows, and hauling spring water from frozen rivers . . .), families pass much of their waking hours around the hearth cooking, eating, drinking, gossiping, and storytelling. (They pass their nonwaking hours there, too, sleeping sprawled out together in the same room.)

Bootleg Bollywood DVDs playing on small secondhand TVs have increasingly supplanted the *meme*'s old stories, but the hearth fires rage on. And Zanskaris, though they are more habituated to it than most, know well that all this burning is dangerous.

When I ask a circle of old men sitting in the sun around the prayer wheel in Kumik if they think burning less fuel would be better for their health, there are a few snorts of laughter, the kind you might expect if I had asked: Does the abbot of Stongde Monastery wear a pointy yellow hat? (The answer: yes, he does, and everyone knows that.)

"Yes, it's bad for our eyes, and it causes coughs," says one man.

Another, Meme Stenzin, gives a loose cough and just spits into his hand. He holds it out to me by way of a cheeky answer. I stare at the phlegmy yellow mass in his palm while he smiles mischievously up at me through rheumy eyes.

Our rotating teams of villagers—ten on a lean day, forty on a busy one—and a few hired Nepali masons have been making steady progress on the solar *lhakhang*. The foundation is completed, the mud brick walls are rising. Must be time to celebrate, with yet another *chang* party in the community hall in Kumik.

I find myself sitting along the wall next to Gara, who is one of just a handful of men who don't have some kind of cash-generating job outside the village. Some Kumikpas work for the Indian army's road-building unit, or lead ponies for trekking groups, or have shops in Padum; eighteen serve as soldiers in the Ladakh Scouts unit of the Indian Army, posted all over India and even to places like Lebanon for peacekeeping. An unusually high number of men work outside of Kumik, much higher than in a typical Zanskari village—testament to the powerful economic constraint of scarce water. But Gara just farms. So I run into him frequently on the footpaths around the village—he's almost always there, in his fields, working on his house, taking his turn as the rotating *rarzepa*, the shepherd.

Gara assiduously keeps my cup full of *chang*. He decides to tell me a short tale and leans in to be heard over the blaring music.

"My parents have six or seven children who died," Gara says. "Then me, a brother, and one sister lived."

When he was an infant, he says, he fell deathly ill, like his lost siblings before him. "One blacksmith came and gave me milk from small spoon."

This was a common practice not long ago in Zanskar and Ladakh, where there is a limited caste system. It's nowhere near as strong or rigid as in Hindu communities in India (for example, there are no "untouchables"), but nonetheless certain communities have lived and eaten separately for centuries. Blacksmiths in particular were regarded as compromised souls by other Tibetan Buddhists, and therefore of lower caste, because they plundered the earth for their raw materials, disturbing the *lha* and the *lhu*, the spirits inhabiting the ground. For some long-forgotten reason, a tradition developed: when an infant was sick and on death's door, a blacksmith, or *gara*, would be called to the house. With a spoon he himself had forged, he would feed the child some fresh milk.

In Gara's case, this trick appeared to have worked. He recovered and survived, so they called him Gara—Zanskari for "blacksmith." (No one ever seems to use his real name, Tundup Gyaltsen, not even his wife.) He now has seven children of his own.

"Like a machine," Stobdan, who has been listening, jokes with a sly grin.

"I have strong pole, yeah?" Gara says, with an elbow in my ribs and a naughty chuckle.

Tsewang Rigzin, who is several years younger than Gara, has a similar story. Around Kumik he is sometimes called *Garshung*—little blacksmith—for the same reason: he is the only one of his parents' six children who survived to adulthood.

Rigzin, as a boy without siblings, would splash around instead with his best friend Stobdan in the irrigation pond. As an only child, he comes home from his government posting and causes a pleased fuss in his grand, dark, old house. For Rigzin has a doting, no-nonsense mother who asks me, whenever I arrive in Kumik after having passed through Leh, Kargil, and Padum, if I've seen Rigzin and when is he coming home to help with the fields? (I relish each chance I get to tease Rigzin about the fact that he, the block agriculture extension officer, never seems to be at home to help with his family's own harvest, weeding, or planting.)

When I ask him how his siblings died, Rigzin replies with a shrug, "Who can say?"

He remembers when he was a boy and researchers from the University of Bristol, named Keith Ball and Jonathan Elford, came to Kumik in the summer of 1981 to conduct a comprehensive health survey. He recalls how they asked everyone endless questions, put the blood pressure cuffs around his arm, and measured the force of villagers' exhalations with a strange device. They stayed in a room just off his family's summer kitchen; Rigzin and other boys would occasionally pilfer apples from the researchers' bags.

Ball and Elford recorded their findings as part of the classic (and only) study of the human and natural history of Zanskar, *Himalayan Buddhist Villages*. They chose Kumik partly because they were based in nearby Stongde with other researchers, and partly because they thought Kumik would be representative of Zanskar as a whole: not too big, not too small a village; not too rich, not too poor; off on its own, but not exceedingly remote from Padum. They found a surprisingly healthy, sturdy bunch of farmers. There was no obesity in Kumik, "nor was there any evidence of malnutrition and the food intake during the summer appeared to be adequate." They observed a high prevalence of conjunctivitis, "probably a result of the dry, dusty summers combined with exposure to domestic smoke during the long, cold winters."

But one thing stood out in their interviews with the villagers. "Nearly one-third of the subjects said they had a cough . . . or had had one during the previous year. About half said their cough was worse, or only occurred in the winter."

One wonders what the Bristol team might have found if they had visited in winter, when snows make access to diagnosis and treatment at the hospital in Padum impossible for most Zanskaris—and even if they do make it, they can't stay, because there is no heat in the wards. During my two winter visits to Kumik, in 2009 and 2011, most of the children seemed to have loose coughs.

"It seems likely that the many complaints of coughing and conjunctivitis were associated with exposure to domestic smoke," they concluded. The men from Bristol were surprised to find no evidence of chronic bronchitis and ventured that "those who survived infancy and early childhood could look forward to an adult life almost completely free of malnutrition or obesity, severe respiratory disease and hypertension."

But how many survived infancy?

Elford and colleagues conducted an extensive demographic survey of Kumik during their 1981 visit. Based on interviews with every household, they found that, of the seventy-two children born alive between 1962 and 1981 in Kumik, thirteen (18 percent) died before the age of one. That's a rate of 180 infant deaths per 1,000 live births (the standard public health metric for infant mortality). India's national rate at the time was 160. They asked mothers and fathers what they thought the cause of each death was: "fever" and "coughs" accounted for eight of the fifteen postneonatal deaths, for which the cause was recorded. "Although these were rather vague categories, it seems likely that these children died from a respiratory infection of some kind." Elford concluded that "measles and respiratory diseases appear to explain most of the post-neonatal and childhood deaths in Kumik, and many of these occur during the cold hard winter"—when the hearth fires were burning longer and harder.

This wasn't an epidemiological study, which would employ randomized sampling and sophisticated statistical techniques to separate out the smoke exposure from other factors, like nutrition, parasites, other sources of infection. (Randomized control studies are the gold standard of the public health world, used to tease out causal relationships between specific risk factors and illnesses, as well as between effective interventions and health outcomes. They are also very expensive and sometimes ethically fraught.) "Such studies are almost nonexistent [in Ladakh and Zanskar]," says Dr. Stenzin.

Based on their own experience—and between them they see every patient who comes through their doors—Dr. Stenzin and his colleague in the Padum hospital, Dr. Nawang Chosdan, say that until fairly recently, child mortality rates in Zanskar were alarmingly high. In the last ten years Dr. Stenzin says that child deaths from pneumonia have declined dramatically as hygiene has improved in some villages. But respiratory problems remain the most common problem the two doctors encounter. "Still hygiene is poor in the rural areas," Stenzin concedes, and consequently, he speculates, infant mortality remains high in the more remote communities.

In many rural places like Zanskar, the cause of children's death remains a mystery to their families—bad luck, random infections, angry gods? And to doctors and public health researchers, a lack of access to adequate health care, malnutrition, respiratory damage, and the combination thereof are all plausible culprits, in terms of environmental factors that make infections both more likely and more deadly.

"I had five children, who all died," Tsewang Rigzin's mother, Rigzin Yangdol, told me matter-of-factly out of the blue one day as we sat quietly and drank tea.

She was sewing pieces of white cloths into small sacks, like local Ziploc bags, to hold barley flour for a nephew, a snack for a monk who was about to travel to Manali. She nodded toward the dimly lit courtyard room, where an ancient mud hearth sat against the far wall.

"We put wood in it to keep them warm. But the smoke hit their eyes." She made a motion toward her own eyes, as though the fire were attacking them.

I asked if she knew why her five children died. She shrugged just like her only son did.

"Who knows? There were no doctors then."

Then she asked if I'd seen Rigzin and if he was coming. She told me to tell him to build a house for her and her two husbands in Marthang. "Just two small rooms is all we need," she said. "In Marthang it's warm, the sun is hitting it all day. Here the sun goes down behind the hill early." That means more time spent gathering fuel, and more burning, and more smoke hitting her aging eyes, which she still needs for sewing. "And all this going up and going down is hard on my knees! *Chi choen?* What to do?"

She said all of this placidly, with small sighs, sincere but also winking at her own complaints, as she performed one household task after another.

In the 1970s there were indeed no doctors, and no postmortems to ascertain the causes of the deaths of Rigzin's and Gara's siblings. Just thirteen long periods of mourning, the arrival of the abbot from Stongde in his pointy yellow hat to perform the proper prayers, and the cremation that followed among the rocks at the edge of the village. Then ashes and smoke.

Even so, many Zanskaris know why some of their children died—and still die today—before their time. In the absence of doctors and multiyear studies, sometimes common sense suffices. Life in Zanskar is *hamo,* they say—difficult. After all these years, even with the comforts and technologies brought by the road, it remains a place of hard choices, just like Csoma's: stay warm or breathe easy?

In 2005, Elford returned with a team from Durham University to conduct a follow-up survey in Kumik. According to Carla Lewis, who led the social research component of the expedition, they interviewed all of the adult women then residing in the village. Between 1995 and 2004, the women reported 69 live births. Of these, 12 children had died in their first year—a rate of 174 deaths per 1,000 live births, for an infant mortality rate of 17 percent. Despite the improvements in nutrition, hygiene, medical facilities, and training since the advent of the road, the researchers were surprised to find that infant mortality in Kumik hadn't improved much since the 1960s and 1970s.

The global average in 2011, according to UNICEF, was 37. In 2007, the infant mortality rate in the United States was 7 deaths per 1,000 live births. In 1915, the year my grandfather was born, infant mortality in the United States was in the neighborhood of 100 deaths per 1,000 live births. At that time pneumonia was the number one killer of all ages. It was 160 around 1900.

So Kumik's infant mortality rate is about where the United States' was well over a century ago, when Americans burned mostly coal and wood to stay warm and cook. Kumik remains stubbornly where the Indian average was sixty-five years ago. If Kumik were a country, it would rank below Afghanistan, below even war-ravaged Sierra Leone, which has the worst infant mortality rate in the world, at 117.

For as long as we've made fires, the prevailing attitude of humanity toward all those fumes has been: "Well, we have no choice, so we'll just put up with it." It's only in the last couple of decades that we have learned just how deadly all that smoke can be.

Take wood smoke, for example. We now know that, far from being the wholesome alternative to coal that John Evelyn thought it was, wood smoke can contain known carcinogens such as formaldehyde, butadiene, benzene, carbon monoxide, and, of course, particulates. Of these various nasty ingredients, concentrations of particulate matter, specifically of those tiny particles labeled PM2.5, are the best single indicator of health risks to those who are exposed to the smoke. And of those particulates, it's the ultra-fine ones such as black carbon that can reach deepest into our lungs and which are therefore the most dangerous.

How much PM2.5 is too much? Under pressure from advocacy groups citing a huge body of peer-reviewed research about the dangers of exposure to these fine particles, the EPA in 2013 lowered its annual mean threshold concentration from 15 to 12 µg/m3, saving an estimated 15,000 lives per year. (Clean air advocates wanted a standard of 11 µg/m3, which would have saved an additional 12,300 lives per year; the Obama administration seems to have split the difference, in deference to industry concerns about the cost of compliance.) The World Health Organization's air quality guidelines recommend 10 µg/m3 as a limit and state that no public microenvironment, indoor or outdoor, should have concentrations of PM2.5 in the air greater than 35 µg/m3 (while still pointing out that *there is no safe threshold*).

And how much black carbon and other ultra-fine particulate matter hovers in Zangmo's and Urgain's and Rigzin's kitchens? No researchers have ever measured concentrations directly in Zanskar. But daily average PM2.5 exposures range from 285 µg/m3 for children, to 337 µg/m3 for women, to 204 µg/m3 for men, respectively, across India.* From many other studies done in rural kitchens across India, we know that fires made in similarly simple stoves using similar fuels can spike as high as 400 µg/m3. That's forty times the maximum threshold recommended by the World Health Organization.

---

*India has the dirtiest air of any country in the world, which probably helps explain why Indians' lung capacity is 38 percent lower than that of North Americans'; why India has the highest mortality from chronic respiratory diseases in the world; why India has more deaths from asthma than any other country; and why half of all visits to doctors in India are because of respiratory problems.

No public health agency tracks the health impacts of black carbon; the focus has always been on PM2.5, of which black carbon is a component. Differentiating the toxicity of a sulfate or an organic carbon particle from that of black carbon is notoriously challenging. It has been assumed that black carbon's health impacts are similar to that of PM2.5 in general; now there are efforts under way to isolate the health effects of black carbon alone, but it's not clear how useful that knowledge would be from a practical standpoint—since every fire produces a whole stew of stuff that is bad for us, not just black carbon.

Still, Joel Schwartz, a public health expert at Harvard, in testimony to Congress on the health impacts of black carbon, cited a study in the Netherlands showing a shortening of life expectancy, "suggesting that these [black carbon] particles, which in Europe and North America are predominately from diesel, are more toxic than average. Getting rid of them has more health benefits than average," relative to all combustion-related particles.

Black carbon itself is damaging as an ultra-fine particle that can penetrate deep into human lungs and other organs, but it might be just as dangerous as a delivery vehicle for its fellow offspring of inefficient fires, the rogue's gallery of particles, volatile organic compounds, gases, and hydrocarbons that are produced along with it and that piggyback onto it in the form of sticky tars and chemical coatings.

Stobdan quit smoking in 2012, after trying to kick the habit for years. (Zangmo once promised him a glass of fresh milk every day if he stopped smoking; after several weeks, when she discovered he had relapsed she scolded him in front of his neighbors: "You have wasted all my milk!" This phrase has now become a punch line guaranteed to generate raucous laughs in any house in Kumik.) But his winter exposure to kitchen smoke in his kitchen means that even though he has finally quit, he's still breathing in dangerous quantities of particles, tars, and more. Zangmo, who has never smoked, might have the equivalent of a few-packs-a-day habit just from burning dung to make bread in the morning.

Kirk Smith, professor of global environmental health at the University of California, Berkeley, thinks we're well past the point of knowing enough to act. On a December morning, in his office on the fifth floor of University Hall, on the edge of Berkeley's sun-drenched campus, I find him hunched at his computer, scrutinizing charts from the new Global Burden of Disease study, a comprehensive review of global health risks that had been released that same morning.

The GBD is an exhaustive report compiled roughly every decade, on everything that kills us or makes us sick. Smith helped lead the household air pollution part of the study. His team's findings are, by almost any measure, shocking.

As of 2010, 4.3 million premature deaths were directly caused by exposure to household air pollution (HAP) from burning solid fuels for cooking. Of those needless deaths, 500,000 were children who succumbed to pneumonia. The rest are adults who were killed by lung cancer, cardiovascular disease, and chronic obstructive pulmonary disease. (And these are just the diseases for which Smith and his colleagues have ironclad evidence of causation by HAP, but there is a growing body of research suggesting it also likely contributes to many other illnesses and negative health impacts, including tuberculosis, adult pneumonia, cervical cancer, low infant birth weight, and cognitive development problems.)

Smith acknowledges there's a need for more research, given the complexity of the problem—but he likens it to cigarette smoking, which we know beyond a shadow of a doubt to be dangerous, even though we haven't pinned down all the mechanisms by which it kills us.

"They've done literally tens of thousands of epidemiological studies, and God knows how much toxicology and animal studies, and they still can't tell what it is in cigarette smoke that's causing the ill health," Smith notes. "But they know it's something, and what do they use? I actually have a pack of cigarettes for this very demonstration. Tar and CO. Well, tar is just small particles, it's just the way they measure it. So I sort of take it, well, we've got to get rid of combustion particles. And sure, maybe one is 1.4 times more damaging than the other. But then what do you mean by damaging? Well, there are a lot of health effects from small particles. There's heart disease, stroke, lung cancer, chronic obstructive pulmonary disease, pneumonia, cataracts, and if you go into the smoking literature, there are hundreds of things caused by smoking. And if you had some huge amount of resources and got rid of ethical constraints and did randomized trials to figure it out, expose people to more black carbon here and less there and work this out, it's not going to be the same for health endpoints. It's going to be different for heart disease than it will be for cataracts or pneumonia or lung cancer."

His eyebrows arch in bemusement and frustration. "So what is the point here exactly? We've just got to get rid of this stuff!"

We now know, as the WHO points out, that there is no safe level of exposure to ultra-fine particles like black carbon. And thanks to Smith, there is now powerful evidence of a link between particulate exposure from household fires and childhood pneumonia. He led the first study to employ the gold standard of scientific inquiry—a randomized controlled trial—to investigate the health effects of installing improved stoves, in order to "better understand the relationship between acute respiratory infections in children and exposure to indoor air pollution." In communities in the western highlands of Guatemala, Smith and his colleagues randomly assigned 534 households to either receive a new wood-burning stove with a chimney or continue cooking in the same way they always had, over an open fire indoors. For eighteen months, health workers visited each household to see if there was any change in the incidence of acute respiratory infections (ARI), including pneumonia, in young children. After two years, every household got a stove.

During the course of the study, from 2003 to 2005, twenty-three children died; nine of them died from pneumonia. The smoke exposure of infants in the control households—the ones cooking the same way they always had—was similar to smoking three to five cigarettes per day; the children in the intervention households were exposed to 50 percent less smoke on average. While the intervention group had a slightly lower incidence of pneumonia than the control group, the researchers found a reduction of about a third in the number of cases of severe pneumonia among households that received a stove. Those who were exposed to the least amount of smoke were 65 to 85 percent less likely to contract severe pneumonia than those who inhaled the most, according to Smith.

"This study is critically important because it provides compelling evidence that reducing household woodsmoke exposure is a public health intervention that is likely on a par with vaccinations and nutrition supplements for reducing severe pneumonia, and is worth investing in," Smith told a reporter in 2011. Introducing even cleaner burning stoves could reduce severe pneumonia cases much more dramatically, he argued.

I ask Smith if there's still an urgent need for more randomized control trial studies.

"From the health sector side, yep. Go to the air pollution community and they tell you, 'What the hell for? The air pollution levels are so high. We know there are health effects at 20 micrograms and you've got 200. What's the deal!' "

Later in our conversation he distills the central lesson from hundreds of peer-reviewed studies and from his own research spanning four decades in the field of public health. It is redolent of your grandmother's common sense.

"Bad combustion—this is something you want to get rid of. And the worst thing you can do is stick the stuff in your mouth." Smoking cigarettes causes 6.3 million premature deaths every year. "But the next worst thing is to have it in your kitchen."

While household air pollution from solid fuels kills over 4 million people worldwide every year, another 3.2 million people die prematurely each year from exposure to outdoor air pollution (including black carbon) from other combustion sources. (There is overlap here, with smoke vented from indoors becoming *outdoor* pollution, where it kills half a million people yearly.) These estimates are from the GBD, which examined sixty-seven health risk factors—malnutrition, tobacco smoking, heart disease, alcohol and drug use, lead exposure, lack of access to clean water, injury, and on and on—for all 7 billion of us. In addition to estimates of mortality, the report offers estimates of another important metric: the disability-adjusted life-year, or DALY. One DALY is equivalent to one lost year of healthy life; add all the

DALYs up from all manner of illness and injury across an entire population, and you get the "burden of disease." It basically captures the difference between reality and the ideal world in which everyone lives to their full life expectancy, free of disease and disability. Using the DALY as a measure, according to the GBD, household air pollution is the "second most important risk factor for women and girls globally."

Household air pollution (HAP) is the leading mortality risk for people in South Asia. In India alone, in 2010 there were 900,000 premature deaths due to household air pollution. The upshot: household air pollution is the "most important single environmental risk factor globally and in poor regions" and the fourth overall worldwide, after high blood pressure, tobacco smoking, and alcohol use.

In richer countries like the United States and the nations of Europe, where relatively clean fuels are predominately used for household cooking and heating, outdoor air pollution steals more lives and healthy years of life than household air pollution. (The same is true in China, where outdoor air pollution ranks fourth among all risk factors.) But Smith points out that a nontrivial amount of that ambient miasma originates indoors: 25 to 30 percent of the outdoor air pollution in India comes from cooking fires. In China it is 15 to 20 percent. Smith has coined the term "secondhand cooking smoke" to capture this contribution of billions of hearths to ambient air pollution. In India in 2010 it killed a staggering 150,000 people.

Perhaps the most heartbreaking impacts aren't even reflected in the GBD numbers. Recent work on infants exposed to cooking smoke suggests serious cognitive development impacts. Startlingly, when Smith and his research colleagues went back five years later to measure cognitive development in children in thirty-nine of the participating households in Guatemala, they found a strong association between wood smoke exposure in the third trimester of pregnancy and impairment of fine motor skills and visual-spatial perception. Other studies, including some on which Smith is a co-investigator, suggest a link between low birth weight and exposure to HAP.

It seems that before they're even born, some children are exposed through their mothers' inhalation of the fine particles—and start out in life way behind their peers growing up in cleaner kitchens.

In 2010, at the end of a keynote speech I heard Smith give to fellow scientists, he quoted an old English proverb: "Wood is the fuel that heats you twice"—when you chop it and when you burn it.

"It's actually four times," he said cheekily, adding to the list the fever from pneumonia and the global warming caused by black carbon and other constituents of wood smoke. He punctured his gloomy assessment with a ray of hope: "But better combustion can get rid of the last two."

Smith has developed a knack for eyebrow-raising imagery in the course of a career spent trying to draw people's attention to an overlooked epidemic. He has experienced previous waves of policy interest in household combustion and energy, and he sees another one cresting right now (mostly because of interest in the climate impacts of black carbon, though, not the health impacts) and is determined not to waste the opportunity.

After so many years of making the case to clean up dirty hearths, citing epidemiological study after study, offering up rigorously vetted but jaw-dropping numbers like the ones in the GBD, he feels he might finally be homing in on a message that is both empirically grounded and credible, and that resonates with people—decision makers, the public, those beyond his field of public health—and that will open people's eyes to the tragic, silent, raging epidemic he has been observing up close, from Guatemala to India to China, for forty years.

Before I leave him to field inquiries piling up in his inbox that morning from colleagues and the media in the wake of the release of the GBD's findings, he waves me over to his computer monitor. He clicks on a PowerPoint that one of his PhD students helped him put together, inspired by his favorite TED Talk of all time, by the statistician Hans Rosling, on the rise of the humble washing machine.

"It's the world cooking, in pictograms," he says, chuckling at the stick-figure humans, like the ones you see on pedestrian crossing lights, each representing 1 billion people. He gestures to one of them. "What do the richest one billion people cook with?"

"Gas or electric stove," he answers. "Plus a whole range of appliances, rice cookers, etc. So four billion people worldwide, roughly, cook with LPG, natural gas, and electricity."

These are the people who have access to market-based options for cooking. They are connected to the electric grid so they can plug in ranges and toaster ovens and microwaves and rice cookers and cook with the flick of a switch. They have propane and LPG delivered so they can fire up a gas burner or oven with a turn of a dial.

"But what about the other three billion? Well, they're using open fires. So I called it the smoking section and the nonsmoking section!"

Smith points out that the battle over cigarette smoking has resulted in a resounding victory, albeit much and tragically delayed, for the public health community. (When was the last time you were in a restaurant or public venue where smoking was allowed?) But nearly half of humanity is still locked in the smoking section, thanks to their dependence on solid cooking fuels, with no apparent way out, no access to affordable alternatives.

"So what do we do about these three billion?" he asks. "Well, these people are generally using unpurchased wood, dung, crop residues. They're gathering it. Maybe some of them—the middle billion—are near and able to join a market. But these people probably not," he says as he points to the single human figure representing the bottom billion. He points to another figure in the middle.

"But these people are already purchasing their fuel, bad fuels: kerosene, charcoal, wood. And some have electricity, actually." These people are more likely to be willing and able to join a market. "So why not move these people here into using electricity and gas and maybe ethanol stoves, fancier kinds of things. Biogas certainly fits in there. Don't try to get them Philips [biomass] stoves; just get them over to gas—the nonsmoking section!" He points back to the other stick figure.

"This is the hard one . . . that bottom billion. And of course that's where most of the health effects are."

So where to begin?

"Well, moving two billion people off biomass is a good start."

Mendo Tamang's hearth certainly doesn't look like a weapon of mass destruction. In fact, she seems proud as she shows it to her visitors. It used to be much worse.

Mendo is an energetic, forty-three-year-old mother of six. She lives in the village of Haku Besi, which clings to a steep mountainside just above the Trisuli River, a day's bus ride north of Kathmandu. On the day I visit her home in the company of Suraj Kumar Sharma, a development worker with an organization working to reduce indoor air pollution around Nepal—where 82 percent of the population uses solid fuels for cooking and heating—Mendo wears a bright gold shirt with floral patterns, a round red cap, and a blue skirt that soon becomes coated with ash as she kneels down on the earthen floor to light a fire.

Mendo leans in and blows air repeatedly over the sticks. For a few minutes, bluish smoke billows back into the room; then it abates. Her stove is simple: two iron rings set into a low-slung, boxlike mud frame, with space below to load the fuel and a wide, open mouth. A soot-caked kettle sits to one side, a pile of light gray ash underneath. Atop the whole setup, at eye level, is a four-foot-wide metal box that tapers to a narrow throat, which then exits between the blackened beams of the ceiling. It's a stove hood, the first that Suraj's organization, Practical Action, helped to install a few years ago.

"We now use half as much firewood as we used to," Mendo says with evident satisfaction. "Most of the smoke used to stay in the kitchen."

Before the fire was in the center of the room, and the family would sit in a circle around it, warming themselves and inhaling an equal share of the particles. There was no chimney, not even a hole in the ceiling. The smoke would spread and find its way out via the gaps in the eaves between the tops of the mud walls and the thin ceiling beams. She says her family used to suffer from frequent eye problems and respiratory infections.

"Our children are now fighting to sit in front of the stove," she says.

The project had three phases: moving the fire up off the floor; installing the smoke hood; and applying mud plaster to leaky walls to better retain heat in the winter months. But after a few years of use the hood is coated with soot, and one side's rivets are coming loose. Mendo is eager to get a new one.

She pokes at the fire as Suraj describes her critical role in installing smoke hoods in forty-six of Haku Besi's fifty households. Mendo was the first person Suraj worked with in this part of Rasuwa District.

"She was very active," he says. "She convinced the people. At first she gets benefit within two or three days. Then other women are coming here to discuss, how do you get benefit, and she explained about the benefit."

Suraj went through eleven different design phases, in close consultation with Mendo and the other women. "Mendo gave us advice to put the hood at a lower level on the front side. So we put the front side lower. The other thing they suggested to us is how to make use of firewood more efficient, so we put the iron rod [to allow better air flow]. These ideas came from the community."

Using a revolving loan fund, each household received a 4,200 rupee loan (about $45) for the smoke hood and repaid it within forty months. Mendo and her husband are now making a separate room for guests away from the kitchen, which still sometimes gets smoky.

Suraj estimates that 80 percent of the smoke is ventilated out of the kitchen by the stove hood. (But while they seem to be marked improvements over the status quo, Kirk Smith is skeptical of the efficacy of chimneys and stove hoods. "It just moves the smoke," he says. "Moving the smoke isn't the answer." He points out further that even if you replace some, but not all, household hearths with cleaner-burning stoves, there is still a huge pollution problem in the village—because the combustion products hang all around outside the homes.) But the bigger improvement might have come from simply getting the fire up off the floor so air could flow and mix with the wood.

Mendo's husband, Karsang Tamang, is also pleased with the incremental change. "There is not too much smoke, or the dirty, the black," he says from where he sits near the doorway on a charpoy. "Now, yes, it's cleaner. Before, like ten or twelve years, it was different. There were health problems before. At that time no chimney hood, just a fire. And no one could save the wood. Now we save some wood, and the smoke goes to one side, not everywhere. No headaches."

"But he is not cooking, I think!" Suraj says, and lets forth what I now regard, after a few days in his infectiously upbeat company, as one of his signature, booming belly laughs. The laugh echoes around the small dark room, as Suraj gestures toward Mendo. "*She* is cooking!"

"I cook a little, sometimes," says the husband sheepishly. "We now have a quiet place, separate. Before we have together the kitchen and sleeping room for people. Now one room for bed, one room kitchen."

"Are people aware of the dangers of smoke?"

"Yes, people know about danger of smoke," he says, nodding slowly and looking at the floor. He pauses and looks up.

"Two girls a long time ago," he says, gesturing toward the house next door, "they had headache from smoke. After a long time we take them to go for a check at the hospital. They were very sick. Four or five years ago, we had to carry them, they were coughing, to Dhunche, to the road." He pauses.

"What happened?"

"The two girls died. They were young women." The room falls silent.

"Now there is less sickness and headache," Tamang continues. But life is still hard. "Everyone who lives here, it's very difficult. Nothing is easy. No ambulance, nothing. Everyone carries, climbing. Very, very difficult here, basic living."

Suraj and I bid farewell and start the steep hike down to the river, weaving through the terraced fields of corn and wheat. We stop to chat with another family on our way out. The mother sings the praises of the smoke hood, and of the revolving loan fund for home improvements and small businesses, that Suraj and his organization helped set up. As Suraj listens and jots down her suggestions for expanding it, I glance into the adjoining hut.

Two young girls stand there, transfixed over a wood fire. One holds her baby brother in her arms. The older girl places a pot on the mud stove. Dark smoke wreathes the girls' heads. The baby boy fusses and squirms. I scan around the room. There is no chimney.

• • •

A few days later I am in the kitchen of the Hotel Yala Peak, a nondescript two-story stone building in the village of Kyanjin Gompa, a three-day walk from Dhunche up the Langtang Valley. It is also the home of Kunzang and her husband Dawa, who open their doors to foreign trekkers for part of the year. The village sits on a windswept plateau at 3,600 meters, roughly the same height as Kumik. Looming above is a glacier called Lirung, which, like the one above Kumik, is dying.

The Kyanjin Gompa villagers have watched the glacier retreat up the mountainside in recent decades. "Before that dark part was all glacier," Kunzang says as she points at the mountain's face, now mottled like a cow's hide, white patches of snow and ice outnumbered by dark blotches of rock. "The black one, no? All that are glacier, full of snow, ice. But all melt now."

She's troubled by these swift changes, but she and her neighbors have more immediate concerns. There are few trees nearby at this altitude, so the villagers have to walk for days down the valley to cut pines for firewood and then haul it up in heavy loads on their backs. Kunzang cradles her infant daughter in her arms as she nods toward her large mud *chulha*. The massive stove burns through their supply of wood quickly, she says. The baby starts to cry, and she murmurs a sing-song series of rhymes that all key off the girl's name.

During the height of the two trekking seasons, in April and November, Kunzang and Dawa have to feed up to thirty guests. They burn some propane for fast-boiling tea and soups, but it's expensive: the cylinders must be carried in from Dhunche by porters like everything else, and by the time they reach Kyanjin Gompa they can cost the princely sum of 1,600 rupees or more (about $20). So they can only afford to use a few cylinders each year. An NGO in Kathmandu has also installed some biogas systems, fueled by cow dung, in Kyanjin Gompa, but they have worked only sporadically in the cold temperatures. The rest of the time Kunzang and Dawa burn local wood, split and stuffed into the maw of their altar-shaped stove.

The girl stops crying, and Kunzang keeps cooing gently. I gesture toward the *chulha*, which is as mottled as the mountain framed in the window next to it, its sand-colored mud plaster spliced with dark sooty patches.

"Does it make a lot of smoke?"

"Of course the smoke is bad," she says, with a note of exasperation.

She looks meaningfully at the black streaks lining the wall and the stove's face. She gives a weary shrug. Then she turns back to me with a defiant expression, chin slightly jutting out, eyes flashing, and articulates the precise lineaments of the damned if they do, damned if they don't situation of the 3 billion human beings who still burn solid fuels for cooking and heating.

"But what to do? We have to eat!"

In China's Yunnan Province, I found evidence of yet another burden of fire, one that is often overlooked, but which has been quietly borne every day, ever since Prometheus gave us his fateful gift. (Or perhaps for the past 1.7 million years, which is the age of blackened mammoth bones found in a cave in Yuanmou, a candidate for the site of the earliest known hearth, situated in Yunnan, between Kunming and Lijiang.)

I boarded a diesel bus in the heart of Lijiang. It took me past the giant statue of Mao Zedong and north out of the city, where I bore witness to the juggernaut that is China in full-throttle development mode, even in its most rural, bucolic reaches. The bus passed through thick upland forests, heading toward the famous first bend of the mighty Yangtze River. We sped past brick kilns, long narrow buildings surrounded by hundreds of rows of stacked bricks, spewing gray smoke into the sky. We got stuck behind small trucks loaded with produce as they pulled heedlessly out into traffic, belching velvety puffs like dark regrets in their wake. Out in the fields I saw peasants piling up corn stalks and setting them alight. I saw dozens of thick tapered columns of smoke pouring off dozens of mounds of crop waste in the fields lining the banks of the Yangtze—fires morphing from bluish to dark to light as they smoldered, then to dark again as they flamed. The bus passed a diesel-powered conveyor belt that carried aggregate gravel destined to become concrete buttresses holding up a flyover for a six-lane highway that will soon slice through these red hills on a journey to nowhere, to the future.

I saw roadside clearings slotted out for the sites of future shops and unspecified coal-fired small industries—the coal already there, waiting for them in head-high piles—and spied thin wisps rising from the lonely breakfast fires of the highway workers where they camped up on the raw red hillside. We passed a road-paving station with its metal barrels full of flaming black paving tar. Young men, just boys really, were shoveling trash into a large fire on the riverbank, in an opening in the side of a nondescript building surmounted by two brick smokestacks, and I wondered what useful thing was being made or disposed of within. And I saw another knot of eight young men clustered around a crude, horizontal paddle cement mixer, powered by a diesel generator putt-putting away into their faces, and nearby another boy with a cigarette dangling from his lip, eyes dead, as he manned another diesel-powered jackhammer on the edge of the road.

Then we arrived in the shabby little town of Qiaotou. I paid a 50-yuan entrance fee to the national park, and disembarked up the road a ways to find three dozen men shouting and straining against heavy timbers as they raised a traditional timber frame, all by hand, no diesel cranes, for a new hotel. I watched them clap each other on the back and set off firecrackers to mark the completion of their work. The smoke from the firecrackers was blue, and it smelled like childhood.

Not much farther on, I reached the setting of a very old story, the deepest gorge in the world. A hunter once chased a tiger there, down into its depths. As the tiger reached the bottom, the world turned into a raging river, and it saw there was only one route of escape. Never breaking stride, it raced toward the boulder-strewn stretch of bank where the river was narrowest. The hunter crouched close behind and put the animal in his sights. The tiger leapt into the mist-filled space above the rapids. The hunter could only watch, dumbstruck, as the tiger landed on the other side—eighty feet in a bound—and softly padded away, safe beneath the dark cliff bands of a towering mountain.

This place now takes its name from that bold maneuver: Hutiaoxia,

Tiger Leaping Gorge. The mountain across the water is Yulong Xueshan, the Jade Dragon Snow Mountain. It is home to the southernmost glaciers of Eurasia. Not long ago there were nineteen glaciers on the flanks of the Jade Dragon; today there are fifteen. The glacier-covered area decreased by almost 27 percent between 1957 and 1999, the largest extent of glacier retreat of any glacial region in China. Its rock face is riven with deep clefts that channel water 10,000 feet down to join the boiling river, known as Jinsha, or Golden Sands, which still tumbles and froths through that narrow defile before it becomes the Yangtze. This is Asia's longest river; it flows 3,900 miles east to the sea, watering the southern half of China on its way. Tiger Leaping Gorge now attracts a different kind of hunter, like the young urban Chinese couples armed with digital cameras and foreign backpackers equipped with carbon-fiber trekking poles.

But on an aimless walk on a rain-spattered morning many centuries after the tiger's leap, you might encounter, as I did, a scene that the hunter once saw: a middle-aged woman making the winding descent down the 50-degree slope above her village, one of a handful in the gorge inhabited by the indigenous Naxi people of southwest China, who grow corn and rice and wheat on terraced fields not unlike those you will find on the other end of the Himalaya, in a village called Kumik. Unlike the hunter, she has secured her quarry. It is slung across her back, an unwieldy load of many long pieces of split pine.

Bent over almost parallel to the ground, she takes halting steps, and rasping breaths. She stops and sit-falls to the ground with a shudder. She moans and mutters a few words in Naxi. Seeing you don't understand, she signals with eloquent hand gestures that her burden is heavy, too heavy. She points down at a small house tucked among the pine trees, just above the narrow road. A thin chimney juts up through its red tile roof. It must be fed. Then she points up the mountain whence she has come and mimes the hauling of yet another load. More wood to be fetched.

Her gaunt face shines with a kind of outrage, and her eyes well with tears. There is always more wood to be fetched.

The woman heaves her small bent form upright, along with her appendage-bundle of pine, together as one needful thing. She staggers under her burden, and for just a brief moment it seems as though she might tumble outward into the yawning void above the river. Then her foot lands safely on the ground. She steadies herself, sets her jaw, and lurches homeward again down the steep footpath.

Wood is the coin of this realm. The pine forests that coat the southern flanks of the gorge—and that, together with the antishadow cast by the frequent mists, make it look like the one true setting for all of those ancient Chinese ink scroll paintings of lonely poets and waterfalls—provide the wood for building homes, fuel for cooking and heating, something to sell in the market for a few yuan. The steep hillsides offer a somewhat less picturesque daily prospect for those who live in this village, called Bendiwan. Securing fuel for them means unremitting toil: hours of climbing up a thousand feet or more, splitting wood, ferrying it back down to the hearth's maw.

This price is extracted almost entirely from the women. "Marry a Naxi woman," an old local saying goes, "and you'll be happy for the rest of your life!"—because you won't have to lift a finger. The Naxi creation myth tells of Nlue and Sse (their Adam and Eve figures) and how they came to live together. The first thing they did was to go to the mountain to cut firewood. Once they had cut it, the wood got up and walked off to the house all by itself.*

Some of the sights on the ride to Tiger Leaping Gorge are unique to the rapidly modernizing China of the early twenty-first century. But the scenes on its mountain paths have changed little since Tu Fu, perhaps China's most famous poet, penned "Ballad of the Firewood Haulers" during a sojourn in other western mountains in the eighth century. "Women bustle in and out," he wrote. "When they return, nine in ten carry firewood—firewood they sell to keep the family going."

The tiger may have escaped ages ago, but the women of the gorge are still bound, like Sisyphus, to its slopes.

---

*"Perhaps a hint at the laziness of the people, but later it was arranged for the women to cut the firewood and carry it down the mountain," wrote Joseph Rock, the eccentric explorer of southwest China, student of Naxi culture and correspondent for *National Geographic* in the 1930s. It is said that his dispatches inspired James Hilton to invent Shangri-La.

• • •

A windless, cloudless, electric blue July day in Kumik—as far as one can get from winter, yet winter planning is in full force. Every house has sent a member up into the high pastures, to fill *tsepo* baskets with dung cakes deposited by grazing yak and sheep and goats.

I walk up the valley, past the willow grove, along the fickle stream. Hundreds of feet above on the mountain I spy tiny figures. They move slowly, stop, stoop, and rise again. When I get closer, I see that every one of them is a woman, doubled in volume by her *tsepo*, tossing round cakes of dung over her shoulder.

There's Ane Dolma, working her way laterally across the mountain above me. There's Tsering Putith, the mother of the ten grown Gonpapa children, with a full basket, already heading down with her second load for the day. I toss one more piece in it for good measure. She smiles and with a "*Jullay!*" is off down the path homeward. There's Ishay Paldan's unmarried daughter, Dolkar, with her long graying pigtails. She's sitting with young Stenzin Nangsal, resting in the noonday sun a stone's throw from a fifteen-foot-long concrete tank embedded in the slope. A water storage tank with two metal outlet pipes. I glance inside. Empty.

In the pounding noonday sun of late July, it's hard to imagine the bitter cold of February. But it's coming, and that's why the women of Kumik, of all ages, from unmarried girls to wizened grandmothers, are out en masse—to lay in a store of fuel for the long winter. Each woman fills her own basket, for her own household, but like everything else in Kumik, this task is performed in the company of neighbors.

The three women rest around the tiny trickle of a hidden spring, cracking jokes and gossiping. (There have been three weddings this summer in Kumik, and everyone keeps a gentle but firm tally of who gave what, who owes what to whom.) They call to a woman far up the hill. "Make yourself look nice!" they tease her, cackling with glee. "Jon is taking your photo!" I wander a bit up the goat path, and then they tease me. "Don't go up there—there are ghosts up there!" they warn me with giggles. "*Yato met-a?*"

They are up on their feet again—"*Hup alahes*!" they say, the universal exhalation of a Zanskari rising with a heavy load, like the grunt of a tennis player propelling a cross-court backhand. I go off to comb for the dried brown chunks of treasure, though I have it easy, of course, without a basket. Soon I have an armful of big flat yak dung cakes and smaller conglomerates of golf-ball-sized sheep dung (which, I have to note, look like little chains of black carbon spherules magnified by an electron microscope).

I intersect with Rigzin's Aunt Dolma and deposit my find in her *tsepo*. "*Shichey rak, skomsa rak!*" she says. "I'm dying of thirst!" Then she presses some *paba* on me—I don't really like *paba*, a tasteless lump of boiled pea and barley flour, but she insists—and keeps combing.

Farther down the hill I cross paths again with Stenzin Nangsal. She is the young wife of one of the men of Pang Kumik. I offer to carry her *tsepo*, which is almost as big as she is—she's well under five feet tall.

"*Oh jullay!*" she agrees.

She watches with a bemused expression as I wrap the straps around my shoulders and awkwardly stumble to my feet. I reckon it weighs eighty pounds. We head down the trail. She bounds; I stumble.

Within a minute circulation in my arms is cut off. Stenzin senses my discomfort.

"Do you want me to take it back?" she asks me.

"No, no, I'm fine. Let's go until the school."

I am not fine. I am barely staying upright. I am afraid I will twist an ankle. I am losing feeling in my fingers. The willow twigs of the woven basket have gouged my back, leaving marks that will last for days. I reach the big *chorten* and lean the load onto it unceremoniously, abandoning all hope of appearing nonchalant.

"That *tsepo* is heavy. How many *tsepo*s of dung do you burn in winter?"

She considers for a moment. "One *tsepo* for one day."

Winter in Zanskar can stretch to almost 200 days. Thus the heavy loads. I shake my head. "Zanskari women are very strong."

She just laughs as I hand her the shoulder straps. Then she smoothly transfers the giant load to her tiny back and starts off again. It's another kilometer down to Pang Kumik, her second run of the day. "*Oh jullay!*" she calls again over her shoulder to thank me with a smile.

I've carried her burden for twenty minutes. She has a whole lifetime of hauling fuel ahead of her.

Reliance on solid fuels imposes some particular, and particularly taxing, burdens, beyond the dangerous exposure to household air pollution. The vast majority of the cooks in rural households—the ones who spend the most time out gathering the fuel and near the stove—are women. Around the world, women and girls spend an average of twenty hours each week just searching for fuel. This causes many girls to miss school entirely, while others go to school exhausted by these chores. In some conflict-ridden places, like Sudan's Darfur region and the Congo, they must walk for hours by themselves, risking rape and other violence, to find increasingly scarce wood to cook for their families.

Even in peaceful valleys like Langtang in Nepal or Tiger Leaping Gorge in Yunnan, women can spend entire days harvesting and carrying loads of pine to burn in their mud stoves. One 2003 survey of over 10,000 rural households in Himachal Pradesh, Uttar Pradesh, and Rajasthan in northern India found that women walked an average of 2.5 kilometers to collect wood; finding, harvesting, and transporting fuel to keep their hearth fires burning took each household about fifty hours per month. Burning this fuel in open fires, or in simple mud stoves, means they need more fuel; therefore women must gather more wood, take more time, and expose themselves to more risk.

Smoke from cooking and heating fires is the second biggest health risk factor globally for women and girls; in South Asia, it's the number one threat to women—the environmental factor that shortens their already difficult lives more than any other. But the soot extracts an indirect price, too. For every molecule in the wood that becomes black carbon instead of combusting completely to provide useful heat energy, women and girls from Kenya to Nepal must spend more time and effort combing the hillsides—time not spent in other productive activities, including education.

When he was a young child, my old friend Sonam Wangchuk, the solar guru, watched his mother crouch near the stove each night and light the dung fire.

"I used to wonder as I watched my mother as a child," Wangchuk recalls, "when she would do this *butpa*." She would bend down near the stove and squeeze the *butpa*, a goatskin bellows, again and again, to force air into the firebox so the flames would take. It was hard work, but if she didn't push enough air into the stove, it wouldn't burn hot and smoke would pour back out into the room.

The boy looked around the room and started thinking of different ways to perform the same function while saving his mother all that trouble battling the smoky stove. He toyed with an idea: "A big stone that I would hang on a pulley, and I would pull it once and then it will run the forced air thing. And as it descends it would keep running it for twenty minutes." The weight would slowly descend from a height, pulling the handle of the bellows as it ratcheted its way down, much like the way old clocks were powered, thus freeing his mother to do other things (like teaching him how to read).

As a boy of five or six he didn't have the means to build his gravity-powered bellows. But Wangchuk credits that moment as a turning point that sparked a lifelong fire for making things that are practical and useful to people and lighten their burden, and a propensity for looking at problems from a different angle than most people do. Today he still smiles at the memory of his hearthside puzzlement.

"I kept watching," he says, and a question kept welling up: "Why should they keep doing this?"

In his freshly painted guest room in his new house on the outskirts of Padum, I have given Dr. Stenzin Namgyal a cruelly tantalizing hypothetical. "If your boss called tomorrow and added one crore to your budget, how would you spend it?"

He shakes his head with a rueful smile. Too painful to entertain such dreams: A crore! Ten million rupees—about $183,000. But he indulges me.

"If I get a budget—and that's very unlikely—I would put almost 90 percent into preventive part. It would be a long-term measure. I will do awareness campaign at every village, and provide [face] masks to every village. And if the budget permits—it's a huge amount, yeah?"

I confirm that, yes, in this fantasy he has been provided with a huge amount.

"Then I would try to improve the ventilation condition of the kitchen in every household. We would do some evaluation of how to ventilate the kitchen, how you can get a better *chulha*, so it can be more effective, less smoke and more combustion."

"So more efficient *chulha*s would have a significant health benefit for people in Zanskar?"

"In Padum you will see nowadays in summer most use LPG [for cooking]. In winter they will be using the *chulha*s, with cow dung, everybody will be using. These LPGs are not enough for minus 25, 30 [degree temperatures], to heat. You have to have fire. So I think that's the most important."

The supply chain for gas cylinders and burners is tenuous: a few summers back the distributor in Kargil had a dispute with their supplier, and no cylinders arrived in Padum for weeks. Empty cylinders piled up in the depot, and people shifted to using kerosene camp stoves to supplement their dung stoves. Even when the deliveries of LPG cylinders are steady, the cost and difficulty of transporting them are as prohibitive for many families in the most rural parts of Zanskar as they are for Kunzang and other residents of the Langtang Valley. So they do almost all of their cooking with dung.

"With time people are improving but there are many people who have the old style. Even me! You will see this is well furnished, everything's okay," he says as he gestures around his new house, "but if you visit my old house, it has been there for forty to fifty years. The ventilation condition is almost the same."

I tell him that I have visited his old house. The year before I went with a friend to deliver a letter to his father, the old king of Padum. We ducked under the five-foot-high thresholds and sat along soot-stained walls and breathed the stale, smoke-laden air. Stenzin smiles, shakes his head. The aged king has just moved into the new house that his son has built, but most of Zanskar remains in the smoking section—where children under five are at heightened risk of dying needlessly from pneumonia and adults of cardiovascular disease, among many other soot-borne scourges.

"So you know I would love to change the way the kitchen is," the doctor concludes with a wistful smile.

With a few resources, the doctor would uphold his oath to heal by working with his neighbors, to help them realize that they have a choice beyond Csoma's options of freezing or choking, or Kunzang's dilemma of going hungry or eating amid a toxic stew of dark particles. He nods slowly, brow knitted, staring at the *choktse*, as he mentally catalogs all that could be done to move his fellow Zanskaris into the nonsmoking section.

"Yes, I would spend most money on improving the condition of the people."

It's another late summer night in old Kumik—so, yes, it must be time for another party. (This seems to be the default setting in the village: it's always better to be together, and to drink together. Or as Ben Franklin put it, "He that drinks his cider alone, let him catch his horse alone.")

So I prepare for the *chang*-over that awaits me in the morning, following the hours of sitting and *sheep-sheep* sipping ahead of me that night, and head over to the community hall. The old men, as always, sit cross-legged in a line by descending age on the southern wall, with Ishay Paldan and the other elders in the corner. Stenzin Konchok sets up the old dusty stereo and pops in a cassette. Children flit in and out in a state of excitement. The women haul in five-gallon buckets of fresh cool *chang* and move down the line to amiably harangue the men, joking and gesticulating with their long-handled ladles, into draining their cups. The *chang* seems to flow in much greater volume than the trickle of a stream fifty yards away in the darkness.

Tonight the villagers opt for some old songs. Twelve dancers spin and rotate slowly around the hall, and the women, some still in work clothes caked with the day's dirt from the ancient fields, sing a song of pride about their village:

"Kumik is like a small flower that grows high in the mountains, and only needs little water!"

The dancers do a lenten circuit, making small mincing steps in counterpoise to the subtle gestures of their twirling wrists.

"This village of Kumik has one spring for drinking and one hero—Ache Lhamo Dolma!"

Another circuit, as the women carve spinning circles around the room's painted pillars.

Later, when I ask around, nobody seems able to explain who exactly Ache Lhamo Dolma was. What is known about her seems tautological: Ache Lhamo Dolma was a legendary figure of some sort, who may or may not have existed and now survives in a folk song. (*Ache* means "older sister"; Lhamo and Dolma are common names.) Apparently it's like asking why we row row row the boat gently down the stream.

And why was she a hero? No one has a good answer for that either.

"That is how the song goes," I am told. "She was the hero of Kumik." And what else would you like to know?

"Oh," I reply, nodding, mystified, but perhaps a bit *chang*-stunned, too.

So, as with much of Zanskar's unrecorded past, the matter is left open to interpretation. I like to imagine that Lhamo Dolma was an indomitable figure, much like the other women in her village, though maybe a bit more so. I further surmise that the heroism of Lhamo Dolma was not of the Gesar Saga's monster-slaying variety—conventionally demonstrated in one swift, courageous feat of epic proportions—but rather that the song slyly nods to the quotidian miracles of creation wrought by Kumik's indefatigable women. The ones who gather the fuel, stoke the fires, milk and tend the animals, maintain the homes, cook the meals, irrigate the fields, who make the world go round. Men do these things, too, of course, but as in much of the world, the burden falls far more heavily on the women and girls. The ones who stay all summer up in the *doksa*, those rough stone huts in the high pastures, grazing the animals and transmuting mountain grasses into rich yogurt and butter. The ones who stream in clusters from every household to scour the high mountain slopes on appointed days for precious dung for winter fuel. The ones who shuttle back and forth on the fields' margins all summer long, with only the bottom halves of their dark *gonchas* and brightly colored *salwar kameez* visible underneath truly mountainous loads of freshly cut fodder grasses, bobbing through the village like shaggy Dr. Seuss characters.

"How on earth do these people survive?" asked John Crook when he first traveled to Zanskar.

I watch the unassuming young women dancing around the room, singing with pride of their dying, drying village, unfurling their calloused hands again and again in delicate gestures pregnant with meaning for those who know the story they tell. *That's how they survive*, I think.

Just like the "one spring for drinking," Ache Lhamo Dolma kindles life from the barren mountain slopes. As inexplicable and necessary as the *chumik* ("water eye," as springs are wonderfully called in these parts) that trickles out of the earth up near the *doksa*, Lhamo Dolma must have performed daily wonders as earthshaking as Gesar's or Gilgamesh's, as shrewd as Odysseus's: she drew water and gathered fuel, fed her creatures and her family, and through endless toil sustained life's glowing embers through the long night of winter. She still does.

6

Water Towers Falling

Some say the world will end in fire,
Some say in ice.
From what I've tasted of desire
I hold with those who favor fire.
But if it had to perish twice,
I think I know enough of hate
To say that for destruction ice
Is also great
And would suffice.

—ROBERT FROST, "FIRE AND ICE," 1920

Chu ma-yong kongdu raks.
(Build the dam before the flood comes.)

—TIBETAN PROVERB

Compassion Incarnate is pulsing with laughter in the heart of Zanskar. His cheery baritone ripples through the bustling fair grounds. The ice-clad peak of Padum Kangri gleams a mile above his famous, tonsured pate as he leans conspiratorially into the mic and jabs his finger into the air.

"The teaching of dependent origination by the Buddha is compatible with science!" His Holiness the Dalai Lama declares to the rapt audience.

The day before, on the first day of the Dalai Lama's teachings, I had run into Dr. Nawang Chosdan, Stenzin Namgyal's colleague at the hospital, on the streets of Padum. As he waited for his lunch order from a

Tibetan restaurant, he told me casually, "High lamas, you know they can change the environment, the weather. Yesterday was hot, so today he brings some cloud, not so hot."

If the Dalai Lama controls the weather, then he wants today to be hot again, for the sky is once again clear and the sun pummels our heads and necks. Urgain Dorjay, dressed in his finest woolen *goncha*, passes through the crowd, stopping here and there to hand out cups of fruit juice to thirsty devotees. It seems like half of Zanskar's men, women, and children are here to listen to the teachings. His Holiness occasionally pauses in his explanations of Buddhist doctrine to remind them to drink, to use their umbrellas, to stay cool. Then he returns to the subject of the day's discourse: interdependence.

Depend on these teachings for nourishment, he tells the crowd, "just as here you depend on the glaciers for water."

The next morning I awake to a clatter, and poke my head into the hallway of Urgain's house.

"There is *chu-rut*," Urgain tells me calmly. A flood.

He explains that a violent flood originating from the vicinity of Padum Glacier came down the stream on the other side of Stagrimo Monastery, just over the hill. He speculates that it has ruptured the water pipe that brings water for irrigation to their fields and another one that brings drinking water. So he is filling pots with water from the spigot in his washroom in anticipation of a cutoff. (Urgain, being something of an innovator, is the only resident of his village who has installed indoor plumbing.)

We walk up the hill to investigate, hopping across empty canals, full of little muddy puddles. "The water is stopped," he says. "This means it is damaged."

We crest the ridge above the monastery and reach the muddy stream tumbling down from the glacier, bouncing off huge boulders. Urgain points to where the "wind" of the water's sudden surge blew spray onto the banks ten feet above, caking the leaves of delicate purple wildflowers.

He sketches his theory of what happened in a diagram on the sand.

There are multiple small bodies of water underneath the different glaciers above, he says, and due to this stretch of hot weather, maybe the ice walls that contain them melted. The resulting surge sped out from under the glacier, down the mountain, and onto the plain below.

"Others are thinking it is finished, but I think maybe water will be high for one week. There is cause for worry," he says, crouching there by the still raging stream, not sounding worried. "This is not good for the future. Glaciers are also finished. It means climate is changing."

I ask if this had ever happened before on this stream.

"Never, never before."

"Have you heard of this kind of glacial flood happening elsewhere in Zanskar?"

"Never," he says emphatically.

We walk on down the bank of the stream until we come to a place where two stone bridges have been washed out by the flood. The three-inch pipe that delivers drinking water to Padum, Pipiting, Shisherak, and several other villages is indeed severed, spewing clear spring water into the frothing stream. A fifty-foot section is missing.

We gaze out onto the plain. If the stream had chosen the other channel, the main bazaar of Padum would now be knee-deep in silty water. The headworks of the canals that carry irrigation water to the fields of Stagrimo, Khanggok, and Padum are also damaged. Rocks now block their intakes, leaving a muddy trickle just a couple of weeks before the harvest starts. Urgain makes a plan of action as we walk back to his house.

"I will submit application to the subdivisional magistrate for repairing the pipe and canal." A bit later he laughs, as though the magnitude of the situation has just struck him. "Oh ho! Drinking water is finished for everybody!"

There is some person who makes *puja* [prayer] around the village, and he is like a magic person, yeah?"

As he starts his tale, Tashi Stobdan fills my cup with *chang*. "Everywhere—also in Kumik. Not now—before."

It is the fourth day of mourning for Tsewang Rigzin's grandmother,

The core of the ancient village of Kumik. The snowfields and glaciers above—the community's only water source—have been steadily shrinking in recent decades. (Photo by Jonathan Mingle)

Dressed in their traditional *gonchas*, Kumik villagers gather to commemorate the completion of a new statue of the Buddha in the partially completed solar *lhakhang*, in August 2012. (Photo by Jonathan Mingle)

Top: The single stream that carries snowmelt to irrigate the fields of Kumik. (PHOTO BY JONATHAN MINGLE)

Left: The concrete canal above Kumik, cracked by harsh winters, left dry by early springs and declining snows. (PHOTO BY JONATHAN MINGLE)

Sitting astride the dry canal, Tsewang Norboo (seated, looking left) leads a group of Kumikpas in discussion of the ongoing drought, while Tashi Stobdan (leaning on car) looks on. (Photo by Jonathan Mingle)

Men from Pang Kumik argue with their neighbors over the fairness of a planned timetable to lay new drinking water pipes in the old village. (Photo by Jonathan Mingle)

Friends since childhood, Tashi Stobdan and Tsewang Rigzin sit on the plain of Marthang, gazing out at the future, in the form of piles of stones that their neighbors have gathered to build new homes and walls. (PHOTO BY JONATHAN MINGLE)

Urgain Dorjay navigates a particularly tricky part of the *chadar*— the "ice way"—on the frozen Zanskar River. (PHOTO BY JONATHAN MINGLE)

Clad in old sheepskins, Kumik's *abi-leys* (grandmothers) walk home after circumambulating the old *lhakhang* in midwinter on the ridge above the village. (PHOTO BY JONATHAN MINGLE)

ABOVE: Tsewang Zangmo lights a fire under a soot caked pot to make tea in the shepherds' huts in Kumik's high pastures. (PHOTO BY JONATHAN MINGLE)

LEFT: The women of Kumik spend entire days of the hot summer preparing for winter, combing the mountains for dung to burn as heating and cooking fuel, and hauling it homeward in their *tsepo* baskets. (PHOTO BY JONATHAN MINGLE)

Sonam Wangchuk makes an experimental section of floor from a mixture of clay, straw, and apricot fibers in his house on the campus of the Students' Educational and Cultural Movement of Ladakh. (PHOTO BY JONATHAN MINGLE)

Women bake bread for a communal gathering on a temporary dung-fired hearth set up in the sheep pen of the *Kharpa* house. (PHOTO BY JONATHAN MINGLE)

A young boy pretends to drive a tractor, the new diesel-powered beast of burden in Zanskar, with Kumik's shrinking glaciers in the distance. (PHOTO BY JONATHAN MINGLE)

In the shadow of Padum Kangri and other receding glaciers of the Great Himalaya, a villager digs a small canal to bring water for making mud bricks for the walls of the new *lhakhang*. (PHOTO BY JONATHAN MINGLE)

A diesel-powered JCB excavating machine digs soil from a collapsed section of cliff out of the canal that brings Lungnak River water to Marthang. (PHOTO BY JONATHAN MINGLE)

Gara and Rigzin (a retired soldier) take a laughing break from work on the new *lhakhang*. (PHOTO BY JONATHAN MINGLE)

The men and women of Kumik haul stones, bucket brigade–style, to fill the foundation of the new *lhakhang*. (PHOTO BY JONATHAN MINGLE)

The day after a fierce political argument in the old village, Kumikpas come together in Marthang for a piece of practical, forward-looking business: assigning pairs of households to keep hungry animals out of their new fields. (PHOTO BY JONATHAN MINGLE)

Tashi Stobdan in his new fields in Marthang. (PHOTO BY JONATHAN MINGLE)

A new concrete *pukka* house rises in the new village of Kumik Yogma, with the old village (right) in the distance. (PHOTO BY JONATHAN MINGLE)

Kumik's children haul precious spring water home in jerry cans. (PHOTO BY NICOLAS VILLAUME)

Two future householders from drought-stricken Kumik on a visit to Marthang—"the red place"—their future home. (PHOTO BY NICOLAS VILLAUME)

The *zing*, one of two storage ponds in Kumik, holds water overnight for irrigation during the day. (Photo by Nicolas Villaume)

Rigzin Yangdol, mother of Tsewang Rigzin, sits flanked by his grandmother and aunt in their ancient home in the heart of Kumik. She lost five children in infancy. "The smoke hit their eyes," she says. (Photo by Nicolas Villaume)

Tsewang Zangmo in animated conversation with her neighbors. (Photo by Nicolas Villaume)

Meme Ishay Paldan, Kumik's eldest citizen, points out his window toward the glaciers and snowfields that he has watched vanish for most of his life. (Photo by Nicolas Villaume)

Dark particles coat the surface of the interior of the Greenland ice sheet, in this photo taken by researcher Jason Box on a trip in the summer of 2014. Box is analyzing samples, and theorizes that they could be a mixture of soot and dust, and microbes that feed on them. (PHOTO BY PROF. JASON BOX, GEOLOGICAL SURVEY OF DENMARK AND GREENLAND)

A satellite image capturing one of humanity's biggest smoke signals, the "atmospheric brown cloud" that forms seasonally over northern India. Composed of black carbon, organic carbon, sulfates, and a stew of other particles—largely produced by the inefficient combustion of fuels across the subcontinent—the cloud blocks some sunlight from reaching agricultural crops, and sends its dark tentacles up into the "abode of snows," the High Himalaya. (COURTESY OF NASA)

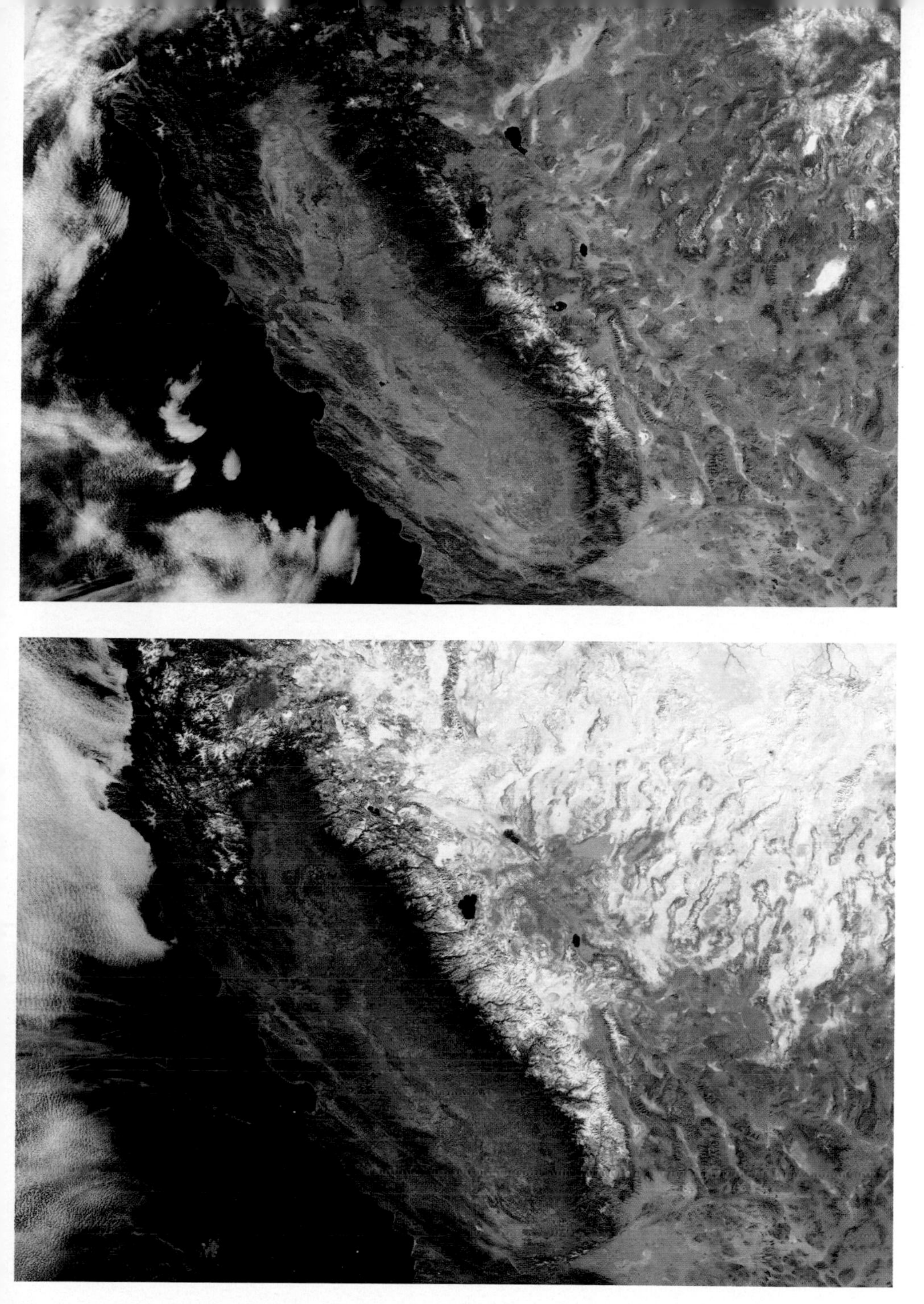

The top satellite image of north/central California and Nevada was taken on January 18, 2014; the bottom on January 18, 2013. California's three-year-long drought had reached extreme levels by early 2014. In January, the snowpack—source of most of the irrigation water in the agricultural breadbasket of the Central Valley—measured at different stations in the Sierra Nevada was between 10 and 30 percent of normal levels. In the expanse of brown at the western foot of the mountains shown in the top photo can be seen the devastating effect of the drought on the farms and orchards of the San Joaquin and Sacramento Valleys. (Courtesy of NASA)

In the top photo, taken by pioneering mountaineer George Mallory in 1921, the main Rongbuk Glacier is a robust river of ice flowing down from the foot of Mount Everest. By 2007, when mountaineer David Breashears took the bottom photo from the same vantage point, the glacier had diminished dramatically. Since Mallory's time, the Rongbuk glacier has lost more than 330 vertical feet and retreated more than half a mile; it retreated 558 feet between 1966 and 1997. (Top photo by George Mallory [1921], Royal Geographical Society; bottom photo by David Breashears [2007], GlacierWorks)

who was in her eighties and had suffered a stroke two years before. (Urgain, working with a local nonprofit, had supervised the construction of a solar-heated room in which she could comfortably pass the long winters.) Red-robed monks are in the courtyard outside making *puja*, clashing small cymbals and chanting. Mourners from Kumik and surrounding villages come and go in the hallway of the house of Lama Thinles, Rigzin's uncle, who is a *geshe*, a senior monk. The mood reminds me of my own grandmother's Irish wake, a celebration of a long full life well lived. Bursts of laughter escape the room next door. A smiling cousin pokes his head in the door and asks if we have enough *chang*, and do we want more rice?

We definitely have enough *chang*. Stobdan tilts a pitcher into Rigzin's cup, then mine, then his own. He looks around, satisfied that all are filled to the brim, nods, and continues.

"And one story goes like this. There is a magic person in Kumik and one is in Rinam."

Rinam is a village about a thousand feet below and northwest from Kumik across the Zanskar River. The "magic person," the *onpo*, is someone who understands the movement of the stars and the seasons; determines the auspicious days for planting, harvest, and marriage; and performs exorcisms of troublesome spirits. He knows the secret ways to placate prickly deities and has special powers and insight into hidden worlds. But he is still human, all too human.

"And behind Kumik there is the glacier, above Stenzin Thinlas's house."

He gestures toward the window, beyond which in the darkness, a couple hundred feet up the spur ridge on the small, 17,000-foot subpeak of Sultan Largo known as Donkey's Ears, two large houses—known as "Gonpapa"—have become squat silhouettes.

"Now there is no glacier," Stobdan says. "[It was] just behind these two houses. This was a long time ago. The Rinam magic person, he, what we say . . ." He rummages for the right word. "Spit! Spit. He put spit to this glacier from Rinam. And the glacier is totally finished."

Stobdan pauses, takes a big slug of *chang*, puts his cup down gently. He answers the question on my face. "The Rinam magic person and the Kumik magic person have"—he puts his palms up apologetically, almost embarrassed, as any right-minded, conflict-averse Zanskari would be, on their behalf—"not a good relation. They are fighting."

Rigzin stretches out parallel to the small table. He leans on his elbow, and wears a slight smile as he listens to his oldest friend hold forth. He is a man of science, the most educated son of Kumik, and a thoroughgoing skeptic on any and all subjects, but he likes these old stories so much that he once asked me to bring him a voice recorder from the United States. He travels widely across Zanskar for his work and wants to interview elders and preserve their old tales before they both disappear forever.

Anyway, the Rinam magic person threw the first punch, so to speak. "He spits from Rinam to here," Stobdan continues, "and it's finished." The vast body of ice just dissolved. "They say that quickly it is finished."

The Kumik *onpo* did not take this lying down. Eager to avenge the loss of his village's glacier, he immediately launched his own saliva bomb into the cerulean. After tracing what must have been a magnificent parabola across the broad valley, it slammed into a spectacular rock formation known as Norbu Todpal—Precious Stone Coming Out of Stomach—for which Rinam was and remains famous.

"There is the rock—and the rock is cut like this," Stobdan says with a chopping motion of his hand. "The rock is very expensive rock. And they destroyed the rock." He laughs a bit mischievously.

Stobdan and Rigzin both clearly relish this story. But they are keen that I not get the impression that they buy into such superstition.

"How we can believe?" Stobdan concludes with a shrug. "Because we didn't see anything. Just stories."

There are other tales of this long-departed glacier (*kangri*, which literally means "ice mountain" and can also refer to "permanent" snowfields or to the mountain itself underneath the snows) and of the snowfields that once almost touched the village. There are many old stories, too, about ancient springs long since dried up and the instances of human ignorance or arrogance that were responsible.

Such legends help explain the Kumikpas' present straits, their psychological adaptation to their challenging environment, and the land's striking features: outlandish layer cakes of red rocks; water frothing improbably out of monolithic cliff faces; huge chunks of gneiss balanced delicately on the lofty ridge that buttresses Stongde Monastery (floated there by an ancient lake that once filled the valley, some say). Some kernels of what they relate may indeed have taken place. But you don't have to take them literally to understand that they have their provenance in some real yet invisible domain, where great tectonic forces—the moral bleeding into the physical—play out. How fast can a glacier die? Who or what has the power to dissolve it?

These are not idle questions. The battle of the magic persons, for instance, had consequences that still ramify. From Gonpapa there is a full view of Stongde, which spreads across the valley floor three kilometers north of Kumik. On its southwestern edge, one can make out the faint edges of brown ovals and ellipses filling the space between the two villages. Kumik's elders say their own grandparents told them of a time long, long ago, when both villages cultivated those fields. They have long since been abandoned. Kumikpas and Stongdepas used to sit down to lunch together under the sun on summer days, so close were the fields they worked—a testament to a time when both communities were flush with water seeping from the spit-dissolved glacier and could farm a wider area beyond their village cores.

There is still another body of stories that seeks to explain these features of the land: the shapes of the valleys, the presence or absence of water, the lone rock or incongruous hill. These are tales told by geologists and glaciologists, and their protagonists are as enigmatic and iconic as the *zbalu* and *onpo*: vast glaciers that bulldozed soil and rock into piles known as moraines, scoured and scored U-shaped valleys like the one above Kumik, and carried massive boulders far from the bands of granitic gneiss where they were born. Echoes and footprints of a race of giants, long passed from the earth. Every part of the demoness's body, every corner of the kingdom, bears these scars and welts. Zanskar is a landscape authored by vanished ice.

"The evidence suggests that there have been two or possibly three major glaciations of progressively smaller extents, when ice occupied respectively the whole, two-thirds, and one-third of the Stod–upper Zangskar valley," Henry Osmaston wrote after his extended study of the region's geology. These are broadly dated to within the last hundred thousand years—it's impossible to assign more precise timing. What we do know is that at different distant points in time throughout the Quaternary period, the whole valley complex was filled with moving ice.

Terminal moraines are perhaps the loudest clue. Like the high-tide watermark on the posts of a beach boardwalk, terminal moraines signal how far a glacier managed to advance before stronger, invisible forces pulled it back.

These singular mounds are some of the most distinctive features of the Zanskar Valley. The oldest lies 100 meters deep at Kilima, near the entrance to the river gorge; atop it now sits the camp of Indian army engineers and laborers, engaged in their decades-long assault on the canyon walls to make the all-weather road to Leh. Old Padum lies on and around another terminal moraine, with the king's old house at its heart. There the geological past elbows its way into people's kitchens: the huge boulders that litter the moraine's slopes form entire, irregular walls of the oldest houses.

But the stories told by such clues aren't complete. "Moraines are great for tracing boundaries and making maps of the glacial world," notes geophysicist David Archer. "But they leave a sketchy record of the comings and goings of ice sheets through time. . . . The most recent, largest ice sheet wipes out the moraines left behind by the others."

So scientists are left to read other signs that capture ancient climate swings—tree rings, lake-floor sediment, ice cores—gathered from around the globe in order to reconstruct ancient glaciation patterns. From this body of evidence we know that an oscillation took place—glaciers advancing, retreating, sloshing in super slow motion across the cold, high basins like the one between the Himalaya and the Zanskar Range—driven by shifts in temperature and precipitation that play out over millennial time scales. (Globally, dating back millions of years, these changes are themselves driven primarily by slight variations in Earth's orbit around the sun known as Milankovitch cycles.) The broad lesson is that masses of ice grow and shrink across the face of the Earth over stretches of time that make even Kumik's long tenure seem as ephemeral as a soap bubble.

Which makes the current pace at which Zanskar's remaining ice is retreating all the more remarkable.

"There are loud indicators that these glaciers are melting," Shakeel Romshoo, the glaciologist at the University of Kashmir, told me. He has studied glaciers in Zanskar and other parts of the state of Jammu and Kashmir since the mid-1980s. "Out of 365 glaciers in the Zanskar region that were there in 1969, about 6 of these glaciers are not there." As in, completely gone. "Similarly we see that the dynamics—length, breadth, thickness—are also changing. I would say, all the glaciers I have seen, they are showing the recession."

Ishay Paldan's lifelong observation of shrinking snow and ice through his window is a data point that fits neatly into larger trends observed by scientists. A recent study of several glaciers in Zanskar found that they lost 18 percent of their area and retreated an average of six to thirty-three meters per year from 1962 until 2001. Ulrich Kamp of the University of Montana measured thirteen glaciers in Zanskar, combining field measurements of glacier topography with thermal imaging and remote sensing data. "Most of the glaciers in the Greater Himalaya Range in Zanskar are receding since at least the 1970s," he and his colleagues concluded, "and it seems likely that this has been caused by a combination of precipitation and temperature."

While every glacier, much like its primary ingredient, the snowflake, is unique, there are a few basic dynamics governing the growth and decline of all glaciers. Glaciers develop as snow accumulates at high elevations and turns to ice under the pressure of successive layers. Over time, they grow as "mass gain," from snowfall or avalanches pouring onto their surfaces, outstrips "mass loss," through melting or calving off of ice.

There is a point on every glacier known as the equilibrium line altitude (ELA), also sometimes roughly equivalent to the "permanent snow line," where the melting theoretically equals accumulation. The ELA is not an actual line so much as a conceptual marker of the boundary between the accumulation zone above and the ablation zone below, where higher temperatures promote mass loss. The ELA is influenced by altitude, latitude, and regional and local climatic patterns; it thus varies widely from place to place. In parts of the Alps the ELA might be 2,800 meters; in the Alaskan panhandle it could be between 1,000 and 1,500 meters. In the Himalaya and Hindu Kush region, the mean ELA ranges from 5,150 to 5,600 meters.

As temperatures are rising across the Himalayas, the ELA is rising along with them. In 1994, Henry Osmaston reported that the ELA on the snowfields above Kumik and Stongde were at 5,500 meters on the north face and 5,800 meters on the south faces. In the main Himalayan range across the valley, in Urgain's backyard, he cited an ELA of just above 5,200 meters.

The ELA offers glaciologists a kind of vital sign, to help them track glacial mass balance. "The ELA is important because, for example, if the glacier has no accumulation area for a period because the ELA is located above the glacier, the glacier is destined to disappear over time," writes glaciologist Ken Fujita.

This seems to be the case for the south-facing glacier atop Sultan Largo. The peak is 5,810 meters high; by now the ELA on that side has gone *above* the top of the mountain. Glance up from the magic person's house toward the ice cap on the mountain and this prognosis won't come as a surprise: it evokes one of those eroding Polynesian atolls losing its last stand against the rising sea. The rising waters are the invisible, implacable ELA.

The double whammy of declining snows and rising temperatures seem to have given Kumik's lone remaining glacier—and the fields it helps to water—a death sentence.

If you stand there by Gonpapa, the magic person's house, and look down toward Stongde, just west of the dried-up fields you can make out a chain of sere, eighty-meter-high hills that stretch across the river to the northern edge of Karsha. This is another terminal moraine, contemporaneous with the bulldozed pile of rock and sediment on which the old town of Padum sits. These features define two edges of a single glaciation event many thousands of years ago that filled the whole central valley of Zanskar. From Karsha a line stretches southward, another legacy of the same glacier. Known as a "kame terrace," it is a long bench of sediment deposited by meltwater that once flowed between the edge of the glacier and the adjacent valley slope. Today the line runs across the lower mountain slopes of central Zanskar like a three-dimensional bathtub ring.

On this gravelly shelf sits the house of Urgain Dorjay and his wife, Sonam Chondol. Rising straight up out of Urgain's backyard are the 20,000-foot peaks of the Great Himalaya Range. On a typical summer day, I can take a fifteen-minute stroll down from Urgain's house into the bazaar to watch a few four-wheel-drive taxis hired from Kargil rumble their way into town, sheathed in dust. Many will be full of *chigyalpa*, trekkers or sightseers from Britain, Germany, France, the Czech Republic. As they take their first walk through Padum's bazaar, working out kinks from the fourteen-hour ride, the new arrivals crane their necks to take in the glistening peaks of Padum Kangri and Kapalungpa, so steep they seem to almost lean out over the town. The tourists gape up at the forbidding glaciated passes that lead to Kishtwar and Paldar, the lands to the south. A few adventurous souls might decide they want to go up those lofty passes for a closer look, so they ask around the market for guides. But, aside from crazy Kumikpas in search of water, not many Zanskaris have the time or inclination to go climbing mountains.

So everyone sends them to Urgain. And though he is always busy and has just a bit of formal training himself (mostly from his old friend and mentor Mohammed Amin, the first licensed mountaineering guide in Zanskar), he usually says yes. The pay isn't great—he does it mostly because Urgain is a good host, because he feels responsible. And he knows those places as well as anyone in Zanskar.

Since he was a boy in the 1970s (around the time researchers say the melting really started to accelerate), Urgain has ranged widely over the ice that remains throughout his kingdom. With long, swinging, sure strides, he has covered a lot of ground as a porter, guide, and expedition leader on long treks, ski tours, and emergency rescues in many of the region's remote corners.

In the course of those travels, Urgain has acquired an intimate knowledge of places that glaciologists have yet to see, much less study. On a searingly hot August day, Urgain and I rest in his cool kitchen, and he reflects on what he has witnessed among the *kangri*, the "ice mountains."

"Many, many changes are there," he says.

In the past fifteen to twenty years, he estimates—"not exact, but guess"—that the glaciers he has traversed among the Great Himalaya Range have lost an average of thirty meters of height. He rattles off their transformations as he might recount changes in the girth and hairlines of friends he meets in the Padum bazaar.

On Shingo-la, the 5,000-meter pass marking the southeastern boundary of Zanskar: "Twenty years before, at that time on top of Shingo-la there is very good snow, and glacier also, in summer. So now after twenty years, on top is totally dry. Totally changing."

On Drang-Drung, the largest glacier in Zanskar, which marks the western boundary of Zanskar: "We went in 2005 for summer skiing. That time we have seen it is not too far from the bottom of the Pensi-la. Now this time, looking, it is going too far from this, pulling [away], even in the last seven years. Oh ho! Totally [different]."

On Omasi-la, a 5,100-meter pass that leads south across the Great Himalaya: "Fifteen years before I am traveling first time with my uncle . . . he wants to go to Paldar. That time we saw on the top, there is good snow and glacier. In early morning we are going and making steps with some tools we have already, like local-made ice ax. In July, early morning, snow is very hard—that time, very dangerous to slip down. On the top there is a dry rocky place, maybe five meters gap. . . . Then five years back I trekked again. At that time the top [edge of glacier] has come down maybe fifty meters—totally dry place from the top to down."

On Kang-la, another high pass leading south: "Kang-la twenty years before, exact glacier length is eighteen kilometers. Now, maybe more than ten or twelve kilometers. Both sides [are retreating]—Kishtwar side and Zanskar side also. Height is also going down. Now it is very difficult to go on the side. There are some rocks. First time we went, there was a stone building, and we have some cup of tea. This time last year it's not possible to go there. The glacier is melting, and on the edge of the glacier is loose and rocky."

"So many, many changing," he concludes, with a rueful laugh.

As a part-time trekking guide, these changes make Urgain's life a bit harder. He has to search out new safe routes and places to rest and camp. There is more rockfall now, as melting ice lubricates the scree and debris around it. And the ever-morphing terrain underfoot wreaks havoc with time estimates for a day of travel. "It is difficult to explain to clients."

But Urgain is keenly aware that the changes pose far greater concerns than mere altered trekking routes.

"Of course in future, very big problem," he says evenly. Zanskar's future—much like its distant past—is likely to be dictated by its now-vanishing ice. And as with the past, it's impossible to know when it all might go. But the exact date matters little. Should current trends of declining winter precipitation and rising temperatures (and with them, rising ELAs) hold, much of the ice will one day melt.

Urgain offers an outline of the dynamics driving the decline of Zanskar's snow and ice that is so clear and concise it could be in the executive summary of the next IPCC report: "So every year what will we see? Every year in wintertime it is snowing very less and less. So, on the glacier, stocking [of snow] is very low. Summer and springtime, the weather is very quick warming, the stock of winter snow is getting very quick melted. So after the fresh snow is melting, the glacier every year is melting more and more."

Declining precipitation, warmer winters, early-arriving springs, reduced reflectivity on the glacier surface—because of less snow lasting for shorter periods—driving even more warming in a feedback loop. A perfect recipe for declining mass balance. As Urgain suggests, it is a simple stock-and-flow problem: turn off the faucet (less snow), open the drain wider (more heat), and the bathtub of ice will empty quickly.

Are the glaciers of the Himalaya in a state of general retreat?

Anecdotal evidence, such as the decades of observations by Urgain and Ishay Paldan and many other mountain dwellers from Bhutan to China to Nepal, certainly suggests the answer is yes.

The overwhelming body of scientific evidence seems to agree. The peer-reviewed literature documents dramatic changes in the extent of snow and ice across High Asia. Taken together, the spate of recent studies on the pace and character of that transformation has the unsettling quality of writing on the wall:

China? Across the entire Tibetan Plateau, over the past forty years glaciers have shrunk more than 2,550 square miles, with the greatest retreat occurring since the mid 1980s. Scientists at China's Institute of Tibetan Plateau Research have been monitoring 612 glaciers across the region. They found that 90 percent were retreating between 1980 and 1990; from 1990 to 2005 the fraction jumped to 95 percent. The season of snow coverage between elevations of 4,000 and 6,000 meters on the Tibetan Plateau has been shortened by twenty-three days on average.

Nepal and Bhutan? An exhaustive survey of satellite images by the International Centre for Integrated Mountain Development has shown that glacier-covered areas of Nepal and Bhutan decreased by 21 and 22 percent, respectively, between 1980 and 2010.

India? A recent study by India's Space Research Organisation of satellite images of 2,190 glaciers across the Indian Himalaya determined that 75 percent are retreating, at an annual rate of 3.5 percent.

To study a glacier over time, a glaciologist uses many different tools: satellite images, infrared cameras, field observations, even lasers. If the goal is to determine whether a glacier is shrinking or growing, there are four different measurements that can be made: mass balance, volume, area, and terminus location. Mass balance is the best, but hard to do, as it requires extensive field measurements. A glacier's area can be determined with remote sensing techniques from satellites far above, but without knowing the vertical profile it's hard to draw firm conclusions about what's happening in terms of mass and volume. The same goes for terminus location. The sheer physical difficulty of getting to and studying Himalayan glaciers—in the highest, most remote, and most dangerous terrain on the planet—means that this data is scarcer than researchers would like.

Still, a team of veteran international glaciologists recently performed a comprehensive review of dozens of scientific studies of glaciers throughout the Himalayan region, seeking to overcome, or at least adjust for, these inherent limits, in search of an answer to the question: What is the evidence for glacial shrinkage across the Himalayas?

The researchers synthesized and analyzed a huge quantity of data on the four measures described above, using a rigorous system for assessing levels of confidence to various reported data from the peer-reviewed literature. The study was laden with caveats, of course. The authors discovered a relative paucity of data prior to 1960, and a sharp increase in the number of studies measuring area and mass balance of glaciers since 1990 (perhaps, they speculate, due to increasing interest in the effects of anthropogenic climate change). The authors spent considerable time detailing the observed complexity of glacier behavior, and the variability of that behavior across different regions and in different climate regimes. But they were still able to draw some robust conclusions. And their findings basically confirm that the trend observed in Zanskar by Urgain, Kamp, Romshoo, and most Zanskaris extends far beyond Zanskar, across the entire region. (If anything, the ice of the central and eastern Himalaya seems to be eroding faster than in the northwest.) In all four measurement categories, they found that shrinkage has been the "predominant change" since 1900, citing "a clear and corroborated picture of rapid mass loss on all glaciers considered by the review."

"The evidence gathered, supported by all measurement approaches, provides an overall picture of glacial shrinkage across the Himalaya, with some regional and local variation," the report concluded. "This suggests future long-term availability of meltwater from glaciers is likely to diminish."*

---

*The debate over the precise date by which the Himalaya will be ice-free managed to dominate a few news cycles in early 2010. That's when attention was retrained on a now infamous passage about the fate of Himalayan glaciers in the IPCC's 2007 *Fourth Assessment Report* for policy makers, in which its Working Group II claimed "the likelihood of them disappearing by the year 2035 and perhaps sooner is very high if the Earth keeps warming at the current rate." In January 2010, several news reports began to point out that this scenario was not just highly *un*likely, but impossible. As glacier expert Graham Cogley later wrote in a review letter to *Science*, the claim "requires a 25-fold greater loss rate from 1999 to 2035 than that estimated for 1960 to 1999. It conflicts with knowledge of glacier-climate relationships and is wrong." A group of leading glaciologists, including some members of the IPCC, issued a joint statement: "This catalogue of errors in Himalayan

• • •

Let us thank whatever *lha* or gods or higher powers we like, of course, that ice and snow *does* melt. That the Colorado River swells every spring, filling the reservoirs of Lake Powell, keeping the lights on and sprinklers running in Los Angeles. That the Sierra Nevada snowpack piles high, and then dribbles all summer long into reservoirs, and from there into Central Valley fields of strawberries and lettuce and wheat and almond orchards. That the little stream surges down toward Kumik, for a while at least, and after watering the barley, joins the Zanskar and the Indus, and then courses through the rice and sugarcane and wheat fields of the Pakistani Punjab—the most extensive, most populated, glacier-fed irrigation network in the world—before it merges with the Arabian Sea.

If we could turn a dial at will—try, as the Chinese did, to manipulate black carbon into flooding our fields and fueling our harvests on *our* terms, rather than its own—dark snows would perhaps be a welcome sight, like the ingeniously engineered causeways that bring water for

---

glaciology . . . has caused much confusion that could have been avoided had the norms of scientific publication, including peer review and concentration upon peer-reviewed work, been respected." The offending paragraph was later found to be the result of an absurd game-of-telephone-style chain of mistakes involving the transposition of the date 2350 to 2035 and the confused citation of a decade-old interview with an Indian glaciologist in a science magazine, among other non-peer-reviewed sources. The IPCC retracted the paragraph and apologized, but the damage had already been done, insofar as the mistake cast a shadow, via hyperbolic media coverage, over the whole body of climate science the IPCC and others had painstakingly amassed. "That's a horrible and egregious mistake the IPCC made," the late climate expert Stephen Schneider later told an interviewer. "They correctly said that glaciers in the Himalayas are melting very rapidly. They correctly said this is socially very threatening because it means floods while they're melting and then droughts after they melt—all that was right. Then they had an illegitimate reference giving a specific date when they'd melt. Now anybody who knows anything at all about complicated science like climate, knows we could *never* know to the precision of one year when a complicated glacier is gonna melt." On the BBC's *OnePlanet* program, glacier expert Tobias Bolch weighed in. "If you overall look at the whole Himalaya there is a clear tendency of mass loss. I've seen some press releases that glaciers in the Himalayas are gaining in mass. This is simply not true. It's only in this specific area in the Karakoram. They are large, but the mass gain is only really slight. There will be glaciers in 2100, they will not vanish before. But what we don't know is exactly how much because it's very uncertain."

miles across desert landscapes, from the *acequias* of New Mexico to the *yura* of Ladakh. But since Shi Yafeng's 1959 attempt (and long before), it's as though the faucet pouring black carbon into the skies has had its handle broken off. We now have a daily upward deluge of these dark materials, which drives the ever faster downward flow of snowmelt—leading to consequences on multiple time scales.

By accelerating the melting of snow and ice on the ground, black carbon threatens farmers in two ways. One, it moves spring forward, starting the melting sooner in the year. This leads to the slow-motion (but accelerating) disaster of Kumik, in which winter is worn down by attrition, nibbled away from year to year, like the *chadar* route being gradually truncated by the new road.

But there is nothing slow about the second threat: deadly glacial-lake outburst floods, or GLOFs, as the experts call them. In remote pockets of the High Himalaya, melting glaciers are feeding fast-growing, unstable lakes that can flood communities downstream in a flash, sweeping away agricultural land, hydroelectric facilities, and lives. Picture the water tower on the edge of your town, that squat white tank perched high up on its spindly metal stilts. Maybe it holds 100,000 gallons. If that tank were to rupture, and its contents to pour out all at once, where would it go? Would it flood Main Street? Pour right into a nearby stream or culvert? Or into your living room?

Glaciers and snowfields are like that: enormous tanks that regulate the storage and discharge of water for huge, heavily populated watersheds, across both Asia and the world. They are buffers, melting just enough at the right time to supplement the seasonal rains to keep the vital arteries of Asia—the Yangtze, the Indus, the Brahmaputra, the Ganges, and others—flowing. The same is true for communities downstream of the Andes, the Alps, and the Rocky Mountains.

But as the ice and snow melt faster, as the water towers rupture, all that water has to go somewhere. In the near-term this leads to a seeming boon: increased water flows (though we're not yet prepared to catch, store, and use all that excess runoff). Then, further in the future, after the frozen stores have melted, their storage function is lost, making the "dry season" that much drier.

Some communities are experiencing the inverse of Kumik's problem.

Whereas Kumik has too little supply late in the summer, as water is doled out over a longer time frame and spring eats into winter, villages in the Imja Khola river valley of Nepal live in the shadow of instantaneous catastrophe: a vast amount of water released in their direction all at once.

In 1960, Imja Tso did not exist. Today this lake covers more than one square kilometer at the foot of Island Peak, a dozen miles south from Rongbuk Glacier, on the other side of Mount Everest.

I reached Imja Tso at the end of a week's walk up the Khumbu Valley, on a cloudy day in mid-June, greeted by a cold rain and shifting clouds garlanding Lhotse, an impossibly immense wall of rock rising straight up to the apex of the sky. The milk-hued Imja Khola braided through the high shelf of a valley, trickling out of Imja Tso's western end through the porous moraine. It was a tortured landscape, moraine ridges all piled cheek by jowl, the silence punctuated only by the frequent clatter of rocks sliding down steep mountain faces above the lake.

The tongues of three melting glaciers hovered just above the eastern end of the lake, dark with debris, occasionally calving off huge chunks into its muddy brown waters. The lake itself was surprisingly long. Lateral moraines towered 200 feet above along its entire length, testifying to the past dimensions of the glacier. I thought of the warning I had heard from Dr. Pradeep Mool, when I visited him in his office in Kathmandu the week before: "Imja will go on growing, and it can burst out at any time."

Mool, an expert on glacial lakes at the International Centre for Integrated Mountain Development in Kathmandu (ICIMOD), had explained the process by which lakes like Imja finally overcome their fetters. "In all the glacial lakes, in the end moraine part there is dead ice. We have found the glacial ice underneath the end moraine dam also are melting slowly. So if the flow of water from the melted glacier and the lake cannot freely flow, and if it is dammed, and the inflow water is more than the outflow, then the lake will rise, then the ice underneath the dam might melt slowly." What follows is the collapse of the end moraine dam and the release of years of pent-up melt. This process is taking places in thousands of lakes across the Himalaya; an ICIMOD inventory suggests that "up to 200 may be potentially dangerous." There is no doubt, Mool told me, that warming temperatures caused by human activity explain the growth of Imja and other glacial lakes in the region.

Mool has been studying glacial lakes across the Himalaya since 1985, and he has found that most glacial-lake outburst floods have a damaging effect up to 100 kilometers downstream. "Up to twenty-five kilometers downstream, it's a very big damaging effect."

Outburst floods have always happened, thanks to the natural advance and retreat of ice. What's unprecedented is the rate at which new lakes are forming and the speed with which they are growing.

"There's more than sixteen hundred glacial lakes in Nepal Himalaya," Mool told me, "and more than twenty thousand from Myanmar to Afghanistan, China to India, Himalayan–Hindu Kush region. It's not only the issue here in Himalaya; in Central Asia also there is the issue of glacial lakes. Kyrgyzstan, Tajikistan, they have a serious problem of this glacial lakes now. So they can learn from the Himalayan experience. If you go look at Latin America—Peru, Ecuador, Bolivia—this problem has been similar since a long time. In the Huaraz event, like in Peru, they have done this mitigation work in the early seventies of fourteen glacial lakes, with community involvement, hydropower projects, early-warning systems." (The wisdom of these preparations was acquired through bitter experience. In Peru, in 1941, an enormous piece of the Palcaraju Glacier fell into the glacial Lake Palcacocha, which burst its banks and killed 5,000 people in the city of Huaraz, destroying a third of the city. Other floods in Peru killed hundreds in 1945 and 1950.)

Mool and his colleagues have kept careful watch over Imja's growth. They helped set up two monitoring stations: one on the lake's shore that sends a steady stream of video images and data to researchers in Kathmandu and one in Tengboche, a village downstream, to provide some advance warning to those who live in its shadow. There on the Imja's banks I spied the small metal tower with climate-monitoring equipment and cameras at the west outlet. The stream spilled down sharply from that point.

I stepped gingerly about the moraine; its loose sediment and rocks didn't inspire much confidence as a stable dam. I tried to picture what the surge might be like when it finally busts through, just like the one that erupted in 1985 out of Dig Tsho, not far off. That flood wiped out the new Namche hydroelectric power plant within a few minutes, ripped up fourteen bridges, took out two dozen homes, and killed five people (the toll would have been much greater if locals hadn't been away from home for a religious celebration that day). Imja Tso's impact is expected to be many times greater than Dig's.

While hiking back down out of the valley I stopped for lunch at the Mandala Lodge in Khumjung, a village several hundred feet above the banks of the Imja Khola river, home to the school Sir Edmund Hillary built for the Sherpas after he and local hero Tenzing Norgay became the first men to reach the summit of Everest in 1952. The proprietor, a short, stolid Sherpa, asked where I had come from.

"Imja Lake."

He looked surprised. Most trekkers head up to the climbers' base camp, or to the rock promontory of Khala Pattar, for dramatic views of Everest and its neighbors. Few are interested in the slate gray expanse of Imja, outside of researchers like Mool. But it weighs heavily on the minds of the local people.

"Not dangerous?" he asked. "Many people here have been talking about the danger from the lake."

"Well, you're safe way up here."

"I hope so." He sounded unconvinced.

Over his shoulder I spied a framed certificate on the wall. A nonprofit in Kathmandu had organized a race ("Beat the GLOF Action Run") in 2010 to raise awareness in the Khumbu region about the dangers of glacial lakes. "Thank you for your concern for climate change!" it read. A trail run seemed like the appropriate fundraising activity: running up hill will be exactly what's required should the alarm sound that Imja has, after more than fifty years of swelling, finally burst its banks.

A half day's walk farther downstream, I reached the Nirvana Guest House in the village of Jorsalle, a cozy establishment in a quiet, lush spot overlooking the Dudh Koshi River, run by Kazi Sherpa and his wife Wangmo. Nirvana is three days' walk downstream of Imja Tso, but it would take mere minutes for a massive churning wall of debris and 1.26 billion cubic feet of water to reach it. And there are other glacial lakes forming in the same watershed that aren't as closely monitored as Imja. Over tea, Kazi told me that a small glacial lake west of Imja had burst the year before and the river quickly rose five meters.

"People were saying it was Imja, but it was another lake. There are many more dangerous lakes around Chola Tso," another large glacial lake. He explained that the greatest risk is from debris, soil, and trees gouged out by the flood getting jammed into the narrow gorge above the rivers' confluence and backing up the river. When the water finally overcomes this temporary bottleneck, an even more destructive surge would be unleashed downstream.

Kazi described how the monsoon has been shifting later: they now receive less snow in winter than when he was a young man. Like Urgain, Kazi used to be a guide and traveled with mountaineering expeditions all over the region at the foot of Everest. And like Rigzin and Stobdan, he has a modern sensibility but still feels the pull of the old stories and beliefs.

"Qomolangma is the mother of the universe, and she is one of five sisters," Kazi said matter-of-factly. He doesn't think people should be climbing up high on the snow and ice, to the peaks of Everest and other sacred mountains. "Too many Sherpas are dying." That spring three climbers died on Everest, including an eighty-two-year-old former foreign minister of Nepal, attempting to be the oldest summiter. "He ate lunch here on the way up," Kazi said soberly, shaking his head. "Stupid."

At the table in his immaculate garden, just above the river, Kazi laid out a map to discuss various scenarios of destruction. He showed me the spot above the community of Phunki Tenga where a small lake burst and flooded, and pointed out other small lakes that he thinks pose even more risk than Imja Tso.

The sun pushed through a low cloud bank. Across the river a small waterfall slid in stages down the steep fir- and pine-clad mountainside. An upstream breeze mingled the scent of wood smoke from a neighbor's chimney with that of the garden's roses. The risks Kazi was describing seemed incongruous juxtaposed with the scene of quiet abundance, the three greenhouses full of ripe tomatoes, cucumber, paprika, and chillies, the fields of tall wheat, the flowering potato plants, apple trees, and lovingly tended roses of every hue. Kazi pointed to a cave in the cliff just above Nirvana, where he plans to build a meditation cell and become a recluse after his kids grow up.

"You can join me, and we'll both retire as monks!" he laughed.

His wife Wangmo smiled patiently at her husband's jokes, poured more tea, then glanced upstream at the picture-postcard view and quietly observed, "It's very dangerous here."

Their home is right in the middle of any future flood's path, but Kazi, being a can-do kind of guy, said he sees an opportunity amidst potential disaster. He pointed to a spot on the map where there is a natural rock wall. "Below that you could use to build a strong dam," he said, to capture future glacier melt and drain water in a controlled way. Just like the old Tibetan proverb: Build the dam before the flood comes.

"We build a drainage system," he said, which could even be harnessed to generate electricity. "But some glacier lakes it is not possible to have this kind of system. Then we have to drain out by hand, or by pumping."

After the inevitable floods, Kazi sees another threat on the distant horizon. "In the future these rivers will be dry," he said, not with despair but matter-of-factly, almost brightly—as if this fact posed an interesting challenge for someone with his energy. Plus there's always the cave to retire to.

"Why do you think this is happening?"

"Some in this valley think the gods are angry," Kazi said, echoing some Kumikpas two thousand miles away to the northwest. But he offered a more prosaic explanation, based on observations made over a lifetime of farming and guiding groups of trekkers in the high mountains.

"I think there is too much pollution in the Himalayas, and maybe this is changing the weather."

Sitting out in her yard on a glorious fall afternoon, Namgay Lam poured freshly threshed rice into an iron pan and stoked the fire underneath. She frowned slightly as she whisked the grains around to roast evenly. Yes, she said, she remembered the morning of October 7, 1994, quite clearly.

"I was at school in Samdingkha," she said. "We heard a very loud noise and rushed down to look at the river. I had never seen it so high."

In western Bhutan's Punakha Valley, the Pho Chu—the Father River—had become a raging torrent. Seven hours earlier, 80 kilometers north and 3,000 meters above where she stood, a fatal tipping point had been reached in the icy darkness. The waters of Luggye Tso, a high-altitude lake fed by melting glaciers near Bhutan's border with Tibet, had risen during the day, building up just enough hydrostatic pressure to overcome the resistance offered by its natural moraine dam. In an instant in the early morning hours, the rock and soil gave way. Within minutes, the lake had poured its contents all at once into the Father River. Gravity took over from there.

I watched silently as Namgay added another log to her fire and squinted through the smoke. I had just come from a conference in the capital, Thimphu, where officials from Bhutan, India, Bangladesh, and Nepal had gathered to discuss ways to adapt to the region's changing climate—and where an ICIMOD colleague of Mool's had forecast that Bhutan would lose half of its glaciated area by 2100. ("I'm actually more pessimistic," Bhutan's famously upbeat prime minister had told me, darkly, when I asked him what that would mean for his agriculture- and hydropower-dependent citizens. "It's scary," he said, shaking his head, and falling uncharacteristically silent for a long minute.)

"The water was very muddy, and full of stones and trees," Namgay recalled. Once the dam burst, 635 million cubic feet of water surged down the valley. First they swept away the bridge linking the 4,100-meter-high villages of Thanza and Tenchey. One hundred meters below, the deluge destroyed houses in Chozo; farther downstream it coated the fields around Ledi village with sand and horse-sized boulders. It caromed off hillsides and stripped them bare of the trees that would fill the river hours later as Namgay stood watch in Samdingkha.

"The waters covered three shops in the village," Namgay said. "They finally went down that afternoon. When the flood was over, there were so many dead fish everywhere, out of the river."

But the worst was yet to come. As Namgay and her schoolmates stood safely on the terraced rice fields of Samdingkha, gaping down at the familiar Pho Chu now transformed into a wrathful demon, the people of Punakha were beginning their day a few kilometers downstream, unaware of the frothing wall of water, trees, rocks, and silt heading their way.

Sitting just a few meters above the river's normal level was the Punakha Dzong—better known to Bhutanese as "the palace of great happiness." Completed in 1638, this imposing fortress had been the seat of Bhutan's government until 1955, hosting royal coronations and generally commanding the picturesque point of land where the Mo Chu (Mother River) meets the Pho Chu, before they flow as one south into India and the great Brahmaputra River.

"At that time the *dzong* was being renovated," Urgyen Karma, a local contractor, told me. He, too, was in school that day. "Many carpenters and painters were staying near the river."

Houses and administrative buildings also dotted the grounds around the *dzong*. Logs and debris jammed the point just above the confluence, causing the waters to rise even higher and flow over the small reinforced bank that had been built after a flood in 1960. When this temporary dam burst, the Pho Chu surged onto dry land, inundating a police compound and prison, uprooting trees and bridges, swallowing cars and homes. By the time it reached the *dzong* it had claimed 1,700 acres of farmland and pasture—in a country where only 7 percent of the land is suitable for cultivation.

The waters claimed lives, too. Twenty-three people died that day. The force of the flood was so great that one victim's body was later found in Bangladesh, hundreds of miles downstream. An eyewitness would later recount, in the pages of the Bhutanese newspaper *Kuensel*, what he saw: "The river had swelled and came rushing down, I saw a man struggling to fight the river, at one moment he looked like he was going to survive, when he managed to catch on to a tree but then we saw a log come down with the flood and hit him on the head, he then disappeared."

Today a marker sits outside the *dzong* memorializing the dead, and sirens sit on nearby hilltops next to ancient shrines, waiting to alert villagers the next time the Pho Chu roars.

Now mostly drained, Luggye Tso is no longer a threat, but it has some bigger, newer neighbors: Raphstreng Tso and Thorthormi Tso. These lakes are growing at a rapid clip, fed by the melting of glaciers in the remote Lunana region. The government of Bhutan is so nervous about Thorthormi that every summer for the past few years it has sent dozens of workers to labor by hand, tossing rocks aside, to slowly drain the lake at a height of over 4,000 meters.

Government officials arrive at least once a year to conduct awareness meetings and remind villagers not to build below the "red zone" marked by poles indicating the height where a lake-burst might bring the waters again. When I asked a clutch of shy schoolgirls in Samdingkha if they know what to do when the siren sounds, they answered quickly and confidently: "We run up the hill where it's safe."

But Pradeep Mool wonders whether these systems, which have yet to perform during the real thing, are adequate. "The community in risk must know how it is working," he told me. "If there is a flood and the warning comes, whether the community at risk really will receive that warning or not, we don't know."

After three decades of researching the question of glacial retreat in the Himalaya, Pradeep Mool echoes other scientists in pointing out that even more research must be done on the changes in glaciers and glacial lakes, and their causes. But, he says, one pollutant in particular clearly bears a large share of the responsibility.

"Black carbon is one of the reasons, because it accelerates the glacier melt," Mool told me. "That's what researchers have found in the Tibetan Plateau and Himalaya. This black carbon is not a long-living aerosol. It can be mitigated quickly. There are already attempts in India, like [Project] Surya, and in Nepal also.

"We have to do something on the black carbon." He waved toward the window in his office. It framed a view of hazy low hills north of Kathmandu. It was spring, just before the monsoon's arrival. "Twenty years ago I can see all the snow-capped mountains from my home, from my office. Now because of all the particulate emissions and dust particles you cannot see the range."

Stobdan tells me one afternoon that he'll be absent for a few days from the construction of the solar *lhakhang*, for an important reason: his niece is getting married, and as the only maternal uncle, he has many duties at the four-day-long wedding. He invites me to join in for an evening.

So Stobdan and I arrive in Shila late one night as part of the bride's party traveling from Stongde. We dismount from the jeep and follow a long string of revelers toward the bridegroom's house, hopping over flush canals in the dark. By the light of my headlamp, the water looks cool, clear, abundant. It even smells sweet.

"Nice water," I say.

"It's our water, yeah?" Stobdan replies waggishly.

We laugh long and loud in the dark.

I only discover the next morning—after a truly prodigious quantity of *chang* has been consumed by what seems like the entire combined villages of Stongde (the bride's) and Shila (the groom's)—that Shila has a *chumchar*. A waterfall. Shila has a goddamn waterfall.

In the early light I go outside to clear my head of its *changnet*—the hangover from last night's revelries. As soon as I spy the cataract, I am pulled into its orbit. So much water collected in one place in Zanskar—whether the limpid lake at Sani where Padmasambhava paused to prophesy or the banks of the churning Lungnak—exerts an irresistible gravity. After all, this is a place so dry that barley grains can keep for thirty years or more in the dark interior rooms of the old houses. A moonscape, as some travelers call it.

But here in the falls' misting backsplash—where I am now closing my eyes, spreading my arms, and tilting back my head to receive its blessing—it could be coastal British Columbia, or the Costa Rican cloud forest, or a TV commercial for Irish Spring soap. When I open my eyes I notice there are fluorescent green bits of moss growing on the cliff walls all around. Moss! In Zanskar! Clover-like plants with tiny star-shaped lavender flowers peek out of clefts in the wet rock. The amphitheater, carved by millennia of coursing water, vibrates with birdsong. It's all a bit much. Just downstream the women of Shila crouch along the stream, washing dishes, washing clothes, and languidly washing their long black hair, in an act of such hydro-decadence as would make Kumikpas faint with delight.

And just downstream of the women are four Nepali laborers who have just killed a goat for more wedding feasting (Zanskari Buddhists outsource the bad karma to their Muslim neighbors or Hindu laborers, but they are happy to eat the meat). They pluck at the fat and sinew around the pink stomach and toss the pieces into the canal. The clear water carries it all away—gristle, clotted sheep blood, shampoo—through the thick fields of *sermo*, a special kind of barley that grows in Shila, and that makes tastier flour and juicier *chang*, other Zanskaris say. So even Shila's *chang* is better!

All thanks to the snow- and ice-melt that by all rights of proximity belong to Kumik, because, well, *kha Kumik*. And *chu Shila*.

Shila, you flush bastards.

But it's hard to bear any hard feelings toward the Shilapas, because they throw a good party. Too good, it seems.

"I have the dysentery," Stobdan groans when I return to the groom's house, where the party is still going strong to a steady beat of *daman* drums. The hot sticky room we shared with two snoring older men, members of the bride's party, is like a college dorm room on a late Sunday morning: slightly rank, the air thick with self-reproach.

Stobdan has been drinking *chang* for four straight days, and I note that he is an alarming shade of gray. As the maternal uncle of the bride, Stobdan has special duties that he cannot sidestep. These include spending large amounts of money on the wedding preparations and getting ritually hazed by *chang*-pot-wielding women from all across central Zanskar.

On such occasions, turning down proffered *chang* is not an option. It is an affront to the individual offering you the *chang* and his or her household and to the spirit of the event—which is actually a wholly solemn ceremony welcoming the bride into her new village and new home, despite the outward appearance of a kegger. (As Urgain once warned me, the older women dispensing the delicious, infernal liquid "will sometimes beat" you with their bronze ladles if you refuse to drink.) And poor Stobdan has several more nieces, yet to marry.

"Too much *chang*," Stobdan groans again with a pitiable weariness, clutching his forehead.

The *chang* was made with water that a few weeks ago was snow and ice visible on the heights above Kumik. I refrain from pointing out this added layer of irony to Stobdan; it seems like the wrong time.

The window frames a scene of abundance: a thick forest of tall dark green poplars rustle in the breeze, among fields of golden barley. Beyond the tent erected for the wedding, in the yard of the groom's house, a tangle of fat-trunked willows stretches down the hillside toward the Lungnak River.

But how long, I wonder, before even the lucky Shilapas would be forced to move? Before their share of the shrinking pie of Zanskar's frozen water would no longer support their thirty households? And what of the rest of Zanskar: Lungnak, Stod, Zangla, Padum? What of Himachal, Kashmir, Uttarakhand, and the other mountain states of India? What of Nepal, Bhutan, Tibet, Pakistan, and western China? What of Peru, Bolivia, California?

"In Zanskar there are many villages which are depending on glaciers, like Kumik," Stenzin Thinles, the head of Gonpapa, perhaps Zanskar's oldest and original household, told me. "Maybe after some years the glaciers are totally finished, and they are getting big water problems. They must build near the river, so they can get river water for a long time."

While you're standing in the cool mist of its waterfall, this outcome seems highly implausible. The oasis of Shila has an air of permanence. But the speed of the observed changes on the roof of the world back up Thinles's prognosis. Some distant day even Shila will be known, like Kumik, as a "waterless village." It's a question of when, not if.

Still, outside of Kumik and a few other water-strapped villages in the remote area of Shun, Zanskaris, like most of the rest of us even farther downstream, are too busy getting ahead, or just getting by, to fret much about it.

"They're not so much worried," says Urgain, describing his neighbors, though he might as well have been describing my own neighbors at the time, in the Bay Area of California, where most of the drinking water being poured into herbal tea in Pacific Heights and shooting out of playground water fountains in west Oakland, comes from snowmelt in the Hetch Hetchy Valley of Yosemite. "They're not thinking about this long-term. They're just busy with short-term thinking. Very few people are thinking like this—mostly old people, because they know." Older folks like Ishay Paldan, who have lived long enough to see the sweep of change and have passed the point where they are much concerned with material advancement. "They see. They're a little bit worried maybe in future something will happen, facing problem."

One day not long after the Shila wedding, I visited Tsewang Rigzin at his office in Padum, just a short boulder's roll down the slope from Urgain's house, where he supervises agricultural programs for the government. As we walked around his experimental vegetable garden, full of thriving spinach and onions and cabbage, I asked him about the future of agriculture in Zanskar. He saw some potential for drip irrigation and tanks and improved canals and other water-saving and water-storing techniques but wondered if these would just delay the reckoning.

Behind him reared the Great Himalaya Range, whose name in Sanskrit means "abode of snows." It was mid-July and their flanks were mostly dark. The hanging glaciers coating the sides of Padum Kangri Mountain were unmistakably smaller from my own first memory of them just nine years before.

"Kumik is not alone," he said, echoing his neighbor, Thinles. Rigzin waved a hand at the mountains. "In twenty or thirty years many people will have problems with not enough water, due to these glaciers melting."

"What do you think it will look like here, in a hundred years?"

"In a hundred years, who can say what will happen? But it's possible farming in Zanskar could be finished."

At four o'clock on the day of the flood, Urgain descends to the cluster of administrative offices on the edge of the Old Padum moraine, bearing a letter requesting funds for immediate repairs to the severed pipe and damaged canals.

He first goes to see the subdivisional magistrate, a Ladakhi civil servant newly posted to Zanskar. The SDM is asleep. He is roused, and after eating his lunch, he emerges to hear Urgain's account.

Then Urgain goes to the station house officer, Zanskar's chief of police, a man from Kargil. He too has been sleeping through the hot afternoon and comes to the door rubbing his eyes.

Then Urgain goes to talk to the executive engineer of the Public Works Department, another Ladakhi civil servant, and finds him coming out of the bathroom.

"He was probably asleep," Urgain says later.

They all decide to form a committee to investigate the matter, and to alert authorities in Kargil. So that evening Urgain shows all the local brass the damage, and they say they will fix the pipe the next day or else bring water tankers to Padum if it's not possible. On his way back he passes a cluster of kids heading to their village of Ubarak. He warns them off. "The bridge is gone," he says, "you'll have to go around the long way."

That night, Urgain stretches out on the floor and shares the story of his day with his wife Chondol, who works at the hospital and who herself had a long day—many children came with their families from Lungnak to see the Dalai Lama and went to get vaccinations before returning to their distant villages.

"It is very dangerous for Mani Ringmo, if water comes in other channel," Urgain reflects as he looks up at the ceiling, in his even way. The other channel is aimed right at the heart of Padum. "Many houses could be damaged."

"With no water, we can't make food, wash clothes, irrigate the fields," says Chondol. "*Skitpo miduk*. No fun."

"Just like Kumik," I say. "*Chu met*. No water."

Chondol shakes her head. "No, Kumik never has water. For us, it's just today." She says every household in Khanggok will send someone up the mountain to clear the headworks of debris and make a new canal section tomorrow.

Feeling like I should pull my weight and earn all the *thukpa* I've been eating at Urgain's house, I volunteer to represent their household in the canal repair. The next morning I shoulder a shovel and walk up the hill with Thinles Tundup, Urgain's nephew and neighbor.

"Without water we have nothing," he says. "*Mausm tsokpo, mi tsokpo*. Bad weather for bad people! God gives no water when we need it and too much when we don't!"

We join fourteen other villagers at the spot a few hundred feet above the monastery, where the flood tore away the headworks and plugged up the canal. Five women and nine men (including one ancient *meme*) wade in and out of the muddy froth, which still looks like warmed-over milk tea. Pants rolled up to their knees, they heave large rocks to and fro, sing short snatches of song, tease each other for being lazy. One guy gives Tundup a hard time for arriving late to the work party. Tundup grabs a spade and runs down the hill toward the next choke-point, to attack another jumble of lodged boulders, issuing a kind of ululating, joyful war cry. He's clearly having fun. "Do you work like this in America?" he asks me, soaked and grinning.

The next day Urgain and I rise before dawn and climb a couple of thousand feet up to the surface of the glacier that gives water and life to his village. It sits in a hanging alpine valley below the soaring rock face of Padum Kangri. We want to test his theory and see where the flood originated.

After an hour of fast climbing under a hot sun, we reach the thirty-foot-high terminus of the glacier. Just as Urgain had supposed, there's a small river flowing out from underneath. We scramble up onto the surface. Wending our way through debris and boulders, we soon come upon a half-acre-sized, thirty-foot-deep hole in the glacier. There are signs everywhere that still water recently filled this empty basin and then drained quite suddenly: incongruous puddles yet to evaporate in the blistering sun, wet mud lining the walls of ice, freshly deposited silt coating stones, bright new scrapes on rocks that must have tumbled down from the lip as the water drained. We climb down to its bottom, and hear rushing water in ice caves below.

"This was where the flood came from," Urgain says.

Clambering about we soon find some other small lakes dotting the mottled surface of the ice. How many more would empty their contents all at once in the days and years to come? As we walk on along the glacier's edge, Urgain starts to sing a trilling melody, "*Trokpo ju-ju, chu-lok ma tang . . .*"

It's an impromptu adaptation of a line from a famous old Zanskari folk song—"Dear wind, please don't throw dust in my lover's face"—to the need of the moment: "Dear stream, please don't give us your floods."

• • •

A few days later it is the end of Sani Nasjal, the festival of masked dances at Zanskar's most ancient monastery. Sani might be the site of the longest-running religious observance in all of India's northernmost reaches. There have been rites and monuments and importuning of the gods at Sani ever since the time of Kanishka, the ruler of the Kushan Empire, which stretched from Iran to Afghanistan to the Ganges in the second century CE. That's when, legend says, a stupa was built at the site of Sani Monastery.

Outside of the stupa monks call out names written on scraps of paper, pulled out of a box at random. It's the *chosphun*, Urgain explains, a sort of spiritual friendship matching. In the next world, he says, you may not meet your mother or father, but you will always be bound to your *chosphun* partner, your help in this life. "And if a good one is matched with a bad one," Urgain says reflectively, "he will help lift him up."

Now the lively mask dances—in which costumed monks reenact the monastery's founding legends—are finished. The tourists have departed with their telephoto lenses. Three ecstatic young friends perch drunkenly on their new motorbike, winners of a lottery to raise funds for a local association. The happy crowd disperses, teenage girls strolling arm in arm and giggling, boys swaggering about in search of opportunities for mischief, mothers chatting and men clapping each other on shoulders—a scene much like any county fair in rural America.

Urgain and his friends from the Kanishka Welfare Society break down the archery target they have set up as a fundraiser—20 rupees for six shots, if you hit the bull's-eye, you win 100 rupees—and comb the fairgrounds for litter. I help out, and my partner is a short, jovial guy, dressed in a smart-looking *goncha*. He wears a stiff new baseball cap that says "Pimps and Hos" on the front. He speaks no English.

He opens his sack, full of plastic bottles and instant-noodles wrappers, losing lottery ticket stubs and foil-faced paper plates streaked with chutney. I dump in my armload and scan the field. It is close-cropped by grazing cattle and as bright green as a golf fairway, fed by the fast waters of the nearby Stod River. Just downstream is the sacred lake of Sani, upon which Guru Rinpoche once gazed as he meditated on the interdependence of all things in his cave halfway up the mountain facing us. "I am from the unborn sphere of all phenomena," Guru Rinpoche declared. "I consume concepts of duality as my diet."

The other Kanishka guys are scurrying around, dragging their own sacks full of trash. There are no landfills in Zanskar.

"Where do you put it all?" I ask.

"In the fire!" cries Pimps and Hos.

He points over to the edge of the archery ground, at a small mountain of garbage. A man piles the bottles and paper and cups together artfully, then douses it with a Coke bottle full of kerosene, and lights the mountain with a match. Flames leap from the twisted plastic; dark smoke roils upward and curls off in the direction of Sultan Largo across the plain.

"Then where does it go?"

He grins widely, and with a wide wave of his arms, declares, *"Lungspo!"* The wind!

"To the sky?"

He nods, gives a little flick of his wrist, and says again: *"Lungspo!"*

Problem solved. The wind would sponge it all up, carry it all away, somewhere up over the mountains.

"What if that smoke is reaching Greenland?"

That was the question that Jason Box says he asked himself as he sat watching CNN at the gate in New York's LaGuardia Airport in June 2012. He was about to board the first of a series of flights that would also take him to Greenland, and was transfixed by images of the record-setting wildfires that were raging across his native Colorado and other parts of the American West that summer.

It's a question that Box is uniquely qualified to try to answer. He has visited and studied Greenland over the course of twenty years and is an expert on the interactions between the climate and the vast ice sheet that blankets the world's largest island. Earlier that year Box had published an analysis of satellite data that showed the reflectivity of the ice sheet had been reduced by 6 percent between 2000 and 2011. This drop coincided with a pronounced melting trend, spreading beyond the lower altitude fringes to reach even the highest interior zones of Greenland. Box and his coauthors concluded, "It is reasonable to expect 100 percent melt area over the ice sheet within another similar decade of warming."

Some people scoffed, but it turned out to be prescient: the very next year brought a season of record-shattering melt in Greenland.

In a crowded briefing room in San Francisco's cavernous Moscone Center, during the annual meeting of the American Geophysical Union, I listened to Box and other polar experts, along with the head of the National Oceanic and Atmospheric Administration (NOAA), describe these dramatic changes in a press conference with the anxious-sounding title "What's going on in the Arctic?"

The tone was sober; the panelists seemed genuinely chastened and baffled by the year's data. They reported that 2012 set all sorts of records: the lowest recorded sea-ice extent, the lowest recorded snow cover extent and duration, and the most extensive melting on the surface of the ice sheet ever recorded, among other milestones.

"I've studied Greenland for twenty years now, I've devoted my career to it," Box had intoned somberly, "and 2012 was an astonishing year. This was the warmest summer in a period of record that's continuous in 170 years."

A day later, as we stood in the hallway outside another AGU session, Box was more animated as he described his LaGuardia epiphany. Goateed and intense, he pulled out his laptop to show me a strange image that might be a clue to help explain the astonishing melting he had observed that summer—less "smoking gun" than "smoking forest"—a blue and yellow pixelated composition from NASA satellite data showing a large plume of smoke in contact with the ice sheet itself. On that day, June 17, 2012, the smoke was coming from a wildfire in Newfoundland.

What struck him most about his summer 2012 trip to Greenland was how the snow at higher elevations—in the ice sheet's cold accumulation zone, where it is fed by falling snows—was darker than ever before. The monthly average reflectivity in June and July dropped below the average for the whole period of 2000 to 2011, even at elevations above 6,500 feet, where fresh snowfall and low rates of melting typically keep the surface relatively free of impurities. What was causing such dramatic changes?

Box has devoted the next phase of his career to figuring that out. He is now focused on those wildfires, which have become larger and more frequent in recent years in the American West and Canada as temperatures rise. He has a theory informed by his decades of observations on the ice: impurities such as dust and soot from fires in the Arctic's neighborhood are darkening its surface, causing it to absorb more "melt energy" from incident sunlight. These particles can lower the ice sheet's albedo in certain areas below 30 percent. As they warm the surface, these tiny dark objects trigger a feedback loop that exposes and concentrates more impurities—leading to runaway melting.

There are other factors at work: as the sun melts them, the edges of ice crystals round. With fewer facets, the crystals reflect less sunlight, lowering albedo. If less fresh snow falls in a given year to cover old snow, and if bare ice gets exposed, albedo will be lower. And the biggest "other factor" of all, of course, is atmospheric heating from carbon dioxide and other greenhouse gases, which unquestionably and inexorably raises air temperatures over the ice sheet, contributing to melting.

But just as in the Himalaya, there seems to be a hidden foot on the accelerator, driving unprecedented melt events like those in the summer of 2012. Estimates vary, but most scientists agree that 2 degrees Celsius of temperature increase over preindustrial levels would pretty much guarantee irreversible melting of the Greenland ice sheet in the coming centuries. And if Box's hypothesis about soot's impact on the reflectivity is correct, then it might take even less warming from greenhouse gases than anticipated to tip the ice sheet into irreversible decline, and we could reach that point of no return even sooner.

So just how fast are black carbon and other by-products of incomplete combustion pushing Greenland's ice into the sea? And how much is due to warming caused by carbon dioxide, or other feedback mechanisms driven by both greenhouse gases *and* aerosols?

The only way to figure out what's going on is to go dig around in the snow and ice in the deep interior. But putting together National Science Foundation grant proposals and funding such an expedition—which is akin to a military expedition, with all the flights and helicopters and equipment that has to be lugged around—can take years. Box has a sense of urgency about this work, fueled by the rapidity of the changes he's seeing. So he has put together the first crowd-sourced Arctic science expedition, called the Dark Snow Project. The goal of the venture, Box says, is to answer one question: "What fraction of albedo change is due to light-absorbing aerosols?" He and his team are sampling ice from Greenland's high-altitude interior to determine the relative concentrations—and contributions to albedo change—of soot and other impurities like dust and algae microbes, which are also dark and which may feed on the soot particles.

These are not just academic questions. If the entire Greenland ice sheet were to melt, it would raise average global sea level by more than twenty-three feet. That would submerge 39 percent of New York City, including most of Manhattan below 34th Street, not to mention most of Florida and most of Dhaka, the capital of Bangladesh. (The seven cities with the most assets at risk due to currently projected coastal flooding, a conservative estimate of just *one* meter of sea-level rise, by 2070 are all in India, China, and the United States: Miami, Guangzhou, New York, Calcutta, Shanghai, Mumbai, Tianjin. New York will have over $2 trillion worth of economic assets exposed to flooding risk. Calcutta, Mumbai, and Dhaka will have a combined 37 million people exposed to dangerous flooding.)

That could take millennia. Or it could take a few hundred years—nobody really knows how fast the process could unfold. Systems as complex as the ice sheet and the weather that influences it are subject to lurches and shifts. And these are exceedingly difficult to predict. But as Jason Box points out, once the melting passes a certain threshold, it will be locked in by feedback mechanisms.

The Greenland ice sheet is thus an example of a "tipping element"—those big, important components of Earth's natural systems that we depend on for our survival, and that can be pushed into very different qualitative states by even small changes.

A tipping point is just some threshold at which that small, incremental change can trigger a transformation of an entire system—a watershed, a forest, a continent's weather patterns. Tipping elements are as inscrutable in their way as the *lha* of Sultan Largo—and just as dangerous. It's the permafrost of Siberia melting and releasing vast quantities of heat-trapping methane, now frozen in the peat beneath it, fueling more runaway atmospheric warming. It's the boreal forests of the far north dying back due to warmer, fire-prone summers and exhaling their carbon irrevocably into the sky.

Think of the ELA climbing up above the summit of Sultan Largo; the glacier is still there, but it's doomed. And there's no going back. Think of the moment when Luggye Tso breached its dam, unobserved, in the middle of the night and started pouring down the Punakha Valley. The system has gone from one stable state, represented by the lake, into another one, represented by scattered boulders and debris and the wreckage of lives, by dispersed water and energy. You can't put the water back in the lake. The whole system has shifted suddenly and irreversibly into a new state of equilibrium.

Charles Zender, a professor of earth system science at the University of California, Irvine, who, like Box, has studied Greenland and the interactions of soot and Arctic climate, has said that "nothing in climate is more aptly described as a tipping point than the zero-degree centigrade boundary that separates frozen from liquid water—the bright, reflective snow and ice from the dark, heat-absorbing ocean." Darker-colored exposed seas absorb a lot more solar energy than ice; thus less sea ice means accelerated warming in the Arctic, which has already warmed faster than any other place on the planet. The consensus is that 1 to 2 degrees Celsius warming over preindustrial levels will trigger the inevitable disappearance of the sea ice. Some scientists think we've already crossed the threshold for the disappearance of both Arctic summer sea ice *and* the Greenland ice sheet. That was why the scientists at that press conference I sat through at the AGU seemed kind of stunned: 2012 was a record for both of these tipping elements. The sea ice in particular is declining much faster than anyone predicted; scientists are now predicting that summers in the Arctic could be ice-free within a few decades.

Those who study these large-scale components of the Earth system have taken to warning the rest of us that climatic changes can be "nonlinear." This means that warming won't just necessarily manifest as a smooth, gradual, incremental rise in air temperature. It means that sudden, potentially catastrophic shifts in natural systems that we've grown quite accustomed to, and indeed depend on for our survival, are not only possible but, under certain scenarios, even probable.

Box's preliminary results from his lean and fast crowd-funded expedition to the interior of the Greenland ice sheet in summer 2013 suggest that a small "perturbation" via black carbon can indeed have a very big effect. Bad weather and logistical hurdles prevented his team from taking all the samples they wanted, but they obtained snow samples and ten ice cores from three different sites. In them they found some tantalizing, if inconclusive, clues: concentrations of black carbon from the 2012 snow layer that they could plug into computer models to try to understand just how much of a push the dark particles gave to the ice sheet's melting. Box's preliminary answer, as of early 2014, is that "summer 2012 black carbon concentrations likely increased cumulative surface net heating by 20–40% for the ice sheet as a whole."

The jury's still out, and Box is careful to point out the complexities of the coupled system that is the ice sheet and Arctic climate, and the need for more rigorous measurements and caution in drawing conclusions. He is returning to investigate the role of algae and other dark impurities in accelerating Greenland's melt, and to get more data on black carbon's role. Asked to summarize his early findings for a newspaper in his hometown in Boulder, Colorado, he said, "You could say that without black carbon—if there were no black carbon—there probably would not have been complete surface melting on the ice sheet in 2012."

Other tipping elements are even trickier to parse. Take the South Asian monsoon.

Between its arrival in June and its end in late September, the monsoon dumps 75 percent of India's annual rainfall total. It is difficult to overstate its importance for the 1.6 billion people of India, Pakistan,

Bangladesh, Nepal, and Sri Lanka. About 60 percent of India's cultivable land is irrigated by this rain, and over 70 percent of Indians depend on agriculture for their livelihoods; a good monsoon thus makes the difference between feast or famine. One of India's former finance ministers famously called the monsoon the country's "real finance minister."

It is such an incredibly complex system that it's difficult to make predictions about how it will behave in a warmer world. But the basic mechanics of the monsoon involve the movement of moist, warm air from over the Indian Ocean rushing in over the landmass of the subcontinent. This is driven by a difference in air pressure from sea to land, which is connected to the temperature differential between the two. Anything that weakens or tweaks those pressure and temperature gradients can destabilize the monsoon. The result could be strong shifts of rainfall in space and time: a movement east or west, north or south, or an intensification of rain earlier or later in the season. Nobody really knows how it will respond. Some parts of India and surrounding countries could be rain-starved, plagued by chronic drought. Others could get deluged in shorter time frames, receiving so much rain in so short a time that it overwhelms the soil's capacity to absorb it.

However it ultimately plays out, there are signs that the disruption is already happening.

Torrential rains fell in the Himalayan state of Uttarakhand in September 2012, and again in June 2013, killing 5,000 people in floods and mudslides. The rains even seem to have triggered a glacial-lake outburst flood that destroyed the ancient Hindu temple of Kedarnath in a rare compounding convergence of two tipping points—Himalayan glacier melt and monsoon perturbations—crossing wires to offer a sneak preview of mayhem that might become routine.

William Lau, a NASA scientist, theorizes that black carbon and other absorbing aerosols are heating air over the Tibetan Plateau, increasing the temperature gradient, and drawing in more moist air from over the Indian Ocean—leading to an intensification of the monsoon. Veerabhadran Ramanathan instead argues that the "surface dimming" caused by black carbon and other aerosols (black carbon stratifies temperature in the atmosphere because it reduces the amount of sunlight that reaches Earth's surface, even as it heats up the atmosphere) increases "stability" of the atmosphere, putting the brakes on the overall hydrological cycle and thus reducing rainfall. Some climate model simulations suggest that increasing black carbon concentrations by two to three times from present-day levels would weaken the monsoon circulation to the point that rainfall decreases by over 25 percent, leading to significantly higher frequency of drought in the region. Others say that, because black carbon heats the troposphere, it will ramp up the circulation and therefore *increase* monsoon rainfall.

Ramanathan isn't too concerned that these various theories are in conflict. "When you read the abstracts it looks like we're all disagreeing with each other, because when you say, 'I agree with Ramanathan,' who's going to publish it? So we have to draw the distinction." He pointed out, though, that given how complex the monsoon system is, it could be a long time before these disagreements are resolved and uncertainties are narrowed. It all misses the point. "If you look at the underlying theme, everyone is saying black carbon disrupts the monsoon. Some say it's going to change the monsoon in Pakistan, some say, 'Ramanathan says central India; it's really Bangladesh.' Who cares, right? It *is* impacting the monsoon. That message is loud and clear."

Chien Wang, of the Massachusetts Institute of Technology, agrees. "The basic conclusion that black carbon aerosol forcing over South Asia is large enough to perturb the monsoon system is reached by all the studies so far," he told the BBC News in 2011, "therefore there is no different opinion here."

A certain amount of variability is natural, of course, and would occur even if the skies were clear of black carbon and other aerosols. But these sneak previews give an indication of the profound costs of a scenario in which disruption becomes the new normal. Should the Indian monsoon tip into a new state, it would be wrenching and dislocating for a sixth of the world's population. Over 70 percent of Indians depend on agriculture for their livelihoods. A radically different monsoon could leave them high and dry, or struggling to keep their heads above water—both literally and figuratively.

Black carbon seems to be altering that temperature pressure gradient, interfering with the so-called Goldilocks system—not too much, not too little rainfall—that people have depended on for millennia to farm and survive in the Indian subcontinent, since the time of the Rig-Vedas. According to some scientists, the monsoon seems highly susceptible to nonlinear shifts, even on the order of less than a decade. In fact, of nine tipping elements identified in an influential study in the *Proceedings of the National Academy of Sciences*, the Indian monsoon is the only one that can turn on a dime, so to speak: dramatic changes are theoretically possible on the order of one year.

Scholars link a weakening of the monsoon in ancient times to the disappearance of the mysterious Harappan civilization that rose between the Indus and Ganges rivers several thousand years ago. For decades archaeologists and historians speculated as to what caused the Harappan people to disappear, much as academics used to wonder what happened to the Anasazi civilization of the American Southwest, which seemed to vanish overnight. (The answer: sudden onset of drought.) A team of researchers recently solved the puzzle: their rain-fed river slowly dried up as the monsoon weakened, eating into their crop surpluses. They had gotten used to a certain amount—and certain timing—of monsoon rains. When the monsoon shifted, their water source vanished up and they were forced to move east, reverting from a sophisticated agricultural society to foragers in the jungle.

One day in a back alley in Leh, I have tea with my old friend P. Namgyal, a solar-building protégé of Sonam Wangchuk.

P. tells me about his experience during the terrible flood that took place on August 5, 2010. A "cloudburst," a freak unprecedented storm, dropped a year's worth of rain on Leh in the space of two hours. Fifteen-foot-high walls of mud and rock and water and debris came down the mountainsides in the middle of the night, filling ravines, sweeping away cars, wiping houses clear off their foundations as families slept. Almost three hundred people were killed.

For days afterward, residents of the city fled to the hills in terror at the first sight of a cloud in the sky. They camped out on the platform of the giant Shanti Stupa, a Buddhist monument that was a gift of the Japanese government, on a hill overlooking the city. Some blamed the government's tree-planting campaign along the Indus for putting more moisture in the air above Leh. Some blamed climate change. Some blamed that season's unusual jet stream patterns, which also brought unprecedented, crippling floods to Pakistan. But no one really knows the cause. Sometimes freak things like this happen.

P. narrowly came through it all unscathed—on a last-minute whim, he had stayed that night at a friend's place. The next day he went back to his own rented room near the bus stand and found it destroyed, filled with mud. He and his friends helped rescue half-buried people in the predawn hours afterward. Speaking in hushed tones, eyes downcast, he still seems a bit shaken when he talks about that night. Now, he is working for a local nonprofit development group, leading a project to build dozens of solar-heated houses for families displaced by the flood. The houses, made from earthquake-safe compressed bricks, will be located on a patch of desert across the Indus River.

P. mentions that his village in Ladakh's Durbuk Block, on the border with Tibet, has been having problems. The springs they have used for generations are drying up. Others blame the angry *lha*. P. blames climate change.

We talk about what his village, and Ladakh and Zanskar in general, might look like fifty or a hundred years hence. P. is not sanguine.

"I think the mountain people will have to move down," he says at one point. "And the people near the sea will have to move up." He pauses a beat.

"And then they will kill each other."

"Jesus, P."

I study his expression. Is he winking? It is always hard to tell when P. is joking, when he's serious, when he's half-joking and half-serious. P. has a gentle demeanor that masks a sly, wry sense of humor. P. has an epic poker face.

"For this reason I will not birth a child," he says evenly. "More tea?"

Black carbon seems to be a feedback-generating machine. Wherever it floats, its light-absorbing prowess translates into a specialty for tipping systems into new equilibrium modes.

Ramanathan and Carmichael, in their groundbreaking 2008 *Nature Geosciences* paper on black carbon, succinctly laid out some troubling scenarios. They posited that "surface cooling occurring simultaneously with lower atmosphere warming (due to BC and dust) can stabilize the boundary layer during the dry season and increase the lifetimes of aerosols in ABCs and increase persistence of soot-filled fog." Black carbon can thus tip certain systems into behaving in its own favor, prolonging its brief life span in the atmosphere.

To take another example: if black carbon loading in the atmosphere causes the monsoon rains to diminish in strength, the very factor that *removes* black carbon from the air—rain, or "wet scavenging," as the scientists say—will be weakened. Less rain means longer tenures and higher concentrations of soot in the air.

Black carbon's influence in the Arctic is of particular concern. The Arctic death spiral is a scenario that elegantly communicates the terrifying compounding potency of these feedback mechanisms. The total volume of ice in the Arctic is in decline. Less ice means more dark ocean exposed, which absorbs more heat, which warms the Arctic further, which melts more ice . . . and on and on. More warming in the Arctic threatens to thaw out the vast permafrost, which contains 1,700 billion tons of carbon—twice the amount of carbon currently in the atmosphere. As its blanket of snow thins and thaws out, the warming tundra will release this locked-up store of carbon in the form of carbon dioxide and methane. Once that genie is out of the bottle, it will be exceedingly difficult to slow down the runaway warming that will result. (Preliminary results from a five-year NASA program to monitor methane emissions in the Arctic found 650 parts per billion methane over the eastern Yukon River—much higher than normal levels. "That's similar to what you might find in a large city," where natural gas pipes seep, said the lead investigator.)

While it is a proven scale-tipper, black carbon is also a path-dependent pollutant—not all of it is created equal. A particle emitted by by a ship burning bunker fuel on its way through the recently opened Northwest Passage in the Arctic will have a very different journey—and different impacts on health and climate—than a particle coughed out of a garbage truck tailpipe in New York. The latter particle will have much higher odds of encountering a pair of human lungs, while the former will have far more opportunities to glom onto snow and ice.

As Bond and her coauthors point out, black carbon concentrations are "spatially and temporally variable." Kind of like a flame—if you watch it for several minutes, it flares, dips, alters direction, waxes, and wanes. If you could track individual chains of the pollutant, you would see that black carbon over the high Arctic has a different story than black carbon traveling over the Pacific, or squatting over New Delhi, or climbing up river valleys from a brick kiln in Uttar Pradesh to the glaciers of northern Nepal. Or black carbon liberated from a pile of burning garbage in Sani, making a beeline for Sultan Largo.

For these reasons, it's difficult to compare black carbon's role in changing global climate to that of carbon dioxide. Understanding black carbon's varied impacts is a lot like real estate: it's all about location, location, location. Everywhere it goes, its fundamental behavior remains the same—turning light into heat—but the chain of events it sets in motion stirs thing up in different ways in different places. This makes modeling its climate impact around the world an exceedingly difficult task, fraught with uncertainty. (The only real certainty is that it's disruptive.) Black carbon provides us with a simple, useful reminder: all global climate change is local, in terms of impacts. But precisely because it's so ubiquitous, because 7 billion of us depend on dirty fires every day—and because it's so potent in absorbing solar energy—black carbon is also a huge contributor to *global* climate change.

There is significant uncertainty around how much radiative forcing is caused by incomplete combustion of fuels, partly because black carbon is never emitted alone. If you tried to clean it up, to reduce the amount belched into the skies from New York to Nairobi, you would also be changing the amount of lighter-colored aerosols pumped into the atmosphere.

Recall the toxic waste factory: wood smoke, for example, contains other aerosols, like organic carbon—sulfates that scatter light in the atmosphere instead of absorbing it. These co-emitted pollutants also influence cloud formation and temperature stratification in the upper atmosphere in complex ways that scientists are still trying to get a handle on. The science is young, and much more research is needed to refine our understanding of the physical interactions between black carbon and other particles produced by incomplete combustion and the climate, especially in terms of its effect on cloud formation. (For example, clouds scatter light back to space; if black carbon is found to aid cloud formation, that would have a cooling effect. If it instead evaporates clouds, that would have a warming impact.)

As Bond explained to Congress in 2010, "Any action to reduce black carbon will also affect any *co-emitted pollutants* from the same source. Any emission source produces warming pollutants (black carbon and some gases) and cooling pollutants (sulfates and organic carbon), and the result is like mixing hot and cold water in a faucet. The mixed water can be very warm, very cold, or in between depending on the amount of each flow. Sources with high emissions of warming pollutants are the most promising targets for reducing black carbon warming." The conclusion is logical—darker smoke means more warming, and more acceleration toward tipping points.*

Likewise, from Greenland to north India, there is uncertainty about the relative contribution of soot borne on the *lungspo*—the wind—in driving these systems past their tipping thresholds. How much of the decline in albedo in Greenland is due to algae or dust versus soot particles? Will light-absorbing aerosols like black carbon shift the monsoon north . . . or cause more rainfall early or late in the season, overwhelming the land's ability to absorb it? In the frozen reaches of

---

*But new research suggests that even lighter-colored smoke from biomass fires may have a net warming impact. Ramanathan's study for the California Air Resources Board also looked at brown carbon, a type of organic carbon emitted in large quantities from forest fires and domestic wood burning in stoves. Previous modeling efforts have assumed that emissions from these sources had a net cooling effect, owing to the light-scattering properties of lighter-colored aerosols. But Ramanathan's recent research suggests that on the spectrum of light absorption, so-called brown carbon may be tipping the scales of these co-emitted pollutants' impacts decisively to the warming side. In light of these new findings, he now thinks that his 2008 study, in which he estimated that dark aerosols were responsible for half of observed snow and glacier retreat in the Himalayan region, is out of date. "I think it could be worse," Ramanathan told me. "That was based on a paper I did the year before in 2007—we flew UAVs into the cloud, measured the heating. I think I've changed my mind, with the addition of the brown carbon. I assumed all the biomass was cooling." When you add brown carbon's light absorption to that of black carbon, the climate argument for cleaning up biofuel and biomass smoke becomes more urgent.

the High Himalaya, where data is difficult and dangerous to collect and complex microclimates abound, there is likewise uncertainty about how much black carbon is deposited, and where.

Researchers will continue to investigate and debate all of these questions. But most concede that the larger mystery has been resolved. There seems to be zero uncertainty among the scientists who study these phenomena about this fact: black carbon is wreaking serious havoc with systems whose stability has fostered the development of human civilization over the past ten thousand years. And if anything, each new study suggests the stuff is even more potent and more abundant in the air than we thought. Ramanathan's Project Surya, for example, measured concentrations of black carbon inside of kitchens and around homes in rural Uttar Pradesh that were much, much higher than they expected to find.

"I never expected to see such massive amount of pollution either inside or outside," he told me. "We found out the concentrations we were measuring outdoor is about a factor of ten larger than what the IPCC models were claiming. It's huge!"

Too much water, then not enough. Melting ice, rising seas, a long tail of drought. Vritra, Indra, and Agni locked in a struggle of ever-increasing intensity, and planetary stakes. We are heading into uncharted waters, from the Central Valley of California to the dusty plain of Marthang—and black carbon is pushing the throttle down for us, leaving us little time to adapt.

Focusing on black carbon thus means focusing on hotspots, on the particular places where it comes from and where it does the most damage, both in terms of natural systems whose stability and reliability we depend on for survival—the Himalayan water towers, the Greenland ice sheet, the Indian monsoon—and in terms of our hearts and lungs and health. It also means focusing our attention on the here and now: the benefits of cleaning it up aren't abstract or far off in the future. Whereas carbon dioxide's impacts strike us as distant in both space and time (though this is not always true), forcing us to consider the welfare of future generations in our choices, black carbon is a threat wherever it's emitted, *right here*, and because it's so potent a warming agent, *right now*. And unlike carbon dioxide and other greenhouse gases, you can see it. It gets in your face. (And in your eyes and lungs.)

"In the climate physics, there are two levers you control," Durwood Zaelke, an expert on environmental law and policy and president of the Institute for Governance and Sustainable Development, says. "CO_2 is the long-term lever—you've got to pull it today. But, like stopping a supertanker, it's going to take a long time to get a response because the memory of the climate system is so long. The other lever is short-lived climate pollutants. You pull it, you get almost an immediate response. Black carbon, you get a response in days to two weeks—really fast."

Drew Shindell, a NASA climate scientist and the lead author of an influential UNEP study examining the varied environmental and health impacts of black carbon and other short-lived climate forcers, agrees. "We will have very little leverage over climate in the next couple of decades if we're just looking at carbon dioxide," he has said. "If we want to try to stop the Arctic summer sea ice from melting completely over the next few decades, we're much better off looking at aerosols and ozone."

Zaelke acknowledges the uncertainties around the climate impacts of inefficient burning, when you consider all of black carbon's co-emitted aerosols. "But if you don't start now trying to reduce black carbon, you don't save those lives, you don't save the glaciers, and you don't save the climate if Ramanathan and [Mark] Jacobson and others are right. Because you can't wait for scientific certainty before you start a policy. It's too late. And time is our enemy here. Things are getting worse so fast, we are just about to break the 400 ppm barrier for CO_2. If we think we're going to have any chance for staying below 2 degrees [warming over preindustrial levels], it's going to require heroic efforts on short-lived pollutants. And we know we can do that.

"We're in a race with accelerating warming and accelerating feedback mechanisms, which feed upon themselves, and technology and policy over here, not moving very fast," he says. From the Himalaya to the Arctic, the clock is ticking. "So you've got to find a way of speeding this up."

Zaelke points out that, in the United States at least (and almost nowhere else in the world), people of different political persuasions disagree on the extent to which climate change is caused by human activities. (NB*:* The scientific debate has long been settled: It is.) "But they converge and agree on another question—do we know how to solve climate change? A lot of people think we don't know how to solve climate change. So we could go from climate denial to climate despair. Despair is the enemy of innovation and action and good policy. Despair is deadly.

"We need to show that we can solve climate change, piece by piece, and we need to build the momentum," Zaelke says. "Short-lived climate pollutants can do this, because we can make progress right now, with known technologies, and we can show results right now. So that's really important. And if you reduce black carbon, your air is cleaned up. You can see it."

The only problem is the problem of human relations," observed Antoine St-Exupéry. And human relations can be as unpredictable, as suddenly consequential, as the monsoon rains. Like the monsoon and the Greenland ice sheet, human societies also happen to be large-scale components of the Earth system that can shift rapidly into new states thanks to the aggregate effects of tiny perturbations—perturbations like human choices and epiphanies, and difficult, pragmatic collective decisions. (Like the decision to rebuild your village.) As such, we are also subject to feedback mechanisms and sudden accelerations. The steam engine is invented and put into a locomotive, and the fabric of time and human relations the day after is forever altered. The Berlin Wall falls one day, and suddenly the cold war that defined half a century of human history ends. In Tunisia, citizens rise up and throw out a dictator who controlled their society for decades. The London Smog of 1952 kills 11,000 people, and Britain as a society decides: never again.

The climate commentator David Roberts finds solace in chaos theory, in the possibility of these sudden shifts in human affairs. "The outcome of the climate crisis depends not just on physical forces," he has written, "but on human beings, complex economic, social, and technological systems, and complex systems are nonlinear. We forget this; our instinct is to think the future will look like the recent past, only more so. We don't anticipate the lateral moves, the lurches, the phase shifts."

You can never discount the possibility of a group of people deciding to go move a mountain without any shoes, or dig a seven-kilometer-long canal, or build a whole new village from scratch on a wind-scoured patch of red dust.

There's hope to be found, too, in this simple, overlooked fact about the physical world: Soot is "what settles." Black carbon may wreak some serious havoc, but it has a life span of only about a week. Stop producing it and the skies will start clearing right away. In the past it's taken decades or centuries to achieve this brightening, but the "transition can be accelerated."

Mark Jacobson, an energy and climate expert at Stanford, believes (along with Ramanathan) that shutting off the black carbon faucet is our only option for slowing down warming in the immediate future. He argues that major cuts in black carbon could buy us a decade or two delay in warming—perhaps enough time to wrap our arms around the longer-term problem of carbon dioxide. Zaelke and Ramanathan have made a similar point: steep, quick reductions in emissions of black carbon, together with targeted actions to reduce methane, ground-level ozone, and HFCs (the short-lived climate forcers), could "slash the rate of global warming by half by midcentury—equivalent to wiping out the warming we have experienced over the last 50 years."

So who knows what might be possible if we decided to stop pouring so many dark particles into the sky? And if we followed through on that decision with some Kumikpa-scale moxy—as Tsewang Rigzin might say, "on war footing"? After a few days, as the air starts to clear, the future could start to look very, very different, even from the recent past, in some surprising, hope-inducing ways.

PART THREE

The Fire Brigade

In general Smoke is a very tractable Thing, easily governed and directed when one knows the Principles, and is well informed of the Circumstances.

—BENJAMIN FRANKLIN, AUGUST 28, 1785,
LETTER TO DR. JAN INGENHOUSZ

7

In Search of Phunsukh Wangdu

Look now at this black substance going up into the atmosphere; there is a regular stream of it. I have provided means to carry off the imperfectly burned part, lest it should annoy you.

—MICHAEL FARADAY, THE ROYAL INSTITUTION, LONDON, CHRISTMAS 1860

There is no saying that, "For the rich oxygen is very important, the poor can do with carbon dioxide." You don't say that! They need oxygen as much as any person on the earth. Clean water is as important for poor as for rich.

—SONAM WANGCHUK, SECMOL, LADAKH, MAY 2006

The Buddha has no pupils. What to do?

Odzer, the master sculptor and painter hired by some of the young Kumik soldiers to make a statue of Sangyas, the meditating Buddha, for the new *lhakhang*, is almost finished after weeks of work. The only spot left to paint is the Buddha's eyes. But he has run out of paint.

Odzer purses his lips and considers for a moment, and then heads into the *lhakhang*'s half-finished kitchen. He rummages among piles of tools and kitchenwares until he finds what he's looking for: a short section of battered stovepipe. He walks outside, props the elbow-shaped pipe up on a couple stones, and leans it against the wall of the *lhakhang*. Under one end he places some small sticks. He pours out a bit of kerosene. He sets the pile alight. Flames lunge at the mud brick wall.

Odzer pokes the stones closer together, reducing the air flowing in from below, and turns the pipe so its outlet points straight at the wall. Soon a dark streak coats the bricks. After ten minutes of burning, Odzer pulls the pipe apart and rubs the inside with his fingers. They come out coated dark with soot.

"This makes good paint," he says. "It lasts a long time. The idea came to me once years ago when my black paint was finished."

Nearby a few soldiers jump into a Tata Mobile to head to Padum for supplies. As they pull away, the jeep coughs out a dense cloud of black carbon.

"Nice *shremok*!" I say. Odzer laughs. "Grab it for your paint!"

Making soot is a pretty straightforward process. Like Odzer, we all seem to know intuitively that if you interfere with the air flowing into a fire, it will produce more of the dark stuff. Well, the inverse holds true, too.

It turns out that we've known how to get rid of black carbon since at least 1860. During his final lecture on the "chemical history of a candle," just before Christmas that year, the self-taught chemist and physicist, Michael Faraday—regarded as the greatest scientist of his era—demonstrated a simple remedy for his audience. If you knew the principles at work, it was quite simple to get rid of those pesky dark particles.

"When the particles are not separated, you get no brightness," he observed. Then he held a piece of wire mesh above the jet of a burner and lit the gas coming out of it, above the mesh.

"It burns with a nonluminous flame, owing to its having plenty of air mixed with it before it burns." He pointed to the pale blue flame emerging from the wire. "There is plenty of carbon in the gas; but because the atmosphere can get to it, and mix with it before it burns," there is no yellow flame.

Why no flame? Because there are no particles.

And why no particles? Because the fuel is mixed with plenty of air.

"The carbon meets with sufficient air to burn it before it gets separated in the flame in a free state. The difference is solely due to the solid particles not being separated before the gas is burned."

You can try this for yourself. Hold a spoon over the tip of a candle flame. It soon becomes covered in a chalky layer of soot. You've impeded the mixing of the air with the molecules of paraffin fuel channeled upward by the wick. The polyaromatic hydrocarbons that are liberated from the fuel in the first stage of combustion are the precursors of black carbon. Note the bluish zone at the base of the candle flame, where it's hottest (this is where precursor gases liberated from the fuel are chemiluminescing, as combustion geeks would say). Just above is a dark core, the oxygen-deficient part of the flame, where soot particles are being formed. And above that is the yellow zone, where even more black carbon is being produced, and incandescing. If the temperature up in the bright tip of the flame is above 1,000 degrees Celsius, the soot is burned off. But if cool air interferes, black carbon escapes unburned into the air.

Simply put, higher temperatures and better mixing of fuel with air lead to more complete combustion. The yellow zone around the edge is where temperatures are lower, and relatively fewer of those particles are burned off.

So that dark chalky layer on the spoon is basically fire, interrupted. The fuel hasn't completed its intended journey—a clean sorting into water vapor, carbon dioxide, and useful heat—because of suboptimal conditions. The fuel needs access to undiluted oxygen as badly as we do.

This suggests a few possibilities for getting rid of black carbon. At the level of the fire itself, you can increase temperatures and improve the availability of oxygen during the combustion reaction.

"Carbonaceous aerosols can be destroyed if the exhaust is kept hot and well mixed with air," Tami Bond and her coauthors wrote 150 years after Faraday's demonstration, making pretty much the same point. Modern, well-operated power plants emit relatively low amounts of black carbon because they achieve these high temperatures and the premixing of fuel and air. (India's coal-fired power plants don't fall in this efficient category, which partly explains why their emissions kill over 115,000 people each year.) Better mixing of fuel and air also explains why modern gasoline engines produce less soot than conventional diesel engines.

If improving combustion further is not a good option, or there are technical limits involved—as in a diesel engine, where the compression-based combustion will always entail at least some fuel-rich pockets of air—the particles that are formed can be caught on their way out of the tailpipe or smokestack. This is the basic strategy of the diesel particulate filter. (The filter is self-cleaning: periodically the soot and chemicals caught on it are automatically burned off at extremely high temperatures.)

And if those two approaches are technically infeasible or too expensive, we can ask ourselves: now that we have a fuller sense of its terrible costs, are we willing to put up with the soot-laden smoke for the benefits of a particular fire? Is there some other way to get the service we want—hot water, a cooked meal, a ride to the office—without burning anything at all?

"It doesn't have to be a stove!" Tami Bond told me. "You just want to cook your food. You want hot food. Why don't we start by saying: what services do we need to provide, and what's the most sensible way of providing it? People want services. We have really clumsy ways of getting those services right now. And so we're right now sitting in our own sewage, atmospherically and otherwise."

Jullay?"

I call out the Ladakhi greeting on the threshold of a nondescript house on a quiet side street in central Kathmandu. The door is slightly ajar. I push it open, climb the stairs, and walk through an empty kitchen. No sign of the guru.

Out on the patio I find several neatly cut square blocks, made of a mixture of dung and straw, still damp to the touch and drying in the sun. Jackpot. Now I know I'm in the right place—there's ample evidence that the guru has been here, up to his old tricks, and perhaps a few new ones.

Sure enough, he soon emerges from his room, where he has been meditating. Barefoot and smiling, looking relaxed in a white undershirt and loose-fitting pants, he greets me with a hug. Though he is as busy as ever, and suffering from a nasty bout of allergies brought on by central Nepal's humid air—so different from the dry climes of his native Ladakh—there is a lightness about him. His time in Nepal is winding down, and a return to his homeland, from a productive self-imposed exile of nearly three years, is imminent.

He ushers me back out onto the patio, eager to describe his latest wild scheme: making self-supporting vaulted roofs for buildings in the cold, earthquake-prone Himalayan regions of Nepal and Ladakh, using panels made of dried sand, dung, and clay.

He picks up and turns over the block of cow manure with a critical eye. "This one is cracking because it's drying too quickly. And that one maybe has too much sand and not enough dung." Then he pauses and fixes me with a sly grin.

"But I'm talking such bullshit!" he puns, followed by his signature high-pitched laugh.

Sonam Wangchuk, my old friend and mentor, the solar guru of the western Himalaya, goes into the kitchen and produces a plate of mangoes, and we sit and proceed to talk nonstop late into the night. It's been over a year since I've seen him, and a thorough update is in order. Over the course of the next several hours ideas pour out of him like water from a burst glacial lake. Some are new refinements on schemes he has long been "playing with," as he puts it—such as a movable, modular housing system set on rail tracks and a high-altitude biogas system design for the school campus he cofounded outside of Leh; others are new ideas—such as a special pneumatic ram for making rammed earth walls in Ladakh and a natural siphon system for draining dangerous glacial lakes. At the end of our marathon session, he circles back to his current obsession: those straw-clay-dung blocks for lightweight insulating roofs.

Wangchuk thinks these straw-clay panels could revolutionize the way people build in Ladakh and Zanskar. They could be fitted in between wooden joists or stacked into masonry-style arched vaults, requiring little to no wood. Traditional roofs use heavy poplar beams and several inches of packed earth; because the poplar is an increasingly pricey commodity in those arid lands, the beams have to be trucked in from Kashmir. The roofs are poorly insulated and frequently let both cold air and water through; in addition, they are dangerously massive in the event of an earthquake.

"India burns 90 million tons of rice straw each year," he points out, producing plumes of smoke that merge into the atmospheric brown cloud over north India. Punjabi farmers set fire to 80 percent of their rice stubble after their fall harvest to get rid of the useless stuff clogging their fields; cattle won't eat it. One study found that 116 million tons of crop waste are burned each year in India, producing 145,000 metric tons of particulate matter. NASA satellite analysts note that these dense straw fires look just like wildfires burning in the American West. Researchers estimate that India produces 600 million tons of agricultural waste each year. A third of it goes unused. It's a huge untapped resource, a major source of black carbon (40 percent globally comes from this kind of open burning), and a serious public health risk.

But Wangchuk wonders: what if all that straw weren't useless?

A handful of power plants that burn biomass like rice straw in Punjab have cropped up in the last few years, creating much-needed electricity while diverting a fraction of this waste stream. (They still produce some particulate emissions, but because the combustion is much more controlled and efficient, substantially less black carbon is produced.) Wangchuk sees opportunity here, and he lays it out for me methodically.

Because Ladakh and Zanskar are part of the disputed state of Jammu and Kashmir, there is a heavy military presence. An endless stream of Indian army trucks brings material over the mountain passes. They carry heavy, low-volume loads like cement and iron that sit on the floor, leaving plenty of space. He wants to buy straw in Punjab, for around 1 rupee per kilogram, and load it on trucks with plenty of room for more lightweight cargo that are heading to Ladakh, where the price of straw is 12 rupees per kilogram (about 10 cents per pound).

Then he wants to set up a facility to mix it with local mud and make materials for earthquake-proof, energy-efficient buildings in Ladakh. The panels would provide "insulation and safety—two things that Ladakh is challenged with," while making use of a waste product that would otherwise be burned, turned to black carbon and organic carbon and other undesirables, and sent aloft. In the process, it would cut down on local pollution from concrete use (diesel-powered mixing and transport) and from the kerosene, coal, and dung burned to heat all those leaky, drafty buildings.

I smile as I listen. It sounds kind of out there, but it also sounds so sensible. In other words, it's classic Wangchuk: take multiple, interconnected problems—air pollution in Punjab ("which surely melts our glaciers," he notes), waste management, high energy heating needs in Ladakh, the risk to safety posed by the combination of earthquakes and heavy traditional earthen roofs, inefficient transport—and try to solve them with one elegant, though politically and logistically daunting, stroke. The guru is not content with mere win-win strategies. As his friends and critics alike will acknowledge, Wangchuk likes to think big: "It's called win-win-win-win-win. Yes!"

His blocks are still experimental, and as with most of his innovations, it will be an uphill push to get Ladakhis interested in his dung-clay-straw panels. They, like most people in India, associate mud and dung with poverty and "backwardness," and materials like steel and cement with *pukka* (literally "ripe," meaning solid) houses, with being modern. Likewise, getting the Indian army, one of the world's most sclerotic bureaucracies, on board will be difficult.

But he is undaunted. He sees ways to clear those hurdles. And Sonam Wangchuk has achieved the impossible before.

When he was a boy in Ladakh, Wangchuk was reticent in school, afraid to answer a single question. His teachers called him a "donkey." These men, posted from outside Ladakh, from Kashmir and other parts of India, mocked and berated him, saying he was slow, stupid, hopeless. The medium of instruction was Urdu, not his mother tongue of Ladakhi, and he could hardly understand the questions in this unfamiliar language. He begged his family to send him to school in Delhi, where he encountered for the first time a sympathetic teacher, who helped develop his self-confidence. He became the top student in his class, went to Srinagar, and got a degree in mechanical engineering.

Upon graduating, he defied expectations yet again. His father, a self-made man and a powerful and famous politician in Ladakh, wanted him to go into government service, the default route for freshly minted engineers and a sure path to financial security and social prestige. But he had other ideas. To earn money for his tuition, he had spent his winter breaks from university tutoring Ladakhi students. So, after graduating, Wangchuk started working full-time with those students who had failed the tenth-class exam, a make-or-break test that is a kind of passport to future opportunity.

Scandalized by the way the government education system was failing legions of students who were bright and desperate to learn, in 1988 Wangchuk and several other university students founded the Students' Educational and Cultural Movement of Ladakh (SECMOL). At the time, 95 percent of Ladakhi students were failing the exam. If teachers showed up at all to school, they made their pupils do rote recitations of formulas and culturally alien stuff like Victorian-era British poems—without understanding a word—and beat them if they got a line wrong.

"One thing I discovered was that these students were so keen to learn," he told me when we first met in 2003. "You tell them the most complex things in trigonometry or physics and they grasped it all. But when it came to writing it, as an answer to a question in examination, they were very poor, because of this language confusion. They were not sufficiently capable in any of the languages. I started thinking that I had to change this system. You know, if 95 percent of the products fail, not just in schools, in any system whether it is a car factory or jam factory, then it is not the product, it is the system . . . that has a defect, and the system has to be changed."

The kind of changes he and his partners envisioned were reflected in the group's name: meaningful educational reform and holistic learning could only be achieved through, and with, a cultural revival in the classroom. The goal was to change the culture of public education throughout the region, while instilling confidence among young Ladakhis, in both themselves and their own culture. Wangchuk and his colleagues diagnosed the obstacles to change and set to work, deliberately and patiently, on a pincer-movement strategy to repair and revitalize the system that had failed them, mobilizing grassroots support for reforms at the village level while persuading political leaders at the top to get behind policies to improve teacher training and accountability and curricula.

SECMOL organized parents to form village education committees in dozens of villages in eastern Ladakh. They published culturally appropriate textbooks in Ladakhi and English and held teacher trainings for student-centered learning at their campus outside of Leh. They published magazines and newsletters calling for accountability for absent and underperforming teachers. The program, labeled Operation New Hope, was so successful and popular that the regional government adopted it as their official policy and took over its administration. His harebrained scheme went from being regarded as impossible to becoming the law of the land.

The success of ONH garnered national recognition for SECMOL and Wangchuk. He was appointed to the national governing council on education in India and advised Ladakh's Autonomous Hill Development Council on education and other issues. He won CNN's "Real Heroes" award, and became an Ashoka Fellow. He became known far and wide in Ladakh and Zanskar as a champion of the people, and of striking a balance between development and technological progress and respect for traditional culture.

"I am surprised when people talk of me sacrificing a career," he told *The Week*, India's version of *Time*, in 2001, when it named him Man of the Year for his education reform work. "Sacrifice is giving up something you enjoy doing. I haven't sacrificed anything."

Having traveled a long way—by train from China's Yunnan Province, through Shi Yafeng's soot-gray city of Lanzhou in desolate Gansu Province, to Lhasa, then across Tibet in a jeep before plunging down from the plateau to Nepal—to find my old friend and mentor, I feel a bit like Raju Rastogi or Farhan Qureshi, on a quest to find their old friend and guru Ranchoddas Shamaldas Chanchad ("Rancho," as everyone calls him).

If you're from India, those names will be familiar. They are the three main characters in *3 Idiots*, the highest-grossing film worldwide produced by Bollywood, the world's biggest film market. When it was released in 2009, *3 Idiots* touched a national nerve. In addition to breaking almost all the box office records, it set off a debate about the pressure-filled higher education system in India and inspired legions of young people to "follow their passion" (while also introducing some memorable, now ubiquitous, new catchphrases and songs to the lexicon of modern India). There are now plans afoot for remakes in China and Hollywood. Loosely adapted from a novel about life in the competitive environment of Indian engineering colleges, the film chronicles the misadventures of three friends. The central character, Rancho, is played with joyful, debonair intensity by Aamir Khan, one of Bollywood's biggest movie stars and India's answer to George Clooney.

Even if you're from India, though, you may not know that the character of Rancho is based, in part, on Sonam Wangchuk.

In the film, Rancho makes it his mission to open his fellow students' eyes, to help them set and pursue their own goals in life, to see their own potential, and to show them the deep flaws in the sausage-grinder education system they are being passed through. Rancho teaches his friends how to let go, to enjoy learning instead of endlessly cramming for exams. Shenanigans ensue. He does "demos" for people to show them what's really valuable in life, how they're being shallow or myopic. He repairs a fellow student's remote-controlled helicopter. He tops his class in spite of his unorthodox methods.

Wangchuk met with Aamir Khan and the rest of the *3 Idiots* team in Mumbai and told them about SECMOL and his work there. This clearly informs certain scenes at the end of the movie. (The film crew first planned to shoot at SECMOL's campus, but later chose a private school near Leh.) Both star and producer, Khan is famous for being involved in every aspect of planning his films, including the screenplays. But how to explain the similarities in the earlier part of the story—not the standard Bollywood ingredients of a romance, dance sequences, and a life-threatening crisis plot point, but the evocation of one maverick's journey through the pressure-filled environment of engineering school? "Some of the similarities are so strong, I'm puzzled by it," Wangchuk admits. "My friends from that time saw it and said, 'It's all about us!' I never talked to them about engineering school. I don't think my friends talked to them. So how could they know? But the stories are just like how it was."

He describes the brutal hazing, or "ragging" as it's known in India, that the seniors would subject the juniors to, barging into their rooms at all hours. In the film Rancho uses his engineering genius and some quick thinking to devise a particularly painful and humiliating tactic to deter a senior student's aggression. "I used to devise different means to avoid that for me and my friends," Wangchuk recalls. "One peaceful method of avoiding it was this: I changed the nameplates on our doors, so it would say 5TH YEAR or 3RD YEAR and they would pass us over." In the film Rancho concocts various clever devices. In real life, Wangchuk did the same. When he was in high school in Delhi, he would sleep right through the chime of his regular alarm clock. So, in order to wake him up at three or four in the morning to start studying, he made his own elaborate Rube Goldberg–style system, whose chain of mechanical events culminated in dumping a cup of cold water on his face while he lay in bed.

Years later, Ladakhis would start telling a joke about Wangchuk: He goes to the tool bazaar in Old Delhi. He goes from shop to shop, picking up this machine, that device, asking question after question of the shopkeeper: how it works, where it comes from, the types of components it contains. Finally, exasperated, the shopkeeper asks, "Well, are you going to buy it?"

"No!" Wangchuk replies, "I'm going to make one!"

After graduating, Rancho vanishes. Seven years later, his buddies from engineering college, Farhan Qureshi and Raju Rastogi, embark on a quest to find him and find out what happened to him. And after many twists and turns (spoiler alert here), the two men, joined by the headmaster's daughter, Rancho's lost love, travel north to find him . . . where?

Why, high in the Himalayan region of Ladakh, of course. There he has founded a most unusual school, where students are taught to think for themselves. There they find children operating a bicycle-powered sheep-shearing device, others a bicycle-powered grain mill. The kids seem to be running the show.*

---

*To anyone who's spent time on SECMOL's student-run campus, this scene of empowered young Ladakhi tinkerers will be quite familiar. On a recent visit, I found a group of students working to make a block of new composting toilets and setting up formwork for a mud vaulted-arch roof; another group of students working under the guidance of a volunteer to make a biogas system that would transform cow dung into cooking gas for the kitchen; and two boys making their twice-daily rounds, walking past solar dryers and a

They finally find their old friend and guru pacing along the edge of a famous Ladakhi lake, piloting a remote-controlled aircraft. He is a scientist now, he tells them, without adding that he is a famous inventor. "But I also teach kids," Rancho explains.

Chatur, his old nemesis from engineering school, mocks Rancho by the banks of Pangong Tso. "Crap is what he gave us! Wanted to change the education system, wanted to change the world!" He mocks him for being a mere schoolteacher. "Finally what does he change? Kids' diapers!"

Chatur then discovers that Rancho is actually the gifted inventor he has been seeking to ink a contract with, for his multinational employer: Rancho's real name all along has been Phunsukh Wangdu. The film closes with Chatur running after the laughing group, desperate and apologetic, pants around his ankles after "mooning" Rancho in derision. And the friend-narrator intones: "His Holiness Guru Ranchoddas had correctly stated: Follow excellence. Success will chase you, pants down."

There's a moment in that final scene, as Rancho/Phunsukh stands on the edge of that vast upland lake, savoring his victory, which another man has just mistaken for defeat. There is a gleam in his eye, neither humble nor proud, just alert and bemused and laced with wry compassion. The look of someone who's looking at the entire chessboard, seeing the third or fourth move out, tabulating the opportunities that lie just around the bend.

Watching the film, I immediately recognized it. It's a look I've seen before on multiple occasions. One version in particular is emblazoned in my mind:

It's 2005. I am with the group of American students I have been teaching, and we're rumbling along in a bus to the Leh airport, leaving the SECMOL campus, where we have been living, working, studying, and playing alongside the Ladakhi staff and students for four months. I look out the window to my left as the sun rises over the dun-colored mountains. There, in a T-shirt and pajama pants, is Sonam Wangchuk,

---

solar water heater on the kitchen roof to check on the 10-kilowatt solar array that powers the entire campus.

head down, looking determined, gripping handlebars, vigorously pedaling a rickety three-wheeled rickshaw next to us on the sandy road leading to the main highway.

He had been up all night, trying to win a bet we had made, the terms of which I now can't recall. In order to win, he had to get the homemade solar-powered rickshaw that we had been working on for the last couple weeks up and running, before we left for the airport at six in the morning. I had checked in on his progress at 3:00 a.m., and found Wangchuk and P. Namgyal and another young Ladakhi student hunched over a mess of gears and charge controllers, hooking the battery up to the motor. A welder had been called in from Leh to put the last touches on converting the old rusty bike frame into a three-wheeler.

"There's no way you'll finish," I said.

Wangchuk just cocked his head a bit, gave me a slight smile, and raised his eyebrows: *We'll see.* I went back to bed.

Now he is accelerating, whooping and gleeful, into a sun-drenched Ladakhi morning.

"You finished it!" I shout from the bus window, amazed, pleased to have lost my bet.

"Well, almost," he shouts back. "There are no brakes!"

And he surges past us down the hill on the sandy track to the main road, crosses the bridge over the stream, then throttles and pedals his way up the other side, childlike grin still plastered on his face. Chasing success, chasing the sun, with his pajama pants on.

Like Rancho, Wangchuk has a reputation for standing up to authority, for stirring up placid waters. "Troublemaker. Rabble-rouser. Certified misfit." These are the terms he winkingly uses when asked to describe himself.

"Gadfly?" I ask.

"What fly? What is that?"

He leans forward and tilts his head, in the intent pose he assumes whenever an opportunity presents itself to learn a useful new word, concept, tool, or fact about the universe. For an engineer, he has a remarkable love of language: he's not a big reader but he loves learning new metaphors, turns of phrase, and, of course, languages. (He can speak ten of them, seven fluently.)

But for a guy whose ideas are so often about resolving contradictions and conflict, finding commonsense solutions, Wangchuk seems to piss a lot of people off. He has tangled with powerful monks and Buddhist rinpoches (incarnate lamas who often wield political power) in order to develop a written version of spoken Ladakhi that students can use in school; they viewed this as heresy, since the only written form of the language is a kind of classical Tibetan used for religious scriptures (and which very few people, including few monks, can read). In public forums, he and his SECMOL colleagues have named corrupt officials and teachers who shirked their responsibilities, sparking a fierce backlash. His crusading against corruption and unresponsive government so angered the deputy commissioner in Leh that the man brought a court case against Wangchuk in 2007, accusing him of land theft and all manner of dastardly crimes, including being a Chinese spy.*

When fellow civil society leaders, fearing reprisals from the deputy commissioner, failed to rally to his defense in the wake of these outlandish accusations, Wangchuk accepted a longstanding invitation to work with an NGO in Nepal, where he spent three years designing and building energy-efficient, earthquake-safe schools in rural areas. It's one thing to be a social critic in a society that values speaking one's mind freely. Ladakh and Zanskar are not such places: given their agrarian-communitarian history, people tend to value averting conflict over social innovation. Gyelong Pandey, a social worker and monk in Leh, says, "After Wangchuk left many people think, 'This is bad for Ladakh,' but they don't come forward." (To be fair to some of his critics, including former colleagues, Wangchuk can indeed be prickly and stubborn, qualities that seem inextricably linked to his deep integrity. He's often the smartest guy in the room and he sometimes lets you know it. His temperament seems perfectly suited to rubbing powerful people the wrong way.)

---

*In 2013, after six years and thirty-six hearings into the absurd allegations, the chief judicial magistrate of Leh cleared Wangchuk of all charges.

Perhaps Wangchuk gets his pugnacious fearless streak from his late father, Sonam Wangyal, who was a powerful politician in Ladakh. Wangyal (who started out "as poor as a beggar," Wangchuk says) advocated for the cancellation of debts and the abolition of the traditional system of forced labor imposed on poor farmers. Late in his career, Wangyal protested his fellow Buddhists' social and economic boycott of Ladakhi Muslims during the 1989 upheaval throughout Jammu and Kashmir, when Kashmiris were agitating for autonomy from India. Other prominent Buddhists and politicians, his former colleagues, imposed the ultimate penalty on him for this temerity: *Chu len me len chaden*. The water and fire connection was cut—a social boycott. Even after the boycott was officially lifted, his political enemies kept it in place. When Wangyal's wife died, he said, "I will carry the body myself!"

So even though the proper comparison for his fellow Ladakhis might be the revolutionary pamphleteer Tom Paine, among all the American founding fathers I tend to think of Wangchuk more as Ladakh's Ben Franklin. "Going about his ordinary affairs, he was more curious than ordinary men and followed up what they only looked at," Carl Van Doren wrote in his biography of Benjamin Franklin. "To warm his house, he devised a new kind of stove. To protect his house, he thought of the lightning rod." And to protect his house and neighborhood, Franklin formed one of the first volunteer fire companies.

The description is equally apt for Wangchuk. Trained as an engineer, by inclination a tinkerer, he is addicted to problem solving. Look at the intelligence and hunger to learn of Ladakhi and Zanskari youth, he says. Why not give them opportunities? And look at all this sunshine and how cold it is up here in the mountains, he says. Why not put those two facts together into an elegant solution like solar heating?

This is Wangchuk's core, driving belief: there's energy and opportunity all around us if we just know where to look and think with imagination about how to harness it. "It's a crime not to do it," he says, in a tone of wonderment.

Wangchuk has long worked to merge his twin passions of education and harnessing the sun's energy. One example is SECMOL's solar-powered and solar-heated campus. Another is his decades-long fight to change the school calendar.

Youth in the Kargil and Leh districts of Jammu and Kashmir go to school through the golden summer and have a two-month vacation during the brutal winter. Years ago, Wangchuk pointed out that this is all backward and kind of crazy. Students should be going to school in the winter, he argued, when there is little work to be done at home or in the fields, and be free in the summers to help their families and learn traditional agricultural and other skills.

Officials in the education department, as inertial as any of Jammu and Kashmir's bureaucracies, replied: Impossible. It's too cold in Ladakh and Kargil and Zanskar for the students to study in the winter. Too expensive to heat the schools. No problem, Wangchuk responded. We can build solar-heated schools, using communal village labor and rammed-earth techniques that are updated for safety but inspired by traditional methods, for a fraction of the cost of building a conventional school building. In the process, he argued, it will save money, provide a warm, comfortable place to learn, improve learning outcomes, and keep young people in touch with their fast-disappearing traditions, all while creating zero pollution. What could be more sensible than that?

Skeptical officials quashed the idea again and again. Some teachers—mostly non-Ladakhis who wanted to keep the vacation schedule the way it was so they could spend winters back at home—resisted the change.

So Wangchuk went ahead and built a school to prove his concept. He and his colleagues and students at SECMOL organized villagers in Durbuk Block and persuaded them to merge their far-flung, poorly staffed schools into one centralized model residential school. Wangchuk designed the school building and the hostels to take advantage of the sun. The villagers contributed labor for months to build its thick rammed earth walls, which can absorb and retain heat for days, like a thermal battery, releasing the heat at night. Natural daylight illuminates classrooms and corridors. Modified Trombe walls (another solar heating strategy) keep the hostel rooms warm at night.

The Lalok School proved to be a resounding success: students and their parents loved the new school and took great pride in its innovative features and, most of all, the fact that they had helped build it. The government never embraced Wangchuk's proposed calendar and curriculum change, but the school still became an example across the region. (On the walls of the guest room in his house in Kumik, Stobdan has taped up four photographs: one of the Dalai Lama; one of the famous Indian social reformer B. R. Ambedkar, who fought the caste system; one of the late prime minister Rajiv Gandhi; and one, taped to the threshold over the door, of the Lalok School.)

The politicians took most of the credit; Wangchuk wasn't even invited to speak at the opening. Powerful men had begun to fear his following with the villagers, the way he related to them, the passion he inspired in Ladakhi youth. They worried he would leverage that grassroots support into a political career like his father's and challenge their hold on power. He shrugs when he recalls it—he's open to running for office but only if it's absolutely necessary to consolidate and advance these hard-won improvements in the education system, but truth be told, he prefers to be on the outside, where he has freedom to operate, freedom to speak his mind, to rabble-rouse, to be a misfit.

The question that he first formulated watching his mother use the goatskin bellows over and over to keep her fire going—"why should they keep doing like this?"—he would keep asking for the next forty years, in different forms.

Why should his fellow Ladakhis accept a government education system characterized by chronically absent teachers, corporal punishment, and a total neglect of their mother tongue? Why should people put up with corrupt officers and their graft and bribes? Why should villagers keep buying or gathering and burning wood and dung to heat their cold, drafty homes when the sun was shining every day, free for anyone to use, with a little ingenuity and modest investment? Likewise, why should the Indian army spend immense time and money to truck vast quantities of gas and kerosene and coke (a fuel made from coal) over the Himalaya to burn in poorly designed tin-roofed barracks for its half-million troops stationed throughout Jammu and Kashmir?

When he poses these questions out loud—whether over tea in his kitchen-cum-sitting-room in his modest house on SECMOL's campus or in a public forum attended by political VIPs and thousands of listeners—he sounds genuinely puzzled. It's a rhetorical strategy—Wangchuk is a shrewd communicator and political operator—that also happens to be sincere. He truly, genuinely can't understand how so many people, if they had all the facts in front of them, could continue to accept such wasteful, inefficient, costly ways of doing things.

Like Tami Bond, Wangchuk is focused on more sensible service provision. He may sometimes have his head in the clouds, but he doesn't deal in the abstract. Lately he's been working on a device he calls the "thirsty crow," an evacuated solar heating tube fixed on a pivot, which would tilt as water inside boils, pour into a thermos, and then refill itself, over and over again. "A household can use it to solar preheat water, and then use fire for the last five percent of cooking." People don't want new devices like solar cookers, he argues, they want hot water and cooked food. So why not focus on how to give it to them, affordably and without hassle?

Wangchuk's various technological and sociopolitical schemes usually have one thing in common: they seek to achieve multiple benefits for human welfare with a single elegant stroke. Many of them have the happy side effect of reducing emissions of black carbon and other combustion-related pollutants—from cooking or heating in high-altitude communities, from transport, from open burning—but that isn't always the explicit objective. The point is the wide range of *savings*—the money you don't have to spend on kerosene, or on medicine or a doctor's visit, or the time you don't have to spend gathering dung—and *benefits*. The objective is always to expand opportunity, increase human welfare, prevent human potential from being squandered, and always in *particular places*.

The concept of "co-benefits" has been around for a while, apparently as long as people have been trying to whack two birds with a single stone. But it's gathered steam in recent years, as researchers and policy makers have become more attuned to the potential for single "interventions" to achieve multiple desired outcomes—and as newspapers' op-ed pages have become rife with dire warnings of purported "tensions" and "trade-offs" between efforts to aggressively address the twin threats of disastrous climate change and the poverty that keeps 2 to 3 billion people without access to clean water or clean energy. The Very

Serious People who pen these admonitions claim that initiatives to alleviate poverty and spur human development in places like India, China, and Africa just can't be pursued simultaneously with an all-hands-on-deck approach to heading off climate chaos. Costly efforts to shift to a low-carbon development model, they say, would fatally impede the slow, steady growth via industrialization that is the only hope for the world's poor.

Focusing on soot gives the lie to this zero-sum way of thinking. Kirk Smith, for example, has helped pioneer the idea of co-benefits in the realm of public health, explaining to a wide range of audiences the overlapping climate, health, and economic gains that could be realized from cleaning up household air pollution. Ramanathan makes the same argument, writing in a *New York Times* op-ed:

> Soot . . . offers an opportunity to marry local interests with the global good. A leading cause of respiratory diseases, soot is responsible for some 1.9 million deaths a year. It also melts ice and snow packs. Thus, sooty emissions from Asia, Europe and North America are helping to thin the Arctic ice. And soot from India, China and a few other countries threatens water supplies fed by the Himalayan-Tibetan glaciers. New air pollution regulations could help reduce soot. Such laws in California have cut diesel-soot emissions in that state by half. In China and India, a program to improve power generation, filter soot from diesel engines, reduce emissions from brick-making kilns and provide more efficient cookstoves could cut the levels of soot in those regions by about two-thirds—and benefit countries downwind as well.

Though he doesn't use the term himself, like these scientists, Sonam Wangchuk is one of the foremost practitioners and proponents of the same co-benefits pursuit. He is quite simply obsessed with eliminating stupid, needless waste. He finds it deeply offensive—almost like a personal affront. "If there is a stone on a path I just can't leave that there," Wangchuk says. "That's not in me. Wherever it is I'll have to move it." He smiles almost apologetically, like he's confessing to some kind of pathological condition. "I want to make things how they should be."

Case in point. One day in Kathmandu he devoted an hour to outlining his strategy for a "green bike brigade" to me. He would start with the staff of NGOs in the city. If young professionals commuted on bikes, choosing pedal power over the aspirational car, people would start to notice and emulate them. Then he'd build a marketing campaign that appealed to Nepalis' patriotism, leveraging their resentment at their behemoth neighbor for the fact that fuel and all other goods imported to Nepal must pass through India, with prices rising along the way: "Don't be dependent on India for fuel! Ride a bike!"*

It doesn't take long battling Kathmandu's maze of smog-choked streets and maddening, multidirectional traffic before you start imagining how sweet it would be if everyone were bicycling around and goods were being delivered by electric vehicles moving at a sane speed (and in their own lane). Of course, most of us snap out of such reveries quickly: *Pie-in-the-sky stuff. The world is the way it is. Current trends will be extrapolated. The laws of human nature and desire cannot be bent.*

Not so Wangchuk. He is the master of the buoyant counterfactual: the burden of proof is on *you*, defender of the status quo, to explain why we should keep walking headfirst into the same brick wall, again and again. Look around at the soot-stained alleys of Kathmandu. *Tell me again why we should keep doing like this.*

After ten years of observing him in action—watching some of his schemes come to fruition, while others are frustrated, prove to be impractical, or are simply left by the wayside—I have come to appreciate

---

*A few years later, in early winter, Wangchuk would pedal his bike 120 kilometers from Kashmir, over the 12,000-foot Zoji-la Pass, and into Ladakh, and post about it on Facebook, to make a similar point to Ladakhi youth and urban residents of Leh: "This was my answer to all those friends who keep telling me, when I suggest bicycles for commuting in Leh city, that it's too cold, too high, too tiring, etc. . . . With my coloured glasses (green) I see a car as a tank, a chemical weapon to be precise, that drops a kilogram of CO_2 and many other deadly poisonous gases roughly every 4 to 5 Km of travel. Apart from contributing to global warming it does a lot of local violence. Kills old people years before their time, gives respiratory diseases to children. . . . And guess who is at the trigger of this deadly weapon? But of course you don't see it as a deadly weapon! You see it as a 'style statement,' you think of its colour and its looks to show the world that you have 'ARRIVED' (on the crime scene! If you ask me)."

the rarity of Wangchuk's gift for imagining alternatives and then doggedly, shrewdly, stubbornly making them sprout to life, just like the apricot trees that now blossom in the patch of scrub desert outside of Leh where he started the SECMOL campus. The model for Rancho/Phunsukh Wangdu is a one-man source of societal nonlinearity, a tipping element unto himself. He has no time for despair, no patience for resignation. Like Ramanathan, he has his sight trained on an opening in the dark clouds obscuring our vision of a better future.

All of his ventures and schemes and innovations can seem radical, if only because they suggest we can be doing things much, much better than we currently are, and at a relatively low cost. But when viewed from a certain light, his efforts seem not even political—just more like plain common sense.

The logic is simple. What people want is the hot water, the cooked meal, the light, the motion—not the combustion products. Whether that end is obtained by burning hydrocarbons or coal or wood or by harnessing the sun doesn't make a big difference to most of us, as long as the service is reliable, convenient, affordable, *clean*. So if you could get the same end result without all the soot and other nasty by-products of combustion for a similar investment, well, why wouldn't you opt for that?

Take bricks, for example. Brick kilns feed the rapid construction growth in cities from Delhi to Chengdu. During his three years in Nepal, Wangchuk introduced the use of compressed stabilized earth bricks to the government's school-building program. Conventional bricks are fired in coal-fired kilns, which are a major source of black carbon; after transport, these kilns are the second biggest contributor to air pollution in the Kathmandu Valley. What's more, they must be hauled long distances on diesel trucks to their point of use. Wangchuk's system uses a mix of sand, clay, soil, and a little cement to produce bricks that have high strength in compression and can be used, with vertical and horizontal reinforcement, for earthquake safety. They also provide thermal mass, to absorb and store solar energy overnight and even out extreme temperature swings inside.

In a pilot program sponsored by the Nepal government, Wangchuk organized villagers in Tharu and Madhesi communities in the Terai of

Nepal to build schools for their own children using these earth bricks in under thirty days. In the wake of those experiences, Wangchuk tells me he wants to experiment with similar community-construction models in villages across the region, to build family houses. His approach would bring back the old ways of community building, with a few modern twists.

Wangchuk spells it out for me one evening. It is beguiling—and so hopelessly romantic, so utopian, and so impossible.

He sees Ladakhi villagers coming together at the end of their workday for two or three hours every evening to build each other's houses. A pilot program, perhaps with support from the government, will purchase the formwork for pouring rammed earth (an ancient building technique in Ladakh that Wangchuk has updated) and the mixing equipment, which will move from job to job. A team of master masons will be trained in the skilled work of setting up the forms and achieving the perfect mix of sand, clay, and cement. They will set up all the materials. Then the villagers will come and ram.

They will use their own soil, their own wood, their own hands, their own labor, to create warm, safe houses of rammed earth. There will be songs, written to guide the cadence of different types of ramming, and dances as they ram the earth into forms together. The dancing and singing will give shape to the work—hard work—and people passing by will have a hard time figuring out whether those people are partying or laboring in a frenzy. And then, every night, the builders will gather and eat together (and drink *chang*, of course), telling each other tales the way people used to. The villager-builders will dance and sing and ram earth until the walls are built. Then a roof. After three weeks, there will be a house. He leans back. "So. What do you think?"

"It will never work."

I am playing devil's advocate, but I am mostly sincere. Even in the space of the mere decade that I have been returning to Ladakh and Zanskar, I have watched the older logic of cooperation that bound villagers together get eclipsed by the ineluctable forces of the modern market economy. When people give up farming, take on paying jobs, whether office or manual labor, and move to the city, those reciprocal arrangements that once made sense, that enabled survival, become obsolete.

Time becomes money. It's a force as irrestistible as gravity, as flowing water. Whether you think it's good or bad or a mix of both, it is what it is. You can't fight it.

He smiles inscrutably as he listens to each of my objections.

"Whether you like it or not, centrifugal forces are pulling the villagers and townspeople of Ladakh and Zanskar away from each other," I continue. People are moving apart, thinking more and more of their nuclear families, less about their neighbors' needs. Like people everywhere, they are growing obsessed with money, with sending their children to private schools, with amassing status. The incentives to cooperate are less and less powerful with each day of global connection. My Ladakhi and Zanskari friends, both young and old, admit as much to me, lament the fact a bit, but also throw up their hands and say it's inevitable—it's the price of progress! Why should someone give up three hours of their busy day when they could be earning money and hiring builders to do what they perceive as drudgery—as something they've fought for generations to escape?

"You're moving against the grain," I conclude, in summation.

As soon as I say this, I am struck by how ridiculous it sounds. The guy has been moving against the grain his whole life.

Wangchuk just nods and smiles his wily fox smile, the same one he wore when I bet him he'd never get the solar rickshaw up and running by dawn's light. It's far from the first time someone has told him he's being impractical, that it simply can't be done.

"I will start in villages in Durbuk, perhaps," he says, unruffled by my litany of obstacles, "where they still build this way. It's still alive there. Everyone comes together to build."

"So that will be your Saspol?" I ask, referring to the first village where he and his fellow SECMOL staff started their education reform campaign over twenty years ago. (After the success of the Village Education Committee that they set up with students' parents in Saspol, 300 more villages requested SECMOL's help, paving the way for their radical idea to become the mainstream government policy.) His eyes light up and he wobbles his head in the Indian way of affirmation.

"Exactly! And then other villages will see, and people will think, 'Why can't we build this way?' "

And from that one village, he says, the practice of solar earthen building will spread. The embers will be carried forth, just like the education reforms. And then, who knows what else might be possible?

It's a lovely picture, isn't it, this Amish-style barn-raising in the High Himalaya, with dancing and *chang* thrown in for good measure. (A nice concession, since Wangchuk doesn't really like *chang*.)

But it really *is* a hopelessly romantic plan—way too late-period John Lennon, too drum-circle to have a chance in this cold hard world of indifferent market forces and time-is-money logic (not to mention the head-spinning amount of corruption in official India that is connected to the flow of building materials and fuel, which any scalable model of this scheme would have to overcome.)

Yet this, too, happens to be one of Wangchuk's gifts. In addition to his mechanical genius, his ability to pick up languages quickly, and his penchant for vivid analogies, he can make the outlandish sound so hypnotically practical. As though it is the only reasonable, sensible way to proceed. It's like some kind of Jedi mind trick. Listen to him long enough, and you will start to notice the staggering amount of waste—of energy, materials, human potential—all around. Then you will quickly pivot, as his mind invariably does, to the scale of the opportunity it implies, and how small tweaks in the way we do things could start to realize them.

You look around and think about how crazy it is that more people aren't riding bikes through Durbar Square in Kathmandu. Or that long-haul truck drivers aren't stopped at the foot of Baralacha-la Pass into Ladakh, as he recommends, and given steaming cups of chai as they wait for government-sponsored mechanics to adjust their fuel injection systems at no charge to account for the thinner air at high altitude, thus reducing the black carbon coughed out almost directly onto the glaciers they trundle past (while saving costly fuel).* Or that Punjabi farm-

---

*Wangchuk has given much thought to the physics of soot formation. As an engineer he knows the technical piece is straightforward; the human element is a bit trickier. Ladakhi weddings are "long and notoriously boring," so when he was younger he and his friends

ers don't sell their useless rice straw to Ladakhi building entrepreneurs and toss it on top of a load of steel rebar that will be heading over the Himalaya to Leh anyway. Or that more than 8,000 liters of diesel are burned each day to power Leh, when the same amount of energy is contained in the sunlight that falls in a day on just 15,000 square meters in the city (equivalent to a few tenths of 1 percent of Leh city's total area). What madness!

When you read about the benefits of mitigating short-lived climate forcers like black carbon and listen to some of their proponents among climate scientists, economists, and public health experts, you start to get the same sensation: Explain to me again why we *wouldn't* do this?

A distinguished team of international scientists recently concluded, in a paper in *Science*, that deploying a suite of just seven technologically proven measures would clean up 75 percent of black carbon emissions, save the world $5.142 trillion by 2030 in health improvements alone, and avoid up to 4.7 million deaths globally every year. This, it's safe to say, would be one of the most cost-effective health interventions anyone has ever proposed—the ultimate co-benefits coup. (Even the researchers I've spoken with who criticize this study as being based on some overly optimistic assumptions about the rate and extent of technology deployment, and black carbon's climate impacts, concede the health benefits alone would be mind-bogglingly high.)

On top of that, the *Science* study authors note, these measures would improve global food security: cleaning up the air over key grain-producing regions increases the amount of sunlight reaching fields, improving photosynthetic efficiency, and avoiding over 50 million tons

---

would amuse themselves by taking soot from the stoves and mixing it with water and painting the faces of those who fell asleep while they were waiting to see and congratulate the couple. At one wedding years ago, he started to give his brothers and friends some technical guidance, as is his wont, on how to achieve the optimal soot mixture. "I was giving a scientific lecture—'You must make the solution the same as the body temperature' and this and that, so that the person won't wake up, it can't be too cold, and so on. And the others were saying, 'Yes, what was that, tell us more,' listening carefully and taking notes. And then I woke up the next day painted all over with soot! And they said, 'You were right, that was good advice, it worked perfectly!' And I became such a joke the whole wedding!"

of crop losses by 2050. Combined with a suite of measures to reduce emissions of methane, a precursor to harmful ground-level ozone and a powerful greenhouse gas in its own right, mitigating black carbon emissions through these measures would prevent an enormous amount of waste and suffering. Oh yeah, and the rate of sea-level rise would be reduced by 18 percent by 2050 and 24 percent by 2100, and the rate of warming in the Arctic would be cut by two-thirds over the next thirty years.

Not too shabby. How, you ask, could this improbably upbeat scenario be achieved? Inventing cold fusion? Hair shirts and raw-food diets for everyone?

Actually, we know how to do all of these things. It's not rocket science. All seven black carbon–reducing measures would entail deploying proven technologies on various fronts to address the four main source categories: requiring the installation of diesel particulate filters for both on-road and off-road vehicles (in compliance with Euro 6/VI emissions standards), as California has done, and eliminating high-emitting, older vehicles; replacing wood heating stoves (mostly in the developed world) with wood pellet stoves and boilers; replacing chunk coal fuel, used for example in China for heating, with coal briquettes; replacing traditional cookstoves with clean-burning biomass stoves in the developing world; upgrading or replacing conventional brick kilns and coke ovens with more efficient vertical-shaft brick kilns and modern recovery ovens; and banning the open burning of agricultural waste. That's most of the problem, right there.

These measures combine the deployment of proven technologies and policies. There is a general consensus that we know how to do most of these things. We just have to decide, collectively, that the up-front investment of political and financial capital is worth it. None of which suggests that it's easy (anyone who's spent time working on the design and dissemination of cleaner-burning cookstoves, for example, will tell you how challenging and multifaceted that particular problem is, as Chapter 8 will demonstrate). But all signs suggest that it will be—overwhelmingly—worth it.

Every major analysis, from the United Nations Environment Programme (UNEP) to the World Bank, shows that the cost of implementing most of these black carbon reduction measures would be substantially less than the total economic value of the benefits to human health, agriculture, and climate. The health savings alone, in terms of productive years of human life saved, constitute the majority of these benefits. But it pays to be skeptical. So let's assume the benefits were half of what the *Science* study's models suggest. These measures to reduce black carbon would still be some of the most cost-effective policy maneuvers available to decision makers from Washington to Beijing to New Delhi, with benefits rippling outward on all fronts of human development: agriculture, health, energy, environmental stability, fiscal savings.

And if the climate benefits don't prove to be as extensive as the study's authors project, will we really regret saving 4.7 million lives each year? Or, assuming that's too rosy a projection, even a half or a third of that, for a relative pittance?

"No matter what people think about global warming, there aren't a lot of fans of dirty snow, poor crops and diseased lungs," as the libertarian columnist, John Tierney, pointed out in the *New York Times*.

On the other hand, if nothing is done, black carbon and its fellow sooty by-products of inefficient fires will continue to kill millions of people each year, and the most potent agent melting the Arctic and the Himalaya will continue running amok. So, tell me again: why wouldn't we do this?

Well, you know, there are the standard political-economy realpolitik objections. Many climate commentators like to point out, correctly, that well-reasoned white papers issued by think tanks or UNEP reports replete with impeccable cost-benefit analyses of low-carbon energy solutions can never substitute for the hard, slow grind of politics. Otherwise we would have licked this problem long ago.

"I am not encouraged by mere reports and events," I. H. Rehman, director of the Social Transformation Division at the Energy and Resources Institute in New Delhi, told me. "The global community has been talking about this, but there has been very little done in terms of any actual headway. We've had reports for a number of years—the problem, to see it in black and white, is it requires action!"

The same is true in the world of public health, where good intentions and even randomized control trial studies of interventions, like hand-washing education campaigns, don't automatically make the clouds part so that political leaders suddenly see the wisdom of taking a chunk of money from their defense budget and putting it into classroom hygiene initiatives or building toilets for everyone. Appeals to enlightened self-interest are a notoriously ineffective answer to powerful entrenched constituencies or to the myopia and inertia that characterizes most institutions' short-term planning.

But without good information, you get nowhere. That's why Kirk Smith has labored for thirty years telling people "you get what you inspect, not what you expect," conducting rigorous measurements of what people are really exposed to, and trying to determine what the health burdens of fire really are—so that public health officials can better understand the opportunities and opportunity costs of investing in efforts to clean up hearths and kitchens. That's why Veerabhadran Ramanathan has been studying aerosols like black carbon in the atmosphere for over thirty years, trying to reduce uncertainty about their effects on the climate.

If you read enough of the skeptics' reactions, though, they tend to blur together into one standard party line, a grand self-congratulatory gesture of circular reasoning: "Things can't change because that's the way things are." This is much like the way energy forecasters predict future trends of consumption: by taking a ruler, laying it against the trend of the past few decades, and drawing a straight line with the same slope. This is status quo bias: the assumption not only that the future will resemble the past but that it *should* resemble the past.

In response to claims about the cost savings of, for example, energy efficiency measures in buildings, economists like to ask: If there's this free twenty-dollar bill laying on the sidewalk, why has no one picked it up? If we could do all these things, with such huge benefits, for such a modest investment, at a *negative* cost—well, why hasn't the market done them already? Surely because it must be harder, less feasible, more expensive than you think?

Fair questions. But these people miss two important points that Sonam Wangchuk understands in his bones. One, change is an unpredictable, nonlinear process. Social systems can shift to new equilibrium states as quickly and dramatically as natural systems like the monsoon.

Two, much like the ramming traditions in the villages of Durbuk, many of these things are already happening. Every day, all over the world, people are experimenting with and perfecting solutions to our daily global hemorrhage of black carbon, and to the scourge of respiratory disease it contributes to.

The answers are all around us. They don't quite require moon shot initiatives or Marshall Plan levels of investment. They do require some tinkering, some careful bets, investment and smart program design, and patient monitoring and follow-through. We just have to decide whether we want to invest to refine them, scale them, and, most importantly, *hot-rod* them. On both the climate and health fronts, time is not on our side.

But as Tami Bond points out, while the march of development may clear the skies in the long run, "this transition can be accelerated." We just have to decide to push the pedal down.

One day Wangchuk loaned me his bike, the one with the I HEART MY BIKE sticker on the little silver bell, so I could sample a range of black carbon solutions already being implemented in one of the world's poorest countries.

As the view from Pradeep Mool's office window suggested, Kathmandu is one of Asia's most polluted cities. Like Los Angeles and Mexico City, the city is surrounded by hills that prevent the wind from just sweeping pollution away, making it prone to particulate-trapping temperature inversions. Much of the pollution comes from vehicles: diesel trucks, passenger vehicles, and "tempos"—small three-wheeled vans that function like buses or taxis—all clog the maze of streets. The valley is also a hotbed of brick production. Dozens of brick kilns, obvious from their dark smokestacks jutting up into the sky, dot the outlying areas. In the villages clinging to the rim of the hills around the city, people mostly burn wood for cooking. In the spring and fall, they burn agricultural waste in the fields. And to top it off, in the past few years smoke from wildfires have reached the city. Every flavor of soot can be found here. It's the world burning, in microcosm.

Kathmandu is also home to a wide spectrum of simple but ingenious technologies—electric vehicles, more efficient brick kilns, biogas systems for cooking—that produce dramatically less black carbon while providing the same, or better, service as the methods involving dirty fires. First I biked west of the city to the offices of the Biogas Sector Partnership. The ride was educational. It took only fifteen minutes before I was clipped on the arm by a bus. But the air, given what I now knew about black carbon's empire of smoke, seemed more dangerous than the threat of a collision. When I arrived, I wiped sweat off my brow with a handkerchief before heading inside. It came away black with soot.

The assistant director, Hariwar Durkheshi, explained to me how the Biogas Sector Partnership, an unusual public-private partnership born of a collaboration between the Nepal, Dutch, and German governments, has installed 260,000 clean-burning biogas systems for cooking and lighting in households in all seventy-five districts of Nepal. The BSP subsidizes the cost of the system and helps set up supply chains and maintenance schedules. Maintenance costs are 300 rupees per year, and they have a twenty-year lifespan if built properly. The cow dung that gets added to the tank every day is free, and so is the (mostly methane) gas produced by anaerobic digestion. The biogas reduces a family's expense on wood, kerosene, and LPG; on average those savings exceed the cost of installing the system within three years.

And there's another measure of the program's success. "In Chitwan District," Durkheshi told me, "if you want to marry, the girls have one question: 'Do you have biogas or not?' "

The next day I biked over to the offices of the Nepal branch of Practical Action and heard about the awareness-raising campaign and improved cookstove installation program the NGO has implemented in rural communities from the Terai to the Himalayan district of Rasuwa (where local activists have made it their mission, as they told me when I visited the region later, to "eliminate indoor air pollution in five years"). Later, I took a taxi to reach the humble garage of Shri Eco Visionary, a six-year-old company that manufactures electric vehicles and converts old Mitsubishi tempos to electric, on site, in the middle of a busy neighborhood in west-central Kathmandu. The company maintains four of the twenty-eight electric vehicle charging stations around the city and owns and leases 150 electric tempos. These are part of a fleet of 650 electric tempos, ferrying 150,000 people around the city who would otherwise require diesel-powered buses to get them to work, to the market, to school. "It's the highest concentration in the world probably of electric public vehicles, with no government subsidy, entirely privately operated," said Bushcon Tukedhar, a schoolteacher and air-quality researcher who was leading a visiting group of students from one of Kathmandu's private schools.

Later, I visited Brick Clean, a vertical-shaft brick kiln west of the city that uses half the amount of coal consumed by conventional kilns and reduces black carbon emissions by 80 percent. The conventional kilns are crude and extremely inefficient and rely on seasonal migrant workers to operate them in dangerous conditions, with inadequate shelter and safety protections. The workers, many of them children, are exposed to the worst of the pollution.

Vertical-stack brick kilns, if operated properly, can dramatically improve on the performance of conventional kilns: VSBKs emit 90 percent less particulate matter, consume 30 to 40 percent less energy—in the form of coal imported from India—and cut carbon dioxide emissions per brick by half. The Brick Clean staff told me that there are twenty-six operating in Nepal, three of them in Kathmandu Valley. One produces 10 million bricks a year; the owner spent three times what he would have to to build a conventional kiln, but the fuel savings over its lifetime will pay him back in a few of years, and subsidies help cover up-front capital costs. Still, the older, heavily polluting kilns vastly outnumber the new improved kilns.

And when I got back home one night, I found Wangchuk ready to light some charcoal briquettes made from waste wood by a small startup company in Kathmandu. We lit them on the gas stove, and they glowed with a steady blue flame shooting from inside their perforations—a sign of good air mixing. They burned for twenty minutes, producing no smoke. We cooked bitter gourds over them. (Wangchuk loves bitter gourds.) I had bought him a copy of *Ratatouille*, the Pixar animated movie, a couple days before. We watched it on his laptop, and he loved it so much he watched it again the next day. As he stirred the vegetables over the smokeless heat of the briquette, he affected a French accent and gleefully repeated the tagline from the movie: "Anyone can cook!"

The solutions to cleaning up the soot coughed out by diesel engines, cookstoves, small industry, and open burning of agricultural waste have arrived. They aren't waiting to be invented by the Ranchos of the future. Anyone with pluck and vision can implement them.

It hasn't always been true, but now it is: anyone can cook, and without soot. But to hot-rod them—scale them up rapidly across the globe—will require sustained institutional focus, grassroots energy, political will, and, of course, money.

On each of these visits around Kathmandu, when I asked what was holding their venture back from scaling up and spreading throughout Nepal, they all had the same answer: "Finance." The guys at Practical Action even asked *me* if I knew of any funding sources. "International climate adaptation and mitigation funds are not contributing much to household energy," Min Malla, a staff member, told me glumly. They had vast experience, promising models, and an extensive grassroots network . . . but there was no money out there for what they were doing.

That summer, I started encountering Phunsukh Wangdu in unlikely places.

When I got a shave and a haircut in Old Leh, I noticed in the mirror that my extremely competent (and competitively priced, at 30 rupees, or 50 cents) barber wore a T-shirt with an image of Aamir Khan and his two fellow "idiots" in the movie. It read: IDIOTS REFUSE TO CHANGE THEIR MINDS.

At the Sani Nasjal festival in Zanskar, I looked on as monks dressed as the eight manifestations of Guru Rinpoche sat and watched as the protector deities—other monks disguised in masks and raiment—rose and swirled in concert. The *tenten talu*, child monks dressed in skeleton outfits, scampered through the crowd and grabbed people's hats, tossing them about on a mission of pointed mischief, reminders of mortality and puncturers of pretension. A young boy sat watching their hijinks intently, wearing a T-shirt that read: I AM ONE OF THE 3 IDIOTS.

A few days later, on the building site for the solar *lhakhang* in Marthang (itself inspired by Sonam Wangchuk's work), two of Kumik's young women mixed mud for brick making in their bare feet, laughing as they splashed water on the rest of us. One of them half sang the chorus of a song from *3 Idiots* (and the catchphrase of Rancho): "All izz welll."

And a couple of months later, I was visiting my friend Sandeep in the rapidly growing satellite city of Gurgaon, south of Delhi, home to dozens of multinational firms and gleaming new shopping malls, all powered by private diesel generators. Sandeep described the new training and team-building program his company designed for a large engineering firm's fresh batch of 150 recruits. He explained with entrepreneurial zeal how the three-month-long program, unlike most such orientations, cultivated creativity instead of stifling it, and how it would culminate in a three-day-long treasure hunt called Wangdu Ko Pakaro—"Catch Phunsukh Wangdu!"—starting in the hill town of Mukteshwar. And how the whole thing ends with the Baba Rancho Dev Innovator Program: the new recruits would have to design and build a rainwater harvesting system in the space of forty-eight hours, in partnership with a nonprofit group working with water-stressed villagers in the foothills of the Kumaon Himalaya.

"Each one is a crazy guy!" Sandeep said, describing his desired outcome of helping these young corporate recruits break out of their mental cages. He was getting fired up. "We want to create one hundred and fifty Phunsukh Wangdus. Innovators!"

In Sitapur District of Uttar Pradesh—in the rural, semi-medieval heart of north India that Aravind Adiga famously described as "The Darkness" in his best-selling novel *The White Tiger*—there are some crazy guys going around from village to village. "*Bati-walli age*," the children shout when they see them: "The lamp men are coming!"

At first the villagers in Kaharan Purwa thought they were tourists, though foreigners had never come to their village. But these men said they wanted to sell them light, at a fair price. No one had ever offered them this before. Their representative in the state assembly had been reelected five times, promising to bring electricity again and again. Nothing ever happened. But then the lamp men came in December 2010 with some strange equipment and this offer: for 25 rupees per week, they would provide solar-powered lights and a mobile phone charger to any household that wanted it.

The villagers were distrustful of the outsiders at first. But after some of the people of Kaharan Purwa agreed to their novel proposal, on the first day they connected thirteen houses.

Once the lights came on—for the first time ever in this rural patch of the Gangetic Plain—everyone else wanted some too, and soon all thirty-four homes were flipping light switches and plugging in their mobile phones (which are a source of information, entertainment, and even crop prices across rural India).

One of those crazy guys was an American named Nikhil Jaisinghani. With his friend Brian Shaad, he started Mera Gao Power in 2010 (*mera gao* means "my village" in Hindi). Their objective was to start a business that would bring solar-powered electricity—and with it, basic services like light—to impoverished parts of India where the slow-moving, dysfunctional government agencies might never reach. Kaharan Purwa was their test case, their pilot village. Once that first solar-powered micro-grid was up and running, they knew they were on to something big.

"People came from other villages—fifty people together, they come," said K. K. Rai, one of MGP's first employees. "They came with connection list, 4,000 rupees. 'Please come to bring the light,' they say."

"It was a huge crowd," Sandeep Pandey, the operations manager, told me. "The very next morning, clients were coming with the 40 rupees [connection fee]. That convinced me to keep going."

The lamp-men didn't want to impose on their customers, so they slept and worked in a grain storeroom that was partly open to the elements. For two months, Brian, Nikhil, Sandeep, and K. K. stayed in that storeroom, sleeping on charpoys, walking for kilometers to even more villages—there are no roads to many of them—explaining what they wanted to do and signing up more and more households like some kind of LED-bearing pied pipers, returning to Kaharan Purwa in the evenings. "Our struggle is successful," K. K. recalled as we walked around

Kaharan Purwa a year and a half later. "We struggle too much—mosquitoes, jackals!"

"Jackals?"

K. K.—who is a slightly more animated Indian version of "the Dude" from *The Big Lebowski,* with his paunch and tank top and flip-flops—walked over to a corner of the storeroom and picked up a bow and arrow from the corner.

"For hunting?" I ask.

"Not for hunting! For defending from animals. The jackals used to come."

The jackals came into the village at night, occasionally making trips out to the fields—the local toilet—hair-raising. Mera Gao's customers are mostly farmers, tailors, small shopkeepers, and day laborers. They can't afford to spend much on electricity and light, let alone lodging crazy *batti-wallah*. So MGP has a policy of not accepting anything from the villagers—no food, or tea, or shelter—because they don't want to tax their limited resources or create a culture of entitlement among their employees. MGP wants to build an ethos of first-class customer service. If a connection comes loose, lights stop working, or a mongoose chews on a wire, the MGP repairmen are dispatched immediately, the same way the power utility in my own country sends out its crews when an ice storm brings down the power lines. MGP's attitude: they may be some of the poorest people in India, in the world even, but why shouldn't they be given the best customer service possible?

This is MGP's true innovation: "Whenever there is a problem, our team will go, and we have committed that we will solve the issue within twenty-four hours. If customer is calling me around ten, my staff will be there around eleven," Pandey says. By comparison, some other villages in this part of Uttar Pradesh that were once connected to the grid by the government have been waiting for years for problems to be fixed.

MGP has produced impressive results, but they got off to a rough start.* They ran out of money after a few months, but a timely grant

*Brian and Nikhil, who don't speak Hindi, were hauling solar panels and batteries around the city of Kanpur, getting nowhere in their attempts to find partners and a place to pilot

from USAID kept them afloat and bought them more time to prove their concept. When I first spoke with Jaisinghani in July 2012, after a year and a half of operation, MGP had installed solar micro-grids in 40 villages, bringing lighting and mobile phone charging services into 1,000 households. By the end of 2013, MGP had brought light to 75,000 people in 15,000 households.

A typical household used to burn four or five liters of kerosene each month for lighting. "Now they are not purchasing [kerosene] at all," says Sandeep Pandey. "They are not taking subsidy, also."

"You can provide lighting super affordably," Nikhil says. He and his partners say their advantage is their low-cost model. "The system is quite simple," Sandeep says. "Nothing complicated." Each distribution line serves ten to fifteen households; because north Indian villages tend to be more clustered, MGP can limit the length of their lines to 100 meters, avoiding transmission losses. A typical MGP system is scaled for thirty to fifty households and costs just over $1,000, covering the solar panels, two sealed lead-acid batteries, wiring, a cabinet for the batteries, a charge controller, and LED lights and mobile phone chargers. MGP can connect a hundred-household village for just $2,500.

All a village needs to be connected is a good, centrally located rooftop (one household hosts the system for the whole village), demand for light, and the capacity to pay. "We go into a village and have a one-hour discussion," Nikhil says. At the end of the meeting, the logistics are settled. "We will charge a 40-rupee connection fee, and 25 percent of them pay the fee at the end of the meeting, 50 percent when we start collection." In a typical village, 80 to 90 percent of the households want to be connected. By the end of the week, the village has light—it takes a team of three installers less than a day to hook up an entire village.

Villagers typically pay 60 to 80 rupees per month to charge their

---

their idea, and about to give up on their whole crazy scheme. They sent a short appeal heavenward: "Just let us find someone with a vehicle who speaks English." It would be a sign for them to keep trying. Right then and there, K. K. pulled up in a small battered white van, asked where they were going in fluent English, and presented them with a card that read "K. K. Rai, Problem Solver."

phones on lead-acid batteries in the market; the batteries are themselves charged by diesel generators. And for light, they use kerosene lamps. MGP's micro-grid replaces both, at a cost of 25 rupees per household per week for two lights and a phone charger (or they can pay 45 rupees for four lights).

The system may be simple, and straightforward to install, but the results can be life-changing. When I asked people in Kaharan Purwa and other villages about the benefits of MGP's systems, they offered a long list. Before MGP arrived, most households used kerosene lamps at night for light. These quickly burn through expensive fuel and produce a sooty smoke; telltale dark streaks rose up the mud walls from the niches where the lamps used to sit. The shopkeeper in Kaharan Purwa told me, "One of my customers, his whole house burned down, because of kerosene lamp. In many places, children are burned by them."

Ramesh, a tenth-class student, told me, "Before when I tried to read, smoke from the lamp would get in my eyes." Now he studies in the smoke-free glow of the LEDs. "First, we save money. Second, we save our houses and children—safety. Third, we can study. In wet season, flame is moving in lamp, but with light no problem."

In the village of Bhitauli, Betti Devi, a young mother, said she no longer uses five liters of kerosene every month, an important savings on her family's meager farming income. But even more important is that "the children can study."

"I am happy," she said with a bashful laugh. "It's easier to prepare food. We are saved from smoke." Her neighbor, village headman Krishna Kumar, said, "Before with kerosene burning, the house was always black. Now there is less smoke and soot. We like that there is less smoke now. And seven hours of light each day."

And then there are the snakes. Another woman described how one night there was a cobra hanging down from the ceiling of her kitchen. Her daughter switched on the LED lights, and her husband saw it just in time. "If there is no light, we are searching for matchbox, searching for the kerosene lamp, and in this time the snake is cutting us! Very long snake!" Her neighbor, Awadh Ram, concurred: "Yes, now I can see snakes, scorpions, jackals!"

Betti Devi pointed out yet another benefit of MGP's lights: it deters thieves, who come in the night to steal the grain they store in their homes—these farmers' only source of wealth. One seventy-year-old woman told Sandeep that for forty years she was unable to see her food every night. "With your light now I'm able see what I'm eating."

"People come and sit and talk because of the light," a shopkeeper in Bhitauli observed. "The tailor is sitting up late, working. The doctor is looking at patients at night." The light has brought people together.

Our enemy is kerosene," Sandeep Pandey told me, with feeling. Some customers want to pay more, so they can run a small TV or a fan, but so far MGP is limiting usage to light and phone charging. "We are here to remove the kerosene oil, the flames, and the health issues. That's our core goal." There are no safety issues with MGP's system, whereas kerosene lamps pose a serious fire hazard and the risk of burning and poisoning for children. "And because they are inhaling all the flames which are coming from the kerosene oil, and lots of issues with their health coming from this. They put a very high value on this."

It is better to light a candle than to curse the darkness—unless that candle is a wick in a kerosene lamp. A recent study led by Nicholas Lam, of Kirk Smith's lab at UC Berkeley, found that kerosene emissions are 99 percent pure black carbon. Kerosene soot is about as dark as it gets. It contains almost zero emissions of "cooling" aerosols like organic carbon or sulfates. So cleaning it up, from a climate perspective, is more bang for the buck because emissions from burning kerosene are entirely warming. Not to mention, kerosene lamps are responsible for countless burns and poisoning cases.*

*New research suggests that kerosene smoke might be even nastier than biomass smoke. "We have a study published from Nepal, Pokhara, where we found that people using kerosene lamps had nine times more tuberculosis," Kirk Smith told me. "Now that's a remarkable finding. One study doesn't prove anything, so we are redoing it in a much larger, more sophisticated way. We just had a paper accepted that showed an effect of biomass smoke too—it doesn't take biomass smoke off the hook. But kerosene smoke is a bad actor for stillbirth, child pneumonia, low birth weight, and so we have several other papers in various stages of review, one just accepted, that are really starting to show that kerosene smoke is bad for you. Per unit of pollution it is worse than biomass smoke." Smith doesn't know whether those effects are caused by the high fraction of ultrafine black car-

"It's only 3 percent or so of global BC emissions," Lam says, "but when you combine with the many co-benefits, the immediate benefits to human welfare, the low uncertainty in terms of its climate impacts—it's difficult to get that with anything [else]." In other words, Lam's research suggests that replacing kerosene lighting with clean alternatives might be the closest thing to a slam dunk as you will find in the realms of international development and public health. The overlapping climate, health, and economic and social benefits make an overwhelming case for replacing soot-producing kerosene lamps with alternative lighting sources. About 40 percent of India's rural households use kerosene lamps for lighting. "They're used in the poorest of the poorest populations in India, South Asia, and sub-Saharan Africa," says Lam.

During 2010–11, India spent over $4 billion on kerosene subsidies for householders, essentially promoting use of a fuel and technology that is slowly killing people. "You can make a really strong argument" for ending the subsidies, says Lam. "Indian subsidies are meant to serve the poor, right? They have to burn something and it shouldn't be wood and it shouldn't be kerosene. And they have to be able to afford it."

Like those in Kumik, villagers in Sitapur are wary of the smoke. They have asked Sandeep if MGP will sell cooking devices to save them fuel and cut down on all the smoke in their homes. "LPG is very expensive, so it's not very easy for the villagers to purchase," Sandeep says. "And dry wood is very expensive. And all the wood is wet right now [during the monsoon]. But we told them we don't have the capacity to do that, we are here just to light your house."

In winter, the villagers burn wood for warmth too, in the kitchen, in the fields, by the roadside. They used to invite MGP staff to come warm themselves by the fire, but the men declined and walked back and forth instead to fight the cold. "Oh my God, it's so smoky," Sandeep says, shaking his head. "Very dangerous."

---

bon in kerosene emissions or by other particles and gases in the smoke. "But again, to me, as a public health person, we know enough already. Just get rid of this stuff! Get to clean combustion."

• • •

When I visited the MGP office in the town of Reusa in the fall of 2012, I found Sandeep Pandey conducting a weekly debriefing-cum-pep-rally with his employees. He inspected his troops like a colonel. As the guy running everything out in the field, overseeing new construction, payment collection, and customer relations, Sandeep was feeling the pressure to maintain and accelerate MGP's rapid growth rate. It was hard for them to keep up with what he characterized as "huge demand." Villagers come to them with cash in their hands, demanding light. They could hardly keep solar panels and batteries in stock fast enough. On a whiteboard on the wall, the team had written down the names of dozens of villages they hoped to reach and connect in the next month.

At that time, at the end of September, MGP had 1,500 customers. "We want twenty-seven hundred by the end of October," Sandeep said, brimming with confidence. "In this Sitapur District only, my target is that one branch can serve up to eight thousand households. If we had twelve hubs we will touch ninety thousand." He shrugged. "It's not a big deal."

Sandeep was deploying three teams of installers to go out and electrify one entire village each in a day, if the weather was good. But the weather hadn't been good—a source of anxiety as they stared at the whiteboard and discussed the ambitious expansion timetable. With water thigh-deep in some villages, the installers couldn't even get to them. "The rains are one and a half months late this year," K. K. said. "Monsoon is coming late every year, the people say. It causes problems for animals, affects the rice planting. They first planted, no water, so they had to plant again, lost their seeds."

The office had an atmosphere that was part forward operating base, part tech startup. Sandeep slept on the floor of the office at night. His desk was a broken packing crate, and he sat in a cracked plastic chair. There were no fans to fight the suffocating summer heat, and the office only got seven hours of light, just like the villages, in a kind of solidarity. (The villagers would happily pay for ten hours of light, Sandeep explained, but that would require bigger and more complex solar systems, and their model is partly predicated on staying simple, fast, and light, so they can spread quickly.)

Nikhil later told me they have plans to expand their army of lamp-men to Nigeria, Indonesia, Bangladesh, Philippines, as well as other parts of India's Darkness, still 400 million strong. He was confident their micro-grids would spread just as fast in these other countries, where there are much higher kerosene costs (thanks to a lack of subsidies) and the economics of their model would be even more compelling. But one thing holds Mera Gao Power back.

"We thought the limitation would be the number of facilities, people hired—but the big limitation is finance. Our model says we need $40 million in debt. We burn through equity fast."

"Why start in rural Uttar Pradesh?" I asked Nikhil.

"Because of the population and magnitude of the problem. Frankly, if we can do this in Uttar Pradesh, which is a basket case, we can do it anywhere. Plus, there's low risk of the grid reaching these areas." He had a point. Uttar Pradesh is perhaps India's most recalcitrantly poor, corrupt, and "backward" (as official India likes to say) state. Many of the villages in Sitapur, where MGP launched, have never seen electric light and aren't likely to get connected to the grid anytime soon. He and his partners saw an opportunity to build a business that would fill this gap, while achieving profitability and scale.

"There are 400 million off-grid Indians," Nikhil said. "It's a two-billion-dollar market."

With more investment, he explained, they could grow quickly. Hire more people, set up other regional offices, and negotiate better prices with suppliers as they order more LEDs, solar panels, charge controllers, and batteries and scale their business up. "My personal dream is to reach each and every house," Sandeep told me one evening. "It's a basic need, and they have a right to electricity. Government can't do [it]. And if they support the private sector, then none will be left without electricity."

MGP's experience illustrates that India's poor are ready and eager to pay for modern services and cleaner fuels, if only they had *access*. And entrepreneurs—crazy, Phunsukh Wangdu–type guys, fending off jackals in the night with bows and arrows and hauling solar panels through malarial jungles—are ready to tackle these problems with innovative approaches to seeming intractable problems, if they can only find the technical and financial support to scale models proven to work on the ground.

Rarely is this type of clarity achieved in the realm of development or poverty alleviation—especially when you mix in climate change considerations. In this case, you get it all. Mitigation: reduction of black carbon emissions near the Himalayan "abode of snows," where it does the most damage. Health: eliminating the risk of child burns and poisonings, as well as reducing exposure to dangerous ultrafine particles in the home. Adaptation: decentralized energy systems make communities more resilient. (When the grid fails, as it did for over 600 million Indians in July 2012 for several days, the lights stay on, kids can keep studying, and you can still see cobras coiled up in the corner of your kitchen.) Economy: the government saves billions of dollars per year on avoided kerosene and diesel subsidies, with knock-on benefits of reducing its trade and foreign currency imbalance, and fiscal deficits. (India imports the majority of its oil and coal from abroad.) Likewise, the government wouldn't have to spend as much on grid infrastructure to reach the "last-mile" villages, avoiding transmission losses and theft of power.

In fact, it would all seem too good to be true if so many hadn't already paid such a steep price for millennia, for both the light and the heat of the flame. And listening to Sandeep Pandey, watching his men in action, it was not hard to picture Mera Gao's solar empire spreading outward from Reusa, just as he had scrawled it out on that whiteboard, beyond Sitapur to other districts of Uttar Pradesh and Bihar that have been burning torches or oil lamps of some kind every night, since the time of the Rig-Veda.

"Nothing is impossible," Sandeep told me at the end of another long day of bringing light to the Darkness. His open, steady gaze and the firmness of his voice all suggested the zeal of the missionary. "I don't know who has said, but nothing is impossible."

Later, after Wangchuk told me about his crazy utopian dancing-ramming vision, it occurred to me that I'd already seen it in action. In Lhasa I once saw teams of young men and women ramming earthen roofs on a monastery, as they danced rhythmically, and in rural Bhutan I have watched villagers passing bowls full of damp earth up to each other as they chanted full-throated traditional songs.

I've also seen a variant of it in Kumik. In Marthang, to be specific.

One day on the site of the solar *lhakhang,* early in the construction process, things were at a standstill. There was no diesel tractor to deliver soil for making mud bricks. We had no water in any case because the new canal had again been blocked by a soil collapse.

So we looked around to see what we could accomplish without fire and without melted ice. Rocks littered the plain all around us. We set to work on filling up the stone perimeter of the foundation. We formed a line, bucket brigade style, and passed stones from hand to hand—men and women, teenage girls and grandparents, from Kumik's largest households and its smallest and poorest—all the way to the end of the line, where Gara heaved each rock into the center of the building. For hours in the afternoon sun, we kept passing rocks down the line, until we had a decent-sized pile to fill the foundation on which the solar *lhakhang* would be built.

There was some dancing too, unchoreographed. Ache Garskit twirled and sashayed in her dust-coated *salwar kameez*, putting on a show for the rest of us. Meme Yunten may have told a bawdy joke or two. They teased me, the lazy *chigyalpa*, when I took off my work gloves to take photos. And gently teased me even while I worked, as they did every day on the construction site: "When will you find a wife? Why don't you bring your own cup for tea? Why don't you bring *yato* with you, Jon, where is your *yato*?"

"You are my *yato*!" I shouted back as I passed the next rock to Gara.

They laughed and shouted in approval. "*Song, song . . .*"

In fact, we shouted and sang and laughed all afternoon in the "red place," under the heavy sun. And the foundation rose, stone by stone.

8

Now We're Cooking with Gas!

We leave it to the Political Arithmetician *to compute, how much Money will be sav'd to a Country, by its spending two thirds less of Fuel; how much Labour sav'd in Cutting and Carriage of it; how much more Land may be clear'd for Cultivation; how great the Profit by the additional Quantity of Work done, in those Trades particularly that do not exercise the Body so much, but that the Workfolks are oblig'd to run frequently to the Fire to warm themselves: And to Physicians to say, how much healthier thick-built Towns and Cities will be, now half suffocated with sulphury Smoke, when so much less of that Smoke shall be made, and the Air breath'd by the Inhabitants be consequently so much purer.*

—BENJAMIN FRANKLIN ON IMPROVED STOVES, *AN ACCOUNT OF THE NEW INVENTED PENNSYLVANIAN FIRE-PLACES*, PHILADELPHIA, 1744

People have cooked over open fires and dirty stoves for all of human history, but the simple fact is they are slowly killing millions of people and polluting the environment. . . . But today, because of technological breakthroughs, new carbon financing tools, and growing private sector engagement, we can finally envision a future in which open fires and dirty stoves are replaced by clean, efficient, and affordable stoves and fuels all over the world.

—SECRETARY OF STATE HILLARY CLINTON ON IMPROVED STOVES, SPEECH ANNOUNCING THE LAUNCH OF THE GLOBAL ALLIANCE FOR CLEAN COOKSTOVES, NEW YORK, 2010

Zangmo has asked me to join her on a walk up to the *doksa*, the cluster of stone shepherds' huts in the high pasture, so I rise before dawn and splash water on my face.

A crescent moon hangs in the southeastern sky over Padum Kangri and the other peaks of the High Himalaya. Kumik is stirring to life, but as usual, Zangmo has already been up for a while, baking bread, milking the cows. She is waiting for me by the front door, cheap plastic sandals on her feet and empty *tsepo* basket on her back, as I lace up my hiking boots. Once we're off, she charges up the hillside in the half-light, vaulting over low stone walls and picking her way unerringly through neighbors' barley fields. After an hour we come into sight of a stone reservoir, the headworks of the village canal. The bottom is bone dry, hoof prints dotting its cracked mud.

We crest the edge of the moraine where the simple stone huts of the *doksa* lie. Round dung cakes are drying on the ground, placed inside circles of small stones. Eight women are already out milking, shouting, and joking loudly, smacking uncooperative cows on the rump. The rough courtyard reeks of cow urine and the shit that is everywhere. Zangmo wastes no time, tosses her basket aside, pulls on some gloves, and starts piling the fresh, liquidy dung in piles around the yard.

She has come for this fuel. A teenage girl helps her load the wet dung into her *tsepo*, and she dumps it in piles fifty yards away, on the drying ground beyond the huts. They squat and fashion the mounds into small discs and spread them out to dry.

A cold wind picks up. After twenty minutes, we retreat to one of the huts. The ceiling is four feet high. The stone shelves are lined with small bags of barley flour and jerry cans of milk and boxes of matches. Zangmo kindles a fire with some dried dung and twigs to make us tea and hot, fresh milk. She blows repeatedly on the incipient flames. There is a six-inch-wide hole in the willow stick, mud, and stone ceiling above us, admitting a thin shaft of light. We both squint and cough through the thick smoke that soon fills the hut.

"Do you ever get ill from all the smoke?"

"Yes, mostly in the winter. And my eyes hurt." She says this in the same even way she might observe that by August water will be low in the storage pond, or that June brings wildflowers like *palimentok* on the hillsides, or any other seasonal fact of life.

She rummages through the contents of a stone shelf that cantilevers out from the rough wall and finds me a cleanish cup for tea, wipes it with the hem of her *salwar kameez* and fills it. Another woman pokes her head into the hut, and she and Zangmo chat for a while, as they both ply me with fresh curd and tea. The woman outside milks a cow as she talks over her shoulder. By 6:30 a.m. their morning work—milking, gathering dung, priming fresh batches of yogurt—is done. The cattle start heading up the valley, climbing the mountainsides to where the knee-high grasses are. Some women go up with them, while some head down the trail to Kumik, where other tasks await them.

Zangmo has more dung cakes to make. But before she exits the smoke-filled *doksa* she turns and fixes me with a shrewd look. She wants to know just one thing, something it hadn't yet occurred to her to ask in the years we'd known each other.

"Do you cook like this in *chigyal*?"

"Not really, *Ache*. We mostly cook with gas."

She nods and fills my cup again. Though over the years I've hauled my fair share of alfalfa bundles and stones and poplar beams and dung, Ache Zangmo still thinks *chigyalpa* like me are too soft for some of the rough-and-ready jobs of a Zanskari's daily round. She might be right.

"Stay here and drink tea," she enjoins me. "I must go and get more dung."

My answer wasn't strictly accurate. I could have told Zangmo that I once rented a house that had a giant woodstove made in the 1880s, a relic from an old Vermont farmhouse. It was a monstrous thing that belonged in the Smithsonian, with five burners for pots, a bread oven, and even an adjustable surface for ironing clothes. When I lost power for three days during a heavy snowstorm while hosting three guests, I cooked each meal on it with maple and ash and birch wood, harvested from the hill across the valley.

But 99 percent of the time I cook with gas or electricity. Just like 99 percent of Americans and 60 percent of the world's population.

Look around your kitchen. Check your cupboards or kitchen counter. If you live in China, you probably have an electric rice cooker. If you live in Indonesia, you probably have gas for cooking. If you live in the United States, you'll probably find a toaster oven, a gas or electric stovetop and oven, a microwave, a coffee maker, maybe a waffle iron or a Crock-Pot. All of these use gas or electricity; virtually nobody in

Europe or the United States cooks with any solid fuel, except when they are grilling outdoors.

Meanwhile, 40 percent of the world cooks pretty much the same way Zangmo and other women in Kumik cook today in the *doksa* and in their winter kitchens—which is the same way people have cooked since before the invention of the alphabet. They burn wood, dung, coal, or agricultural waste in open hearths or simple stoves.

More than half of those who depend on biomass for cooking live in India and China. In India, over two-thirds of all households burn cow dung and wood in their *chulha*s. In rural China, farmers heat pig feed on wood- and coal-fired stoves, and many burn coal for heat, too. In many countries of sub-Saharan Africa, 80 to 90 percent of the population burns biomass. In Burma, 95 percent cook with solid fuels; even in Burmese cities, 88 percent of people burn wood and charcoal for cooking, just 3 percent use electricity, and less than 1 percent use gas.

So why do these 3 billion people still live in the smoking section?

"It's incredibly easy to say how awful this is in a sentence and a half," says Jacob Moss, the director of cookstove initiatives at the U.S. State Department. "It's incredibly hard to solve it."

To recap the awful part: the Global Burden of Disease study released in 2012 estimated that 3.5 million people die prematurely from respiratory illness, cardiac disease, cancer, and other illnesses due to exposure to indoor air pollution, while another half a million die every year from exposure to smoke that escapes kitchens into the outside ambient air. In China, those particles kill 1.2 million people every year; in India, at least 1 million prematurely.

To solve it would mean, more or less, to help those 3 billion people get access to clean cooking—to help them cook like the other 60 percent. The GBD study used a counterfactual of 7 µg/m3 to estimate the number of premature deaths from cardiovascular disease, lung cancer, and other illnesses due to exposure to household air pollution. That's roughly the mean PM2.5 concentration you'll find in a kitchen where meals are being cooked with gas or electricity. That's clean. That's the target.

"We know that 60 percent of the world cooks with gas or electricity," Kirk Smith told me. "It works! It's clean, easy, idiot-proof. So we know what works. The question is: how do we get the other 40 percent?"

Moss offers some plausible reasons for why so little progress has been made to clean up cooking around the world. For one, it hasn't been on decision makers' radar. "Historically, I think somehow the issue gets marginalized," he says. It's only in the last couple decades, thanks to the work of Smith and many others, that awareness has grown about the severity and extent of the health risks caused by household air pollution. And it's only quite recently that researchers have developed more complete and concrete estimates of the economic benefits—through healthier lives, enhanced productivity, increased school attendance, and other knock-on effects rippling through the broader economy—that would flow from cleaning up smoky kitchens from Ghana to India to Yunnan.*

But perhaps the simplest answer is that most of those still in the smoking section don't have any alternatives. Many live in places where they don't have access to reliable electricity or to gas and LPG distribution networks. And those who do may not be able to afford gas stoves and fuel or electric appliances. So instead they rely on whatever biomass they can gather from the fields and forests because it's free (but for the considerable cost of their time and exertion to find, chop, load, carry, split, and store it). Likewise, instead of purchasing stainless steel

---

*Kirk Smith and colleagues published a study in *The Lancet* estimating that a ten-year program to disseminate low-emission cookstoves to 150 million households across India would prevent 1.8 million premature deaths due to COPD and ischemic heart disease by 2020, as well as 240,000 ALRI deaths (acute lower-respiratory infections, basically pneumonia) in children under five years of age. They found that a coordinated effort to introduce cleaner-burning stoves to 90 percent of India's biomass-burning households by 2020 would save 55.5 million years of human life. (And then there are the climate benefits: in the same 2008 paper in which Ramanathan and Carmichael estimated that black carbon had as big an effect on Himalayan glaciers as CO_2, they estimated the effect of switching from biomass cooking fuels to clean, black carbon–free alternatives, such as solar cookers, biogas, and natural gas, in South and East Asia: "over South Asia, a 70 to 80 percent reduction in BC heating; and in East Asia, a 20 to 40 percent reduction.") The World Bank recently estimated that global replacement with fan-assisted cookstoves would save over 1 million lives annually in 2030. It estimates that 743,000 premature deaths could be prevented each year in the Himalayan region by switching to cleaner-burning cookstoves. And the true number would be much higher, because this study only looked at impacts from ambient pollution. Swapping stoves makes sense in wealthy places, too: replacing wood-burning heating stoves with efficient pellet stoves in places like the United States and Europe would prevent another 120,000 deaths each year.

gas ranges, they make their own simple box-shaped stoves from mud and masonry or buy cheap ones made locally of sheet metal. Some forgo stoves entirely and simply place three stones together, set the pot atop them, and build an open fire underneath—as Zangmo does in her storeroom and in the *doksa*.

These people urgently need stoves that burn biomass more cleanly. They also need to be able to afford to buy them. "You require a technology which is really cost-effective and it has to compete with zero monetized cost," says I. H. Rehman, who leads the household energy program of The Energy and Resources Institute (TERI) in New Delhi. "At the moment what people are spending is zero."

Many studies have shown that adoption and sustained use of improved stoves boils down to a few key factors. How much does it cost? Is it easy to use and will it cook traditional dishes? Will it save fuel compared to the current method of cooking? Will there be technical and maintenance support to the user after the sale when the stove needs repair or the combustion chamber needs replacement? Does it require a significant behavior change to use it properly? As Kirk Smith points out, if you don't use a "clean-burning" solid-fuel cookstove as it's intended to be used, it can quickly become a very dirty stove. And poorer consumers can be quite sophisticated and discriminating in their decision making—they can't afford *not* to be.

But now that all this information is out there in front of policy makers, Jacob Moss still seems mystified by the discrepancy between the scope of the opportunity and the resources that have historically flowed from on high to "try to develop those interventions at a really serious scale." Moss has led the effort across a dozen U.S. government agencies to muster action and funding for clean cooking solutions. As of mid-2013, $117 million in commitments had been made from agencies like the Department of Energy, the EPA, the Centers for Disease Control, and USAID. (A year later, the figure rose to $125 million.) This is a significant amount of money, Moss says.

But relative to the scale and scope of the problem, which he characterizes as an overlooked epidemic, it seems paltry. "Even the U.S. commitment, which on the research side is about $60 million, that's still a pittance."

In short, in development and public health circles, household air pollution just hasn't been a priority. Even though it robs more years of healthy human life than any other environmental factor, it's still just on the edge of the radar screen as a major development issue. It mostly affects women and children in the developing world, who don't figure much into economists' calculations. These tend to be the most marginalized people to begin with. Their contribution to the formal economy is not captured or quantified by our conventional measures of GDP, imports, exports, and the like. So if the true scale and depth of the impacts aren't properly accounted for, the potential benefits of addressing the problem aren't accurately tallied either.

And then there's the near universal tendency to emphasize treatment over preventive measures. For example, to treat child pneumonia, the standard approach is prescribing a course of antibiotics. But the environmental factors contributing to those high rates of illness don't receive much medical or policy attention, which is the reason Dr. Stenzin doesn't have a budget for educating Zanskaris about the dangers in their kitchens. Health ministries and aid agencies in countries like India have to figure out how to allocate scarce resources to solve a host of daunting problems: HIV/AIDS, water-borne diseases like typhoid, malaria, chronic malnutrition, and on and on. A dollar spent to subsidize or support cleaner stoves or fuels is a dollar not spent on bed nets or tuberculosis drug distribution. Careful trade-offs must be made, so the public health policy community needs a clear picture of the costs and benefits of a given intervention. Though policy makers are becoming increasingly aware of the stakes, they are still surprised when they learn that pneumonia is the number one killer of children and that household air pollution is implicated in half of those cases. Or that globally this pollution is the second biggest risk to women. *It's just smoke, right*? This attitude partly explains why governments have largely neglected the issue. The private sector has historically had little interest in it, as well. To date, not many corporations or entrepreneurs have figured out how to profit from selling cookstoves that burn dung and wood to poor people in Ghana, India, and Guatemala.

Moss has had a front-row seat to this conundrum for almost fifteen years. To answer these structural challenges, he helped create the Global

Alliance for Clean Cookstoves. The alliance, launched under the umbrella of the UN Foundation and with critical support from then secretary of state Hillary Clinton in 2010, has a goal of bringing clean cooking to 100 million households worldwide by 2020. Its focus is on building a global market for clean cooking solutions by building supply chains and supporting entrepreneurs. Moss describes it as an "incubator," making strategic grants to different fledgling businesses and partners to help them refine their product, build their capacity to get it to the consumer, and survive the start-up phase so they can eventually scale up their operations and be sustainable on their own.

The alliance also has to walk a fine line between helping businesses that are likely to succeed by targeting customers who have some capacity to pay and helping ventures and programs designed to reach those who can't. "If you're trying to build a market, you probably don't go after the hardest to reach poorest customers [first]," Moss points out. "Does that mean there's no room for humanitarian aid to help the poorest of the poor or help refugees? No, but for the piece that's trying to develop the market, it means you're a little ridiculous as a business if you're trying to go after the poorest customer because it's going to be the hardest to succeed."

That's where government programs come into play. The alliance has studied these past efforts closely, drawing lessons from both their moderate successes, and, more often, epic failures. In the 1980s and 1990s, the Chinese government implemented a successful nationwide program to introduce new cookstoves to rural homes. At its close, 180 million households had adopted the stoves, which were designed to save fuel, though not necessarily to reduce smoke in the home.

Meanwhile, the Indian government had its own national "improved *chulha*" program over roughly the same period. In terms of widespread penetration, it was largely a failure. Ten years after the program started, it accounted for less than 7 percent of all cookstoves in use in the country. The program's stoves were poorly designed and poorly made, with short lifetimes and features that didn't reflect how people actually cooked. The beneficiaries reported little fuel savings, high breakdown rates, and little reduction or removal of smoke.

"Smokeless *chulha*s, they called them," Kirk Smith recalls. "They were never smokeless. At best they had a chimney and the chimneys didn't work. China's stoves lasted fifteen years. They're still there. A five-dollar stove that lasts six months: is that cheaper than a fifty-dollar stove that lasts twelve years? I don't think so."

He went to India to investigate the results of its stove program. In one rural village, he attended a monthly meeting of the local workers tasked with implementing it. "And they saw me sitting there, all the women, and one of them clearly asked, 'Who's the gringo here?' And they explained that he was *chulha-wallah*." A stove man.

"And she stood up and started shaking her fist at me, and finally I got a translation: "Oh, goddamnit, get rid of this stupid program! Nobody wants these stoves here. They were a waste of time, we've got better things to do. Go back and tell them that!' "

He throws his hands up, laughs, and summarizes his lesson: "Top-down programs don't have to fail. But, like anything else, you've got to be careful."

He attributes the success of the Chinese program to three factors. They started with regions where strong local capacity raised the likelihood of initial success and adoption. Province-level officials sent inspectors out to see installed stoves before releasing all of the program funds to county-level officials. And they didn't try to get away with super cheap stoves. "They were thinking of it as an asset, not a consumable. It was changing your life, something you hand down to your children, you're proud of it, making your home clean and so forth."

All these experiences were the origin of Smith's mantra—"You get what you inspect, not what you expect"—now famous among cookstove and household air pollution researchers. He has become a leading advocate for rigorous independent monitoring and evaluation, for going back to verify the actual results of a program to introduce better cooking technologies, to see if the health and other benefits claimed are actually materializing. He has learned to be wary of claims of higher performance. It's one thing to get promising results in the laboratory. It's a whole different ball game out in the world, in Mendo Tamang's mud hut in Haku Besi or in Zangmo's high-altitude kitchen.

Throughout all this work, Smith has reflected, too, on the persistent gap between the scale of the opportunity and the technology that has actually been developed and deployed by national programs, nonprofits, universities, and small enterprises. Many stove prototypes have been cooked up in labs by well-intentioned engineers over the last few decades, but very few, if any, of them burn biomass cleanly enough to meet World Health Organization standards for particulate concentration exposure.

And those few that do approach that level of performance haven't really excited potential customers in villages in India and Africa because of various performance and functionality problems. Smith observes that if a new stove only performs a little better or saves a bit more fuel than the old method, then "women may understandably be unwilling to make even small shifts in behavior for relatively small benefits." On this reading, the failure of these stoves to improve indoor air quality is mirrored by their failure to markedly improve the lives of the cooks, who are mostly busy women like Tsewang Zangmo.

After thinking longer and harder about these questions than just about anyone in the world—why have so few truly *clean* stoves been developed and disseminated by public and private institutions and adopted by cooks in the developing world as an alternative to their smoky hearths?—Kirk Smith had another epiphany in rural India.

One day a few years ago he was visiting villages in Haryana, an hour south of New Delhi, where he and colleagues were doing a pilot study. Before they introduced a certain new cookstove to households in the village, they wanted the *anganwadi* workers to try it out, since they would be responsible for distributing the stoves. (*Anganwadi* workers run small government-sponsored child and maternal care programs in every village, often out of their own homes.) At one meeting, a young woman *anganwadi* worker offered some pointed feedback that, after thirty years of working on stoves, left a profound impression on the world's foremost household air pollution expert.

"You know I've been in this business a long time. And she turned my head around in a few minutes, in ways that changed my thinking. She said, first of all, 'Don't call it "improved"! Everything's called "improved"! It doesn't mean anything.'

"And I thought, you go across the street here to the drugstore, and I went, and about a third of the products are called 'improved.' And what does it mean? It means nothing! And I looked it up in the marketing literature. It says, 'well-known gimmick,' you actually use it when it's *not* improved. You use it when your warehouses fill with the old product and you've got a new one coming out.

"And then she said something more profound. She said, 'Don't try to distance yourself'—I'm paraphrasing—'don't try to distance yourself from the bad thing. Try to associate yourself with the good thing. Everybody wants gas. They see gas in the movies, they see their rich neighbors have gas. They know gas is the modern, good thing to do. So you tell them it's like gas.' "

Gas, he points out, has an aspirational quality, certain status associations—it says to your neighbors that you've 'arrived.' So why not sell a biomass stove that appeals to those impulses? But the young woman went a step further.

"And then she pointed to the stove and she said, 'You can do that. This is *better* than gas. It's cheaper than gas to run. It's safer than gas, it won't explode.' "

The woman ticked off all the other advantages the biomass stove had over gas: "It's faster than gas. It's controllable like gas, because it has the knob on it. But the high end of the power is actually higher power than the typical gas stove. It actually can cook faster than gas. So she went down this list of things, [saying] 'This is like a gas stove.' And I thought, "Well, I can say with a straight face that it's *almos*t as clean as gas—I can't say it's *as* clean as gas—I can say it's almost. But there aren't many stoves where with a straight face I can do that, honestly. Maybe two others, based on measurements."

Smith is now working with colleagues in India on a multiyear study to determine the effects of cookstove interventions on the health of pregnant women and newborns. During the pilot phase of this study, the researchers distributed new cookstoves, measured concentrations of pollutants in the participating kitchens, and conducted a survey of the recipients. They distributed two kinds of stoves with small fans built in to improve air mixing with the fuel, resulting in more complete combustion. Then they measured concentrations throughout the day of PM2.5 and carbon monoxide in the kitchens, and, with a personal monitor, the pollution exposure of the person cooking. Cooks in twenty of thirty-two households had LPG but said they preferred cooking on *chulhas* because it was cheaper, and the meals tasted better. Smith and his colleagues used monitors developed at UC Berkeley to measure pollution levels during traditional cooking; they found mean PM2.5 concentrations of 468 mg/m3 in the kitchen and personal exposures of 718. Outside wasn't much better: 400. (These levels, in addition to being alarmingly high, are comparable to what many other studies of Indian cooking conditions have found.)

Most participants preferred a stove made by the conglomerate Philips, because they liked its portability, which enabled them to cook outside at times. Like Urgain over his LPG burner back in 1996, they marveled at the speed and efficiency, and the fact that the fan-driven biomass stove produced much less smoke than the traditional *chulha*. In other words, they weren't cooking *with* gas, but they were cooking *like* gas, only using familiar, renewable fuels that they could gather for free.

As it happens, the key to savvier marketing and wider stove adoption is also the key to avoiding 4 million unnecessary deaths each year. The lesson seems to be that just better combustion isn't good enough—to truly make a dent in the health impacts, you need *really* good combustion. It's faster, more convenient, *and* cleaner.

Smith and many others point out that if the goal is to improve people's health, to save lives—and that should be the goal—then just moving the smoke outside or halving the fuel used for cooking a meal is not good enough. You need to dramatically cut down the amount of smoke coming out of the fire. And more specifically, you need to dramatically reduce the very fine particles, like black carbon.

To pull 3 billion people out of the smoking section, the *anganwadi* worker had it right: you need biomass stoves that make you feel like you're cooking with gas.

Now we're cooking with gas!"

My grandfather raises an eyebrow, eyes twinkling, as he says this, and rubs his hands together. It's a look that says, "Game on!'

We're not actually cooking with gas. We're hooking up a DVD player to my grandfather's television. But we're being efficient, getting things done. The phrase has come full circle. When he was a boy, it had literal significance.

The saying first started cropping up in magazine ads and catalogs in the late nineteenth century. Gas stoves started displacing wood and coal stoves in significant numbers around the time of my grandfather's birth, in 1915. By 1930 there were twice as many gas stoves as coal- or wood-burning stoves in use. Around that time, Bob Hope's comedy writers repurposed the slogan for use in his popular radio show as a kind of catchphrase; it became a cultural touchstone, and the gas industry then appropriated it all over again, to push for an even wider adoption of gas in American households. By the 1950s, the vast majority of Americans were either cooking with gas or electricity, or both, and the slogan took on a new life and meaning beyond the literal association with cooking: *Now we're being modern, smooth, effective!*

My grandfather, James B. O'Neil, was born in Brooklyn. For part of his childhood he lived with his father, mother, brothers, aunts, and grandparents in a house on East 12th Street that was heated by a coal furnace. In the winters his father would call the aptly named Burns Company to deliver pea coal right to their house. The Burns man would fill a wooden barrel from his truck, then roll it down the alley and dump its contents down a chute straight to the basement. There my great-grandfather would load the dark chunks onto the grate and set them alight each night.

"It was a nuisance, as you can imagine," my grandfather recalled. "You had to shake the grate." If you didn't shake it the coal would plug the gaps and prevent enough air from coming through. The grate also let the ashes fall through; he remembers trundling up the basement stairs each night with a heavy can full of coal ash. "And I'd roll it out to the street. It would be picked up the next morning by the city department and they'd throw it, I imagine, in the fill somewhere."

His grandmother and aunts and mother did all the cooking with another fuel. "On one side of the kitchen, [there was] the coal stove," he said, "and the other side facing it they had a gas stove. They didn't use the gas to heat the house; they only used the gas for the oven and cooking."

"But by the 1930s were people mostly cooking with gas?"

"I'd say even before that. I don't remember seeing a coal stove in any house that I ever visited!" His family had one that was only rarely used, for heating a kettle of water. Before he was born, though, the family had cooked most meals on it.

"So people switched because it was cleaner?"

"Oh, it's a job!" Despite all the hassle—of ordering the coal, then washing it, loading it, shaking it, occasionally having the smoke blow back in your face when you lit the pea-sized rocks, shaking it again and again as it burned, then disposing of the ashes the next day—this was the least bad method for heating homes in the northeastern United States at the time. Wood, which had been American's primary heat source since settlers arrived three centuries prior, and well before, was even more cumbersome, especially in urban areas. It was also much less energy-dense—you needed twice as much of it, per pound, to get the same amount of heat coal would release. So his family relied on coal for heat through his youth, despite the fact that around that time, in 1932, an engineer described a coal heater as "simply a device for stewing off tars and vapors of inconceivable variety as to composition, odor and filth for the effective work of polluting the atmosphere."

To supplement the coal heat, in certain rooms they had another device. "A kerosene heater: have you ever seen one?" he asked.

I told him I had indeed, recalling the one my friend Urgain Dorjay set up, with a sawed-off plastic jug full of kerosene hanging from the ceiling feeding it through a plastic tube, in his guest room on a zero-degree night, afraid I would freeze.

"It would be about so high," he said, holding his hand a couple feet off the ground, "round, and burned like a lamp. There'd be a wick in there. It would draw the kerosene up and it would burn like a candle."

When he was fourteen, his parents moved them to a house on East 22nd Street, which had a huge coal steam furnace. Grandpa remembered quite clearly the day his father hired some men to pull the grate out of the old coal boiler to convert it to burn gas. "They installed a pipe about four foot long and put it in the door and connected it with the gas pipe—and that heated the house! We had steam. And it heated the house for a number of years." Thinking of that day took him even further back.

"I'll tell you what, you're gonna think I'm spoofing you," he said with a laugh. "We were at the East 12th Street house, number 1446. And they converted the house to electricity. It was candlelight until then. I would guess I was four or five at that time. And I can remember sitting on the stairs while the electrician tore up the floor above me. The floor was on my eye level. He was tearing up the floor to run the VX cable. And I can honestly tell you I can remember that. I bet if I went in the house I could tell you about where the VX cable is in the wood floor of the house right now."

It's not a bet I would take: my grandfather had a truly prodigious memory (that I don't seem to have inherited). This was one of his earliest memories—and one that perhaps helped spark a lifelong fascination with tools, machines, and appliances, with useful things and how they work. He went on to spend almost half a century with the same hardware wholesale supply company, working his way up from warehouse stock boy to vice president, becoming an expert on all kinds of household devices.

"What were the main uses?"

"Lighting was the most important. Then toasters came—the first electric appliance, I would say it was." He smiled as he arched a bushy eyebrow. "Can you picture, my grandmother said, 'James, will you get my bag?' And her pocketbook would be on the doorknob. And I would bring my grandmother her pocketbook, and she'd hand me two quarters, maybe one quarter. And that automatically meant go down in the cellar and put a quarter in the meter. When they were cooking, sometimes the meter went dry."

"Like a parking meter?"

"Same thing."

"When the time went up it would close the valve and no more gas?"

"No more!"

"So if you wanted to take a hot shower?"

"Better be sure there's money in the bank!"

All the same, dropping a quarter in the slot was a lot easier than the Burns Company, grate-shaking routine. Listening to my grandfather talk about the comparative ease of heating and cooking with gas, instead of the coal furnace his father wrestled with each night, or of using an electric toaster, rather than the coal stove he saw his grandmother labor over as a child, reminded me of Urgain Dorjay describing the day in 1996 when he and Chondol first cooked with gas.

Diesel trucks laden with bright red cylinders full of LPG first started rumbling over the Pensi-la Pass into Zanskar about twenty years ago. Before that, every meal Urgain Dorjay had eaten in his family's home in the shadow of Padum Glacier was cooked on an open fire or in a metal dung-burning stove. Urgain vividly recalls the day when he hauled one of them up the long hill to his mud brick house, hooked up the plastic line to the regulator, turned the dial, and put a match to the burner.

"Oh ho! When we first use we are feeling . . ." His voice trailed off, and he paused. He smiled and shook his head at the memory, as though the initial astonishment had yet to wear off. "It's so easy, yeah? Very fast!"

The gas burner was not only much cleaner but much more convenient. You could turn it on and change the heat of the flame with the flick of a dial. You could hand over a few hundred rupees to the depot agent to get your government-subsidized cylinder instead of combing the mountainsides for patties with which to boil your butter tea. Urgain now goes through ten or twelve cylinders a year. But his wife Chondol still spends hours each week combing the hillside for dung, which they still use for half of their cooking and all of their space heating.

My grandfather's other earliest memory, also around the age of four, was of the funeral of his own grandfather, John McCole, a captain in the local fire company. He recalled watching the hearse leaving his grandparents' house and moving slowly up Jay Street, right in the shadow of the Manhattan Bridge. It was a somber, powerful moment, but two details in particular stayed with him. They were already rare sights, and soon to become as anachronistic in New York City as the coal stove: the hearse was pulled by horses, and the road was dirt.

Within a few years, of course, automobiles had thoroughly displaced horses from the streets of Brooklyn, which were fast becoming paved. From mud tracks to paved roads, horse-drawn carriages to cars, coal to gas, candles to electric lights, and the later rise of oil and all its manifestations, my grandfather was witness to a dizzying sequence of technological transitions in his ninety-eight years. He saw the spread of the radio, the television, the computer. But in terms of the exigencies of daily life—staying warm, fed, and washed and getting from points A to B—none of these momentous developments have transformed our lives in the past century as dramatically as the two shifts that produced my grandfather's earliest memories: the household energy transition, from burning solid fuels like coal and wood for heating, cooking, and lighting to cleaner ones like gas and electricity; and the rise of the internal combustion engine as the primary mode of moving people and things across the city and around the world. For pure time and labor savings, these two miracles are unparalleled. These transitions profoundly changed Americans' lives—and from North America and Europe, they spread across the world to eventually reach the most remote parts of the planet, even to distant Shangri-las like Zanskar.

Though they are marked by single, vividly recalled days in my grandfather's life, these two transitions took decades to be completed across the country. And by the time they were complete, they were taken for granted. The devices faded into the background of daily life. For most of us alive today, there was never a time when you couldn't walk into a room and turn the lights on by flipping a switch or head down to the hardware store by hopping into your car.

For people in Kaharan Purwa or Kumik, this possibility is still overwhelmingly novel. Perhaps that's why the aged woman receiving the army's gift of an electrical appliance at the Farmer's Mela was brought to tears.

On October 29, 2012, a fourteen-foot storm surge poured into Battery Park in lower Manhattan, driven by Sandy, the most powerful hurricane of the 2012 season. The storm ravaged communities from the Jersey Shore to Far Rockaway on Long Island, racking up $68 billion in damages along its path up the eastern seaboard.* In the wake of the storm, much of Manhattan below Thirty-fourth Street was without power for days. Dazed New Yorkers wandered around, looking for a place to recharge their cell phones to tell friends and family they were okay.

Some were drawn to a table set up in Washington Square Park, and another one later deployed in City Hall Plaza. In those locations a few volunteers had set up some curious-looking devices, offering cryptically to charge people's phones and boil tea for them at the same time. Soon a grateful cluster had gathered, marveling at the unlikely technology that was restoring their iPhones to life—until the cops came and, citing the fire code, shut down the whole operation. It turns out you can't light open fires in the middle of Manhattan.

The very popular people staffing those folding tables were employees from an unusual business located at 68 Jay Street, across the East River in Dumbo, the Brooklyn neighborhood that lies in the shadow of the Brooklyn and Manhattan bridges. There, in a fourth-floor loft of a nondescript brick building, is the nerve center of BioLite, a plucky young startup company with a bold agenda: to bring clean cooking to the 3 billion people still stuck in the smoking section.

As it happens, BioLite's office is just a block away from 113 Jay Street, the site of my great-great-grandparents' house (long since torn down to make way for an MTA subway station and ventilation facility). That was where my great-great-grandmother, Margaret McCole, and her fire captain husband, John, had made the switch from cooking and heating with coal to "cooking with gas," sometime around the turn of the twentieth century. (Before they made that switch, Margaret gave birth to eleven children. Seven died before they reached adulthood. At that time tuberculosis was the number one killer in United States, but pneumonia wasn't far behind.)

Earlier in 2012, on an unseasonably warm March afternoon, I had visited BioLite's offices and found five employees bustling about with the quiet, pent-up energy you'd find backstage at a theater before the

---

*Incidentally, Jason Box calculated that the summer melting of Greenland's ice sheet in 2012 added a millimeter of global sea level—including to Sandy's devastating storm surge.

first curtain rises. The scene was pure start-up: road bikes stacked by the door, a workbench covered with tiny circuit boards, shelves lined with a bewildering array of metal and ceramic cylinders, desks littered with sheaves of schematics. Tacked on a bulletin board, under a note that reads "Emergency Provisions," was a packet of freeze-dried ice cream—a wry nod to the tastes of engineers on a deadline.

BioLite's cofounder and CEO, Jonathan Cedar, hunched over a thirteen-page memo just in from the company's lawyers and fielded a call from his board chairman—two of a dozen details that needed his attention, with less than a month to go before their first products started to roll off the assembly line in China. Once they did, Cedar hoped the act of cooking might never be the same.

Cedar and his colleagues have developed a compact, portable cooking unit that they claim uses less than half the wood required to cook over an open fire and reduces black carbon and other particulate emissions by over 90 percent. If these early performance results are borne out and verified, his feat, as any combustion engineer will tell you, is impressive—but equally revolutionary is a feature that has already generated buzz on outdoor-gear blogs and in rural Asian and African kitchens alike: the stove turns excess thermal energy into electricity to charge a cell phone, LED light, or other small devices through a built-in USB port. It's a stove that will charge your iPhone with fire.

In 2006, Cedar and Alec Drummond, BioLite's cofounder and chief technology officer, were both working at Smart Design, a consultancy in Manhattan. Drummond had been playing around with the idea of using Peltier thermoelectric generators—little devices that use the difference in temperature between two metals to create electric current (the same devices that power those fans put on top of woodstoves to blow hot air around the room)—to make a better camping stove that could run on scavenged small wood instead of liquid fossil fuels. "I did nothing about it for years until Cedar and I were having beers at Smart Design," says Drummond. "He got excited, and we said, let's try to throw something together."

"We basically said, 'How do you use wood as a modern fuel?' " Cedar recalls. After some research and afterhours tinkering, they soon hit on the answer: "You need to use a fan."

The fan brings more oxygen in to mix more completely with the fuel, resulting in hotter, more complete combustion. But fans need a power source, and batteries are heavy, clunky, and expensive. After many more months of experimenting, Cedar and Drummond hit on BioLite's core innovation: they placed a thermoelectric generating "peg" inside, rather than up against the surface of, the burn chamber.

The result—after many more prototypes—was the CampStove, which BioLite started selling through outdoor equipment retailers like REI. Small pieces of wood are placed in the top of the small cylindrical chamber, which is about the size of a water bottle. Once the twigs start burning hot enough, the blower kicks on, dramatically improved combustion efficiency kicks in, and the resulting fire is almost smoke-free. The excess power generated in the peg can charge small electronic devices of up to 4 watts through the USB connector (twenty minutes of charging provides sixty minutes of talk time on an iPhone). "The whole idea is to jettison the fuel supply chain," Cedar says.

Since the rollout in early 2012, the CampStove has generated a lot of interest, but mostly in tech circles (people are understandably struck by the novelty of charging your iPhone out in the woods by just burning a bunch of twigs). But BioLite is looking beyond this niche of well-heeled recreators to a much bigger opportunity.

"Once we got the stove working, we started to see this had a lot more potential than simply a camp stove," says Drummond. The proof came in 2009, when they took an early prototype to the Ethos Conference in Kirkland, Washington, an annual gathering of cookstove experts focused on developing-world applications. Competing against stoves designed by engineers with decades of experience, they won the award for cleanest-burning stove. "We kind of wowed everybody. They were all really well intentioned but tended to be engineers and tinkerers. Our background was in product industrial design." Philips, the giant Dutch electronics firm, had been trying for years to develop an affordable, clean-burning cookstove for sale in places like India. "They saw ours, and I saw one of the [Philips] guys almost hit the side of his head!"

BioLite's big bet is on its HomeStove, a larger unit that uses the same technology, a versatile biomass cooking device designed to appeal to a household in Africa or rural Asia. "The expectation is that the CampStove will now generate enough revenue both to support future product development for the CampStove as well as market establishment costs for the HomeStove," Cedar explains. "The big opportunity here on all sides is the HomeStove. It's a $500 million home market if you get it right."

BioLite is far from the only outfit trying to corner this market covering almost half the world. First Energy, in Pune, India, makers of the Oorja Stove, and TERI, the nonprofit institute in New Delhi, are both developing forced-draft stoves. Envirofit in Boulder, Colorado, has probably sold more clean cookstoves than any other company, and Philips, according to Kirk Smith, sells the cleanest stove currently on the market. (The HomeStove isn't for sale yet.)

"I think if you look at all the products out there right now, they rely on a push strategy for success," says fellow cofounder Jonathan den Hartog. "You have to convince people of health benefits, that they'll save money on fuel expenses. It takes education. The return on investment takes a year or longer. You really have to push the product out there. That's why I feel they haven't been that successful—something on the order of a couple hundred thousand stoves worldwide. Our system has a hook on it: the electrical charging capability. For the user it's a really enabling capability. People are demanding cell phones, lights, and radios, and a whole array of other products."

The stove is also *designed*: with its silver anodized finish and the colorful cladding housing the blower components, the stove looks sleek and modern. But the phone-charging function in particular could be a game changer. "It transforms the cookstove into a modern appliance in a way that none of these other guys have," Cedar says. "And we've got hypotheses that probably it engages men more, which lowers some of the barriers for women to get them."

"One of the challenges we see is that the person who controls family finances isn't the one who does the cooking," den Hartog explains. The husband controls the money, while the wife suffers the brunt of the health impacts. The husband also typically controls the cell phone and will see value in the charging function. "They'll see more of a benefit."

When I ask Cedar if there's a model he looks to as an example of successful operation in a vast, complex market such as India, he offers an ambitious answer that illuminates his own evolution from designer to CEO: "Coca-Cola. They give people what they want. They get their message out, and people choose to buy it. Everyone makes enough money so it propagates on its own, in contrast to the aid model."

This thinking is a radical departure from the past three decades of stove dissemination conventional wisdom, which has depended on engineer-centric designs, government subsidies, and a push rather than market-based pull approach to demand. BioLite aims to sell the HomeStove to 1 million homes across India and sub-Saharan Africa by 2018. When we first spoke in 2012, Cedar acknowledged that many larger hurdles and potential pitfalls loom. "Are we effective from a health standpoint? Are we commercially valuable? And can we access carbon subsidies to help reduce consumer prices? I think those are the big three questions."

To get there, the small start-up will need to attract a lot more financing, refine the product, and get more feedback from users in the field. To that end, in 2013 and 2014 BioLite was conducting several trials in Ghana, Uganda, and Orissa, one of the India's poorest states, to understand what cooks are willing to pay and what features they value the most.

Cedar anticipates framing the HomeStove as something that "keeps your house a lot cleaner and more comfortable. For women it's going to be about time savings, and it's going to be about modernity and electricity. I think that's what customers are going to care about."

After several years working in industrial design, Cedar hankered to do something more meaningful. "We all came into this because we want to feel like the things we're making have an impact. We spent a long time remaking potato peelers that were cool. They were great, helpful, but people already had potato peelers!" His primary motivation is making a dent in the global epidemic caused by exposure to cooking smoke.

"Do these things actually affect health? Because that's ultimately, in my view, the reason to be doing this. I think environment is a nice co-benefit, economics are a nice co-benefit potentially, but the biggest story here is 2 million people die" every year from breathing in smoke from traditional cooking methods. (He was citing WHO figures that predate the latest GBD estimate of 3.7 million premature deaths.)

"Hopefully, a lot of this comes back pretty positive and says, yeah, we are effective for health, these things are durable, people like them and want to pay enough money, and that difference can be made up with some kind of sustainable subsidy. And, hopefully, that's enough information to say, okay, this scales from ten thousand units to as far as we can go."

In the two years since my first visit to BioLite's Jay St. office-cum-laboratory, demand for CampStoves has far outstripped supply. But the biggest challenge remains the HomeStove. It is now in its fourth iteration, undergoing extensive field trials and fine-tuning before it hits the market. BioLite claims in its marketing materials that its HomeStove reduces smoke by 95 percent, reduces wood consumption by 50 percent, and almost eliminates black carbon. (Cedar told me that it's not yet quite as good as the smaller CampStove at reducing black carbon, but they're still working to refine it.)

The early feedback from cooks in the villages and roadside *dhaba*s of eastern India has been encouraging. Many people observed that the HomeStove "cooked like gas."

Stamped on the BioLite CampStove is the phrase, "Designed in Brooklyn, Made in China." BioLite has some competitors in the race to serve the 500 million home market for clean cooking. Take the Boss, for instance, whose stoves are both made *and* designed in China.

I find the Boss, aka Zeng Juhong, one morning at his office, which lies down a quiet alley on the west side of Kunming, the main city of Yunnan Province. Dressed in slacks and a windbreaker, he sits at a desk that is bare but for a desktop computer. He is a lean, mustachioed guy who radiates pent-up energy. My companions, two energy scientists who are his collaborators at Kunming Normal University, in deference to my lack of Chinese, simply call him The Boss.

He is the founder and CEO of Kunming Rongxia Stove Company, which manufactures biomass cooking and heating stoves mostly based on Juhong's own designs. The Boss has just returned from Beijing, where the ministry of agriculture agreed to help deploy his cooking and heating stoves throughout China—"as one of best stoves made in China," he says.

His employees have the day off for the May Day holiday, so the Boss leaps out of his chair in his sparsely furnished office to show me around his warehouse and fabrication plant. He takes us to a large building, where hundreds of rectangular metal stoves are stacked on the concrete floor, waiting to be shipped. His stoves burn wood, agricultural waste, dried dung, pellets; Juhong claims a 65 to 85 percent improvement in fuel efficiency over his competitors' stoves. One popular model has two burners, vents for heating, a box for baking, a hatch for loading the fuel, a grate to promote better air flow, and a tray for easy ash removal. It costs 900 yuan to make and sells for 1,000 yuan (about $150); the government fully subsidizes it. He pours pellets into the hopper and lights the stove; it burns with a blue flame, indistinguishable from any gas range.

Then he shows us a large stainless steel stove, marketed for restaurants. It cooks just as well as gas, he says, but burns pellets made from waste wood, shavings, and other woody biomass, even weeds. He wanders outside and plucks a weed from the fringes of the property. He comes back in waving it. "Even crofton weed you can burn!" He pours some pellets in and lights them with some wood shavings. There is some light smoke at first. But after a couple of minutes a huge wok of water is boiling, there is a steady flame, and I am surprised to see no smoke coming out of the pipe. A small household version operates on the same principle, with the same fuel, at greater than 50 percent efficiency. He sells chimneys for all these stoves, but he says they're not really necessary; his stoves produce close to zero smoke emissions. The Boss pantomimes making stir-fry in the wok.

"Makes me hungry!" I say, intending a lame, jokey compliment.

Taking this literally, he quickly ushers us into his car. His son (his wife and children all work for Rongxia) drives us a short distance through quiet western Kunming to a local restaurant to continue the conversation. The Boss orders for all four of us, and soon there are a dozen dishes placed on a lazy susan at our circular table: mushroom noodles, a whole fish, veggies, tomato soup, a shrimp stew, beans and cabbage, and a pile of roasted chicken drumsticks and a mysterious mound of flesh. The Boss also orders a bottle of *baijou*, rice liquor. He pours one out for me. The Boss is pleased that I have agreed to drink rice liquor with him.

"I met another American recently," he says with a frown. "He would not drink rice liquor with me. But you will drink rice liquor with me!" It is eleven in the morning and we are drinking rice liquor.

"Where did the idea for your stove designs come from?"

The Boss just taps the side of his head. He used to live in Guangdong, ground zero of China's industrial rise, where he worked in a heater production factory until it closed a decade ago. Then he sold his small house and some land and came to Kunming to start over from scratch. Now he has 100 workers, and they are moving to a new factory to accommodate the company's growth. He has applied for over 100 patents. The Boss is convinced that there is a huge untapped market in China for his stoves. So many people depend on biomass and waste agricultural residues for fuel—it's a bigger market than for cars, he thinks. (A bold statement, since China has by far the biggest auto market in the world; 55,000 new cars were sold each day in 2013. But he's right: over 55 percent of Chinese, 715 million people, currently cook with solid fuels like wood or coal.)

"And compared to China, there's more opportunity in India," he says between mouthfuls.

I point out that, because incomes are lower in India, his stoves would need to be significantly cheaper. The Boss is undaunted.

"We focus on the market, so we can change our product for India!" The Boss is now speaking with great passion. It occurs to me that he would make a fantastic car salesman back home. "People are important, and idea is very important! We are focused on what you need—we provide to you!"

"So you're saying, the customer is king—the boss," I say.

"WE CAN DO IT!" The Boss is practically shouting now. Professors Yin and Zhang smile and nod at the Boss's exhortations.

"We don't think on Chinese conditions, if we focus on Indian market we can provide more suitable product for India. The market is very important. If India has the interest, we can do it. In the future, maybe

India has the intention for good product, we can provide them: different product, high price, low price."

I mention that I am on my way to Tibet. Does he make stoves for use in such cold, high climes?

"You will see my stoves there!" the Boss crows with delight.

It turns out that Rongxia is one of a very few companies that make stoves designed to burn dung for space heating in high-altitude zones. (This sector is largely neglected by development agencies and nonprofits, despite the fact that tens of millions live in cold mountain areas, burning biomass fuels near snow and ice throughout the region.) He says his stoves are used from Yunnan to Lhasa. Indeed, a few days later, I will find one of them in a simple dining room in a guesthouse in Tiger Leaping Gorge, and a couple I meet there will tell me they have just seen the same one in a remote guesthouse in the mountain town of Dali.

Meanwhile the Boss loads my plate with the mysterious, grayish meat. "It's dog!"

"Dog?" I turn to Professor Yin for confirmation. She smiles and nods encouragingly.

"IT'S DOG!" the Boss repeats. "You must try it, like this." The Boss assembles a dog burrito for me. I try it and immediately understand why dog is not a common dish in most of the world.

"Eat this, then no cancer!" he says.

"Well, if you eat this, and cook with that stove," I say, rather ingratiatingly, "then maybe no cancer, right?"

Polite laughter. The Boss wants me to guess how old he is.

"Let's see. Maybe fifty?"

"He is over sixty!" says Yin Fang.

"No!" I am genuinely surprised. The Boss looks like a much younger man.

"He is very healthful man, very healthful!" Professor Yin says. "You see he's very strong. He works with the stove and smoke, but he's very strong."

The Boss pounds his chest three times. He leans into me, rolling up his shirt and flexing his right arm and points to his forearm, requesting that I appreciate his sinew. It's a fine forearm, and I hope my nod conveys this sentiment. But the Boss wants me to touch his arm.

"Touch it! See how strong!"

I comply. It does seem pretty strong. His point is this: he has spent over a decade tinkering with his stoves, exposed to each prototype's emissions. The proof of how cleanly they burn wood, dung, crofton weed, pellets, and whatever else—the proof of their smokeless performance is in his own vigor and health. And the proof of his health is in his ropily sinewed arm.

This, of course, is not the most scientific verification of his products' quality and cleanliness. So, in case that isn't evidence enough, he shoves some test protocol documents into my hands. "No particles in the smoke, the emissions of the smoke is near zero. You can see the test data. Compared with petrol combustion, this biogas is more clean than petrol combustion. . . . Chinese government give more support for this stove product because it's the same energy but it protects the environment." (At the moment I have to take his word on this: the documents are in Mandarin.) But seriously, the Boss wants to know, if his stoves produced smoke, could he have such sinewy power in his arms, such a youthful visage?

I have no good answer to this. I throw up my hands. We drink more *baijou*. The Boss is so enthusiastic as he lays out his plans to spread his smokeless stoves across Asia that he knocks over his glass of *baijou*.

"I like you," the Boss tells me again as he spins the lazy susan so the fish head arrives in front of him and the boiled dog meat in front of me. "Because you drink *baijou* with me in the morning!"

The Boss's high spirits have to do with more than just the *baijou*. His company has been growing fast the last few years. He has plans to get even bigger.

"He has intention," says Professor Yin, "more money, more money, more big factories, also his products can go to Southeast Asia, also other places in the world. All over the world. He can expand."

"How many stoves do you think you can sell?"

"Maybe quantity of production can reach 1 million stoves per year. Also the value of the products can reach 500 million yuan."

That's $80 million—pretty bold, given that Kunming Rongxia's production in 2009 was 20,000 combustion chambers and 8,000 stoves.

He says the demand is overwhelming, the technology is proven and ready. There are just two things holding the Boss back from selling his stoves, for every conceivable climate, fuel type, and usage, from Yunnan to Gansu to Tibet to Tamil Nadu—who knows, maybe even to restaurants in Brooklyn.

"Investment, and human resource," he says. The demand is there, the technology is ready, but he needs capital to expand. His is a sector neglected by traditional investors. And he needs more skilled workers. "Limited money. Also limited human resource, so I cannot develop my factory in India. But in future when money is not problem, if enough money, enough human resources, I can build my factory in India."

The Boss's message to the world, in sum, is that he is ready to hot-rod this thing. If all of his stoves perform like the ones he showed me—and users achieve the same performance he gets in the factory and test labs (historically a very big if, but not an insurmountable obstacle), which seemed to run with almost no smoke—then maybe, just maybe, the Boss's little concern could punch a decent-sized hole, or rather many, many little holes, in the haze blanketing the western part of his country and drifting toward the glaciers of Tibet.

So, are we ready to move *everybody* out of the smoking section?

"Sure, sounds good," you say. "Kumbaya, bro. But what'll it cost?"

I'm glad you asked. Because if mere appeals to enlightened self-interest and articulation of co-benefits were enough, we'd have moved everyone to the nonsmoking section already, of course. The challenge is summed up by Kirk Smith's thirty-five years of being a voice in a relative policy wilderness, pointing to his rigorously gathered and analyzed data and waving his arms in futility at the world's most deadly and overlooked epidemic. The challenge is also summed up by this reaction from Congressman Joe Barton during a session on clean cookstoves for the developing world: "My constituents don't care about no clean cookin'!" he declared, according to one audience member. Then Barton got up and left.*

* During a September 11, 2012, energy subcommittee hearing, Barton complained about an EPA grant of $200,000 "to study something called 'clean cooking' in Ethiopia" and

The Boss and BioLite notwithstanding, Barton's reaction is worth noting: no one has seen a way to get rich, or get elected, by providing poor households with access to cleaner cooking options. That may change, but even if it doesn't, the rise of organizations like the Global Alliance for Clean Cookstoves demonstrate that we may finally be waking up to the gains we have been leaving on the table all this time, so to speak.

What would it take to meet this challenge? Well, yes, money, for one thing. Smith has noted that "there are 500 million stoves out there we want to replace, so even if they cost $20 each that's $10 billion right there. It's not money at the scale of nuclear power plants, but it's not small amounts either, so we need to bring ourselves up to speed in terms of meeting people's needs."

The development group Practical Action (formerly Intermediate Technology Development Group, ITDG) has estimated that providing clean cooking and heating solutions to 3 billion people would cost somewhere on the order of $2.5 billion annually for twelve years. "To kick-start an effective market in distributing low-cost smoke solutions, it is estimated that government spending and international development aid would be in the region of 20 percent of this total, around $500 million a year." To put this amount of money in perspective, it is less than 1 percent of total aid spending by western nations. The International Energy Agency has a higher but still far from earth-shattering estimate of $4.5 billion annual investment, which would come in the form of grants from multilateral aid agencies, government agencies, donors, and foundations.

Or think of it this way. In an interview, Kirk Smith once estimated that "since the beginning of this whole thing, less than $20 million has been spent on research." Consider that figure next to this one: a UNEP study has estimated the potential health benefits of a global program to fight black carbon emissions by 2030 at $3.7 trillion annually, in terms of health savings; roughly half of those savings come from replacing dirty cookstoves. It's a benefit-to-cost ratio of several hundred to one.

---

observed that during his town hall meetings in his home district of Texas, "not once did I have a constituent stand up and tell me to spend more money on EPA grants overseas."

Moss says the U.S. Department of Energy "would estimate a $50 million to $100 million commitment is really needed to develop the next generation of clean techs and fuels, and all the devices needed to go along with them, on monitoring, on fans, on thermoelectrics."

Meanwhile, Americans spent over $55 billion in 2013 alone taking care of their pets. That's seven times more *in one day* than has probably been spent in the entire history of clean cookstoves research. Americans spent $4.4 billion alone on pet boarding (hotels, kennels, etc.). So, for roughly the same price Americans pay yearly to lodge their pets, the world could almost completely pull half of its people into the non-smoking section, saving millions of lives a year, a huge amount of both fuel and labor, and trillions of dollars' worth of health care and other costs, to boot. Seems a crime not to do it, eh?

But let's say the money were to come through. There's still that core obstacle: the technology is not quite ready for prime time. And scaling up nifty devices to thousands of households, let alone hundreds of millions of them across many different cultures, and then maintaining them, is still a "wicked problem"—as BioLite and The Boss are in the process of discovering.

When I asked Kirk Smith what he thought about the BioLite stove, he pointed out one big problem. "It's not available yet." He wants to test-drive it; he's not one to succumb to marketing hype or company brochures. Remember: it's what you *inspect* that counts. "As a colleague said, really clean biomass stoves are like unicorns," Smith laughs. "There have been sightings, but you can't find any in the stables."

Patrick Kinney of Columbia University, who is coleading a three-year-long study in rural Ghana, has been learning this the hard way. He and his colleagues plan to give a clean stove to half of the rural pregnant women in his study, and half will continue cooking as they always have. "And in both cases we'll be measuring their PM2.5 exposures over the course of pregnancy, so we'll be able to tie the exposure measures directly to what we see." They will measure birth weights, and follow the babies through the first year of life to assess the incidence of pneumonia, and see if there's a correlation with exposure to PM2.5 from cooking fires. Kinney's whole objective is to quantify the benefits of an intervention, to obtain rock-solid evidence to bring to decision makers. "A lot of senior people in the field think that we need a randomized trial with a really clean, truly clean fuel," he says. "To then show the larger world how big an effect it really is, as opposed to getting null results because your stove's no good."

But first, to run their study, Kinney and colleagues needed a truly clean cooking invention. That proved to be harder than they thought. At first they planned to partner with BioLite. "We were hoping to launch our study in the fall [2013] with their stove, and then we had to rethink it," Kinney told me. The HomeStove wasn't quite ready, as Jonathan Cedar and company wanted to refine it further, fix some functionality problems. "So now we're faced with the fact that there's not a clean stove available on the market that we know is going to work." The Philips stove is the best out there, he acknowledges, "but in Ghana it's not going to work that well because it requires being plugged into the grid, whereas BioLite generated its own [power]."

Then they looked at using LPG. But they need a reliable supply throughout the course of the study, and in that part of Ghana "weeks go by and there's no LPG, and people revert to other methods. So we'd have to buy a big tank, and have it for six months so we know we have the stuff." They even talked to a guy who wants to capture waste methane flare gas from oil production fields in Ghana and Nigeria and compress it for household use. "We met him a couple times, interesting guy. Sort of out there kind of idea, but we're open to anything that will provide clean fuel." There's also a biogas demonstration plant—but it's three hours from their study site.

These hiccups in running just one study with a couple of hundred households speak to the larger challenge of reaching 100 million households, GACC's target, with clean cooking solutions—let alone 3 billion people.

"It's probable that a lot of interventions will also want interventions that really get black carbon benefits," Moss says.* "And how many are

*The future seems to be fan-assisted stoves. These deliver more oxygen to mix more completely with the fuel, resulting in hotter, more efficient, more complete combustion. More complete combustion, as Michael Faraday demonstrated 150 years ago with his candles, generally leads to less soot and other unwanted byproducts. But some "improved" stoves

out there? Not many. So BioLite is now getting manufactured, Philips is being manufactured, you've got fuels like ethanol, obviously LPG, methane, and methanol in some places. There's not a huge supply chain of clean cooking solutions other than LPG in the world."

In the world of cookstoves, a one-size-fits-all approach doesn't work. Even from one part of Himachal Pradesh to another, let alone between Guatemala, south Sudan, and north India, cooking customs and fuel use vary. Stove solutions will need to be tailored to specific cultural contexts and resource bases. But there is one cooking technology that is pretty much universally embraced, wherever it's available. Sometimes "cooking with gas" actually means cooking with gas.

Smith once wrote an editorial in the journal *Science* arguing that petroleum gas—a term encompassing LPG, propane, butane, and natural gas—was a "one-time gift from nature" and that its highest use would be to provide clean cooking throughout the developing world, since it would save hundreds of millions of years of human life and potential. "Rather than excluding petroleum, some of this one-time gift from nature ought actually to be reserved to help fulfill our obligation to bring the health and welfare of all people to a reasonable level: an essential goal of sustainable development, no matter how defined."

"I got more hate mail for that, blaming me for the Iraq War," he told me with a short laugh.

Even LPG has its unique challenges: in India, for example, the fuel has to be imported, liquefied, cryogenically stored, transported, regasified, and then distributed and stored safely at the point of sale. But Smith's point, at the time, was that there were no biomass cookstoves currently on the market that were nearly as clean as the standard

---

can actually produce *more* soot and carbon monoxide and other dangerous emissions than do simple three-stone fires: as the fire burns hotter, the fuel is consumed more completely, fewer large-diameter particles are produced, and relatively more fine particles are emitted. (You get what you *inspect.*) With better combustion, fewer of these fine particles attach to the big ones. And the fine particles are the most dangerous—the ones that penetrate our nasal defenses, make it deep into our lungs, and even cross into our bloodstreams. So, paradoxically, in some cases, merely somewhat better combustion can have worse health outcomes. You've got to go all the way, achieving truly *clean* combustion that resembles gas burning.

run-of-the-mill gas burner. Smith has written that "statistics show clearly that biomass use for cooking is nearly 100% associated with poverty. What can this imply, but that every woman in the world will switch to gas when she can afford it—no matter what her mother used." In his presentations to fellow scientists and policy makers alike, Smith notes that three-fifths of the world uses gas or electricity to cook just about every kind of cuisine. Cultural practices and behavior change are obstacles, but they haven't stopped the inexorable spread of washing machines (Urgain got his two years ago) and electric rice cookers and other labor-saving devices. When people encounter a truly empowering technology, they tend to alter their habits to accommodate it.

Some skeptics point out that cooking is a very entrenched behavior and incredibly difficult to change. Fuel choice affects the flavor of tortillas in Guatemala and Mexico, chapatis in India, sticky rice in China. But Smith has seen how quickly cooking customs can change. "You know, most of us don't want to fool with our energy systems. We just want to turn it on and turn it off."

He points out that Indonesia, with crucial political leadership, shifted 40 percent of its population—50 million households—to LPG from cooking with kerosene in under five years. It can be done—and fast. (And at huge savings: within one year the capital investment of $1.4 billion led to savings in subsidy payments of $5.54 billion.) This sort of transition can be accelerated.

There are even companies out there developing cooking stoves and electricity-generating units that produce biochar, a charcoal-like substance (of which black carbon is a component) resulting from the *intentionally* incomplete combustion (in a process called pyrolysis) of various kinds of biomass. Biochar improves soil fertility for agriculture and sequesters carbon in the soil, as the mysterious makers of the *terra preta* in the Amazon discovered centuries ago. Because black carbon is so extraordinarily stable, once it gets carried into the ocean by runoff, some of it stays there for thousands of years, until it gets buried in deep ocean sediments. This technology could actually turn black carbon into a carefully controlled asset—one that might eventually help thin the dangerous atmospheric army of its fellow warming agent, carbon dioxide, and take a whole lot of carbon out of global circulation. Johannes

Lehmann, a researcher at Cornell, and his colleagues have calculated that by 2100, biochar produced from projected biomass energy supplies could sequester more carbon per year than our current yearly emissions from fossil fuel burning.

"The three things we have to do" on the climate front, Durwood Zaelke says, are "short-lived pollutants, carbon dioxide, and [carbon] removal. There's a cookstove that does all three: a cookstove that makes biochar. Capture black carbon, reduce carbon dioxide because it's more efficient, and turn the black carbon into biochar." He acknowledges that these are early-stage technologies that need much more refinement and investment, but Zaelke—like Sonam Wangchuk, like his friend and colleague Veerabhadran Ramanathan—exudes optimism. "There's even a backyard barbecuer that makes biochar. You can do these things!"

Because of the surge in attention to the risks of cooking with solid fuels, Kirk Smith finally sees some glimmers of progress—flames he is keen to kindle further. Entrepreneurs, nonprofits, and government programs have shifted their focus from "improved" stoves, which may have improved fuel efficiency but did little to improve health outcomes, to truly "clean" stoves that dramatically cut down on products of incomplete combustion. But getting those devices to market, and then into the hands of the people who need them most, remains a huge hurdle.

In 1981, in Gujarat, Smith and his Indian colleagues performed the first measurements anyone had ever done of smoke exposure from biomass cookstoves. In October 2013, he returned to the same cluster of villages, searching for a woman who appeared in a photo he has used in hundreds of presentations on air pollution and cooking he had given over the course of his career. He went around asking people if they could identify the woman in the photos; he soon found her.

"Diwaliben is now seventy years old and a bit frail physically but clearly remembers the study and that she was the first in the village to wear the equipment," Smith wrote in an email soon after the trip. "I told her that her photo had been seen by millions of people around the world and that hundreds of more studies like the first one had been done in India and elsewhere since 1981. It was a bit emotional for both of us, as you can imagine.

"Although there was some noticeable change in the villages—reliable electricity, piped water, cell phones, TV sat dishes, a few refrigerators and LPG stoves, a pressure cooker in every house, and a larger fraction of 'pukka' housing and schools—made of permanent materials—much has not changed, including nearly but not all households still using open biomass *chulha*s for most of their cooking. Memories varied, but three to five 'improved' stove programs had swept through the villages since the 1970s . . . but no one could think of any improved stoves that even were still in existence in the villages in the area, let alone being used. Today, India has about 700 million people relying primarily on open chulhas for cooking. In 1981 when we did the first study, there were about 700 million people in the entire country! Clearly, very little progress has been made in terms of the absolute health burden from household air pollution here, which has recently been estimated at about one million premature deaths annually, the largest single health risk factor found for Indian women and girls."

Smith pointed out that there have been many dozens of studies measuring exposure in Indian kitchens since that first one, and yet the concentration of pollutants in kitchens today is more or less in the same range, over thirty years later.

It's only in the last few years that "cooking like gas" has become a viable option for those who only have wood and dung to burn. A few hundred thousand years since our ancestors kindled the first cooking fire, we are finally upgrading to Hearth 2.0.

When I ask Jacob Moss to name the biggest success stories in the history of clean cooking solutions, he cites the China program and Indonesia's rapid conversion to gas. "Today by far the biggest success story in the world is Indonesia switching from kerosene to LPG. So that's huge, and shows you can do really innovative things at a big scale and it can succeed."

"Actually the most successful program was the Industrial Revolution," Moss says. "That wasn't a targeted intervention, but it's true. If your stove breaks, my stove breaks, we don't have an open fire in the middle of the living room. We go get a new stove. It's as simple as that. And we go eat out until we get it fixed. And that's not the case for the rest of the world, and it's just insane."

The transition in the arc of burning that Tami Bond described, and that my grandfather experienced over the course of his long life, is indeed the most powerful soot-scrubbing agent there is. But it takes a while. And given black carbon's ubiquity and potency as a lung and climate saboteur, we don't have the luxury of waiting. The case for cleaning up black carbon is strongest closest to home—in the hearth. And we aren't accelerating that transition nearly as fast as we could or need to be doing.

Moss, the man who leads this effort across the entire U.S. federal government, shakes his head. "And the fact that it's not appreciated generally as an insanity," he says, "and that we should be putting every resource we can into figuring out how to solve it is a constant source of wonder to me."

When I ask him what Marthang will look like in twenty years' time, Tashi Phuntsog, a twenty-nine-year-old Kumikpa and a son of a large Gonpapa house, doesn't hesitate: "Chandigarh."

Chandigarh! The planned city in Punjab designed by the Swiss modernist architect Le Corbusier and built under Prime Minister Jawaharlal Nehru's government in the 1950s. Several Kumik men have spent time there as soldiers on posting. A couple of Kumikpas even studied there. The young generation holds it up as a model to aspire to. By the standards of most chaotic, teeming Indian cities, its orderly grid of wide, clean streets practically shouts *pukka*, organized, modern, at least to the people of these distant mountains. So when they point out the clean, rectilinear layout of Lower Kumik, the "State and Main" intersection of its two future roads—so different from the tight, organic cluster of earth-and-stone homes in their ancient village—you can sense the pride and anticipation.

When it's finished, they will have *arrived*. They will be modern. They will be cooking with gas.

"Cooking with gas," after all, means more than just cooking. It means the whole package: clean homes, modern conveniences, lights that come on with the flip of a switch, easily accessible drinking water, proximity to services and goods, mobility. Just like Chandigarh, or Jay Street circa 2013. The fruits of the soot-drenched Industrial Revolution: now that they've had a taste, it's what the new residents of Lower Kumik understandably want.

We are sitting in the guest room of Tashi and his wife Chorol's new house, the first to be completed in the new village. It is covered with a fresh coat of concrete plaster, so fresh it crumbles a bit when I lean back against the wall. Their other guests, their neighbors Phuntsog and Tashi's brother Nyimbum and his wife Yangdon, all nod in a chorus of approval at this invocation of the paragon city.

"Slowly, slowly, this will happen," Tashi says. "Last year there wasn't even a single plant here, or tree. Now look at it this year!"

"So if Marthang is going to look like Chandigarh, what is the most important work that has to begin now?"

They debate for a while until they reach a consensus.

"First is house," says Phuntsog. "Then drinking water."

"Here drinking is a big problem—main problem," says Tashi. "Water and light. We need hand pumps." In winter, while the men are away, the women have to haul ice from the river and melt it on the dung stove for drinking and cooking.

"Then light," Phuntsog says, using their shorthand term for electricity. "Then trees." Being green, in a world of sere browns, that's cooking with gas, too.

Tashi points out that last year the local government was ready to provide some funds for things like installing hand pumps in Marthang. But the yura committee, including Stobdan's father and the two lamas, whose members supported a different political party, didn't accept the support. There was disagreement in the village. "People should get united and vote for the same candidate, then we will get funds" to develop Marthang further. He shakes his head a bit ruefully. "But people are not united. That's the problem."

As a soldier and member of the Ladakh Scouts, Tashi spends nine months of each year posted to different parts of India. Last year it was

Chandigarh. This year he goes back to the Nubra Valley, India's northernmost. He has served on the Siachen Glacier, the world's highest battleground, where India and Pakistan lobbed shells and took potshots at each other until recently, but where over 90 percent of the casualties are due to avalanches and altitude. Other soldiers from Kumik have served on peacekeeping missions abroad. The wives stay behind in Kumik, managing the household and the animals and raising the children.

"People kept on talking about shifting here," Tashi says, "but nobody shifted. So we take the first step. And now the people are coming down, looking at us and feeling jealous," he says with not a little satisfaction. 'They have gone and planted trees, built up a good house,' they say."

When Tashi heads back to Nubra in September, his wife Chorol will face the long winter with her son and a few neighbors, mostly other young wives, and Bura Tantar and his wife—the Marthang vanguard. "It can get a little lonely," Chorol admits in a soft voice, eyes downcast. "Sometimes lonely, sometimes happy in Marthang. People keep coming, going."

But she keeps coming back to an optimistic theme: before long more people will come to join them. As the first family to finish and occupy their house in Marthang year-round, she feels like she and Tashi and their young son are helping people realize what's possible. She is excited when she talks about a community greenhouse that will be managed by the *ama tsogspa*, the local women's alliance group.

"Slowly they will come down," she says. "The *ama tsogspa* will do good work. We will look after the greenhouse turn by turn, and if one doesn't come to work their turn, we will charge some fine. It could be very profitable. Strawberries grow here, cucumber, watermelon, pumpkin. For the village work, we cannot say no to the work. They should come; otherwise there will be a penalty." She laughs. "That's why people will come and work—if they feel jealousy, it will only be in the heart, not out loud!"

"And what will you use for cooking and heating?"

"We will use dung the same as before," Chorol says. "We can find plenty of dung here."

"If people have enough money and electricity is good," Tashi adds, "then they will get [electric] heating system."

"And where will the electricity come from?" I ask.

Electricity hasn't reached Marthang and probably won't for several years (depending on politics, and on whether the village can "unite").

"From the Karsha diesel generator," Tashi says.

Right across the river, a genset provides intermittent power to Karsha, which has a magnificent view of Kumik's dying snowfields. So the local government's current plan is to add transmission lines to link that power source up to Marthang—to trade the ancient soot emitter for the modern one. (A cheaper option, in narrow economic terms, than installing a new solar electric plant for the new village.) Which will improve the burden in Kumikpas' kitchens but won't do much to spare the snow and ice that remains in Zanskar. The standard "modern" way of doing things isn't *always* "cooking with gas."

Though they started building their new house more than five years ago, Stobdan and Zangmo haven't joined the vanguard in Marthang just yet. They want to move, but their neighbors won't let them.

"They say, 'Who will give us *chang*? You can't leave your house empty. If you shift down to Marthang, don't come back.' "*

Stobdan has been itching to go for years. Zangmo, too, often says life will be better in Marthang: it will be easier to get to the fields, and there will be fewer of them. And no constant anxiety about having enough water. "But big problem for us to shift to Marthang," Stobdan says. "Others have someone to stay in their house. But we don't."

Their daughter, Dolkar, is away in Leh, on scholarship studying at the Lamdon School. Their son Lhargyal is finishing higher secondary school in distant Jammu. And their youngest, the mischievous Dorjay, a good-hearted rogue, has just gone off to Dharamsala to become a monk in the monastery of the Dalai Lama. Stobdan's father, Rigzin Tantar, lives in the *khangbu*, the little house, with his two sisters and brother-in-law, though lately he has taken to staying in

*One household made this fateful move in 2014. The owner locked the door and shifted to Marthang, for good. As the price of letting him go, the villagers insisted that he consign his share of water to other households, permanently.

their unfinished house in Marthang and visiting the new solar *lhakhang* daily.

Other houses have bigger families—like Gonpapa, with its seven brothers and three sisters, plus all of the wives and children—and can spread out to cover all the various tasks of building and planting and guarding new saplings in Marthang while also keeping up with their communal responsibilities in Kumik.

But Kharpa, Stobdan's house, is a lean operation. Zangmo does almost all of the domestic and field work, tending to the cattle and sheep and goats by herself. So Stobdan and Zangmo negotiated a deal with the other villagers, a kind of trial period of not providing *chang*, the liquid currency and fuel for so many group functions, meetings, and rituals (making *chang* an excellent proxy measure for a household's communal responsibilities). "We sign a contract with eighteen houses—no *chang* this year. We will see if it works. Then we are free, yeah? Then we shift."

"Now is another form of competition, the young generation," Stobdan continues, echoing Tashi. "Before they don't have idea, now they are in house, looking nice. Now competition to construct. The [women] say to husband, 'Push!' It depends on water. If there is no snow in two winters then all the persons go. Then no water, so they will shift. Marthang is for future, yeah?"

In the morning Ache Zangmo will be up by four o'clock; she will work in her neighbors' fields until seven, in the ancient practice of *bes*, communal labor-sharing. Stobdan tells her to rest more. He's worried about her health. "She is always working. You see she is getting weak, *zumo*. But she says to me, 'I will work until I die.' "

In Marthang, they both agree they will plant less barley for *chang*, and more vegetables and peas. Zangmo, like her husband, sees a bright future in Marthang. "In winter it's just me to feed the cattle, to go get water from above the school. The road is far, there are many stones. There is no water, the valley is long. Marthang is nice because there is good water! In Kumik there is too much work, it's difficult."

Like this, she counters each devil's advocate objection I offer: Marthang is hot, dry, lonely, desolate.

"In Marthang this year, there are many saplings. After two, three, four years it will look nice. Then Lhargyal will bring a wife. Dust—with enough water, there will be trees, more crops, then not so much dust." Less *chang* will be made. "The soldiers drink it less than the old men."

"With alfalfa and trees getting big Marthang will look so nice," she says. She smiles and sticks out her tongue, as she does when she's cracking wise, or expressing surprise, or about to say something she regards as slightly naughty. "Then you will bring your pretty wife to see!"

"In Marthang there is so much water people will be able to irrigate by themselves," she continues, "no need for help." When the world's hidden forces seem to be pulling you apart, this is not a trivial consideration.

"And there will be plenty of dung for fuel. Marthang has a lot of fuel. People from Ufti and Pipiting bring their cattle, and they eat the grass." This is important, Zangmo notes. The solar heating will help, and the gas will still be used for some cooking. But Zangmo assumes they will keep burning the way they always have.

I had asked Jonathan Cedar, the co-inventor of the BioLite stove, if it could burn yak dung. "I don't know," he had replied, "but why not? You should try it."

So I took his CampStove to Ladakh and Zanskar. I showed it to Stobdan and Zangmo one evening in their kitchen. We loaded it up with willow twigs, *burtse*, and yak dung and set it alight. It smoked a bit at first, but once it got going it boiled water for tea quickly.

"It cooks very fast," Zangmo noted. "And it looks nice." Then she got down to brass tacks: "But how much does it cost?"

I told them, and they whistled in surprise: Too much!

So I asked what they would pay for it. They say about 1,500 rupees, max. Thirty bucks.

One day, back in New York, I showed my grandfather the BioLite, the stove that was born just a stone's throw from his own grandfather's house.

My grandfather was a buyer of tools, appliances, household goods, and hardware for Masback and Sons. He was a hard guy to impress, a famously critical but fair judge of quality. (One time a guy came selling "unbreakable" plates and bowls that he wanted Masback to stock. "Unbreakable, eh?" my grandfather asked him. "You bet!" the man said. So my grandpa calmly pushed all the demo wares off of his desk and onto the concrete floor. They shattered into a thousand pieces.) He hefted the water-bottle-sized stove and peered down through his glasses perched on the end of his nose. He wants to know how it works. I explained. I could tell he was impressed from the way he arched his bushy eyebrows, pursed his lips, and nodded at me. "It's well made," he said as he turned it over and over.

When I went to SECMOL's campus for a visit, Wangchuk—also a man of discriminating taste when it comes to household devices—had a similar reaction. I set it up in the kitchen, where giant solar reflectors cook most of the meals but they still use some LPG. Wangchuk and I dropped some pieces of dried cow dung down on top of the burning sticks. It lit immediately, and the flames turned from yellow to greenish blue. We boiled water for milk tea in about a minute.

"Amazing!" he exclaimed. Wangchuk's eyes lit up with that familiar intense look that I know so well. He wanted to figure out how it worked and make a bigger version of his own for the campus. The elegant, simple genius of the thermoelectric generator—using the fire's own excess heat to power a fan to improve combustion in a feedback loop that burns off all the dark particles and turns them to current, then light in another form—appealed to his engineer's practical mind-set and his aversion to waste in all its incarnations.

He cradled the stove in his hands like a child on Christmas morning. "This is development that I *like*!"

I had also showed the stove to Urgain, a veteran tinkerer. He was impressed enough to deploy his trademark, all-purpose exclamation (for approval, dismay, surprise): "Oh ho!"

"Wood gas! Very interesting. And you can find wood everywhere," he said, unwittingly echoing BioLite's slogan, "Energy everywhere." "Technology is very fast!"

He examined it keenly. Before Jonathan Cedar or The Boss expand their respective empires to Zanskar, with stoves tailored to the locals' needs, Urgain Dorjay might beat them to the punch. He has long wanted to start a stove-making business in Padum. But, at a much smaller scale, he has the same basic problems they do: designing and prototyping a product that meets users' needs, finding good employees, and finding capital.

Urgain and Chondol applied for a 50,000-rupee loan for the venture, but they were turned down by the local bank. He's given up on that route for now. "Loan is too much trouble, maybe, we see."

But if he had a little start-up finance, he has lots of ideas. Now he's thinking that if he saves some money from his contract work to pave the road to Shila (the only thing those lucky Shilapas had lacked), and a little of Chondol's salary as a health worker, he might have enough to start the business. He would have to buy some tools and a stock of sheet metal and rent a small workspace near Mani Ringmo.

"Still nobody has this kind of idea, nobody start this. Maybe this is good. But maybe after two, three years some people are doing like this, then [opportunity] is finished!"

Zanskaris now buy metal stoves from distant Manali or Leh. Urgain has some ideas to improve on those conventional designs and tailor them to Zanskaris' needs. In a book that I once left in his house some years ago, Urgain saw some schematics for a stove that heats water. He wants to make a stove that can capture waste heat from the combustion gases in the chimney to heat water for various uses. "In the middle pipe I have house pipe, fix the water pipe. Then one drum up there behind *chulha*, which is filled with water. Water goes through the pipe inside the pipe, then at end, put a tap. With the smoke, it's getting hot."

Several people have already told Urgain they would buy such a system. "I am looking right now at what people can need: lot of hot water!" Urgain has also been thinking about how to cut down on all the smoke, by partially covering the openings on the front of standard stoves, which lower the temperature of the combustion chamber, reduce the efficiency of the combustion, and give the resulting smoke a way to escape into the kitchen. He would also add a grate or stand at the bottom of the stove, improving air flow and enabling users to easily remove the ash. And one more design improvement, from the man who once trekked for two weeks on a frozen river just to get matches: "At bottom here I make a rack to keep matches!"

Urgain is a busy guy. He runs a small travel agency, leading treks on *chadar* and in the summer. He works with Chondol to maintain their small farming operation, like most Zanskaris. He is the manager of the Zanskar Ski School, the first cross-country ski school in the Himalaya, teaching skiing for transport, communication, and recreation to six-year-old kids and middle-aged police chiefs alike. For many years, as a leader of Kanishka Welfare Society, he has been called on to lead emergency rescues and responses (kind of like a volunteer fire company chief) in Zanskar. He is the *panch* for the fifteen households of Khanggok (kind of like a mayor or city councilman). For four years he worked part-time for the Ladakh Ecological Development Group, based in Leh, to co-lead an EU-funded project to introduce passive solar heating techniques in Zanskar. He moonlights as a contractor. He makes a mean pizza.

But he thinks in April and May he would have time to make a lot of stoves. He sees it as a way to give people a better product to rely on every day as well as to give local youth meaningful skills and employment. "Then they can make also good *chulha*. Then they can make also good business. Many, many boys are just without work." They hang out in the Mani Ringmo, strolling around with nothing to do, caught with zero-value high school and university diplomas between the agricultural livelihoods of their parents and the white-collar jobs that haven't arrived in Zanskar and probably never will. (There are only a handful of government positions.)

This is how Urgain thinks about problems, like a home-schooled version of Wangchuk. If he had access to some finance and some technical resources to help him come up with a good, efficient design appropriate to Zanskar, his simple little stove business could make a dent in multiple challenges: make less smoke, clean up kitchens, save labor, conserve scarce fuel (and the time it takes to gather it), keep money circulating in Zanskar, give local youth confidence and employment.

"Produce some job for others, also," Urgain concludes with a smile and shrug. Then he'd be really cooking with gas.

"Let's see," he says. "Without plan, nothing happens, yeah?"

9

The View from Sultan Largo

The trouble with him was that he was without imagination.

—JACK LONDON, "TO BUILD A FIRE"

"We have fields here, there and there, and I'm just one!"

Tsewang Zangmo sweeps her arm across the tableau of late-summer Kumik to indicate the scope of her task. Today it's her turn to irrigate her household's fields of barley, peas, and wheat. Normally she would have help. But Stobdan is off in town for a government teacher training course. Her three teenage children are at distant schools, outside of the kingdom, in Leh, Jammu, Dharamsala. And her neighbors are all busy working in their own fields.

Zangmo grabs a shovel and walks briskly along the narrow footpaths between the ovals of ripening barley and small triangular patches of alfalfa. The sun is sinking fast toward the spine of the Great Himalaya. She walks over to talk to Nawang Phuntsog, who is crouched down at the edge of the small *zing* pond, trying to dislodge stones that block the outlet. After a few minutes of probing and prying with a long stick, he succeeds; water rushes out into a stone-lined canal and then veins outward through the fields of Kumik.

Now the clock is ticking. As is the custom, Zangmo has just two hours to irrigate her fields, using the snowmelt stored in the pond from the night before. She heads off toward their fields near the southern section of the *zbalu*'s wall. There she sets to work, spading aside clumps of dark earth. The heavy barley grain heads droop and rustle in the breeze. Zangmo crouches down among them, tossing aside heavy stones to let water burble into each *nang*, or "room," of the field.

Irrigation in Kumik is a well-choreographed sequence of openings and closings that exploit topographic opportunities. Zangmo knows the precise amount of water that she wants to give to each *nang* on this mid-August evening, and in what order. She gauges the volume of water in the canal by sight and measures out the liquid dosage in units of minutes. Some loss is inevitable—much water seeps away into the sandy soil of the canals—but of what flows on the surface, little is wasted. After twenty minutes of this dance, reorganizing soil and stones with deft moves of her spade, Zangmo pauses and surveys the scene with a clinical eye. She makes a quick calculation.

"There's not much water. Today we will only water half of a field."

She gives a loose cough. She leans on the handle of her shovel and closes her eyes. She snatches a brief moment of rest, just a minute or so, before she heads off to open another channel. And after that she will head home with a load of cut grass to dry on the roof. Then milk the cow, polish the offering cups, make bread for tomorrow's work. . . . Time is short and there is much work yet to do.

Two weeks later, just before the harvest starts, the Congresspas come to town. It's a few days before the councilor elections, which only happen every five years in Zanskar. The Congress Party candidate for executive councilor and his fellow party activists have come to ask for Kumikpas' votes. They walk by Stobdan's house and call to him through the open window: "Where is the *numbirdar*'s house?"

Stobdan points out the home of Tashi Dawa, and they go to notify him of their arrival, following pre-election protocol. (They are not allowed to visit individual homes, take tea and the like, as it's caused too many problems in the past—modern party politics exploiting the delicate dance of Zanskari hospitality.)

The Congress candidate is a college graduate from Shisherak, Zangmo's home village. "He is a good man," she tells me before we go to hear his pitch, "very educated." His name is Skalzang, but everyone calls him Gara. Like Kumik's own two Garas, I guess he must have lost some siblings as a child.

After darkness falls, villagers from every household fill the main room in the community hall. They listen patiently as this Gara's companions give stilted speeches in praise of the Congress Party (India's oldest, the party of the ruling Gandhi family) and denounce the incumbent chief executive councilor (CEC), a young man from a prosperous family in Padum, who had been affiliated with the Ladakh Union Territory Front in the past.

Then it's the young candidate's turn. He is a bit nervous but deliberate in laying out his priorities, outlining the benefits he would secure for the village. He seems uncomfortable as he delivers the expected pro forma criticism of the incumbent, hinting at his corruption and back-door deal making. He closes by reminding the Kumikpas, "We are only here for today, for a few minutes. You must live with each other, from morning tea to evening tea. After we leave, you must not quarrel. *Tsangma chik choste*. All must work together as one."

Speeches concluded, someone produces a knot of ceremonial white scarves. Tashi Dawa pulls them apart and drapes one around each visitor's neck, as is the custom for honored guests. The men shake hands and walk out to their vehicle.

The room is subdued. Some are still sitting on the floor, but most people get up to go back home. I get to my feet and start chatting with Lundup and some other young guys.

Then the room explodes.

Out of nowhere, the usually soft-spoken Tsering Motup (a younger man, not the *bodyik* teacher) starts shouting, gesticulating, and pounding his fists into his hand, angrily addressing the space two feet in front of his face.

Soon Tsewang Falgyon stands up and does the same. Their shouts mingle in the windowless space of the concrete community hall, which Falgyon, a contractor, built the year before with funds provided by the incumbent executive councilor. Their voices overlap and bounce off the mud brick walls.

"The incumbent has brought great benefits to Zanskar and to Kumik," they declare. "These Congresspas have no right to enter people's houses when they campaign."

Both men have clearly been drinking. The *chang* and *arak* fuel their pique, move them to say things to their neighbors they would likely never say in public in the sober light of day. Soon their twin monologues devolve into a vicious shouting match with Rangdol, a young soldier on summer leave, and a couple other young men who also support the Congress Party candidates.

The rest of Kumik looks on in stunned silence. The older women worry the hem of their frayed woolen *gonchas*, looking upset. The younger women flit about, embarrassed, occupying themselves with the young children. Several *abi-leys*, grandmothers, come up to me with tears streaming down their face as the men scream at each other. *Tsokpo duk*, they say, shaking their heads mournfully, "This is not good." The children, meanwhile, hop around with glee: this violent disagreement is like witnessing a solar eclipse. Something so rare and so seismic is downright exciting.

One of the pro-incumbent men taunts Rangdol, alluding to his absence from the village nine months out of every year: "What do the soldiers do for Kumik?"

This enrages Rangdol. "Do you feed my family?" he screams. "Do you put food in their mouths?" Their faces are inches apart.

Meanwhile Falgyon has drifted over to the *choktse* in the corner, where Tashi Stobdan has remained seated through it all, staring straight ahead, impassive. Falgyon leans down as he hurls abuse and shakes his finger at Stobdan. (Stobdan, as a government employee, says he is "officially neutral," but everyone knows he is a Congress Party supporter.)

Zangmo, seated next to her husband, leans across him and gives Falgyon an earful of her own. Her mother died just a week ago in Shishcrak. ("She is a very good mother," Stobdan had said quietly of his mother-in-law on the day she took ill. "All the people respect her.") Zangmo has been consumed with grief, weeping from dawn to dusk, even as she milked the cows and pressed salt tea on her visiting well-wishers. Now, as she admonishes Falgyon to behave properly, she keeps fingering her prayer beads. Her lips quiver with the silent mantras of mourning, even as Falgyon replies in kind, finger wagging in the air between them. Finally Stobdan has had enough.

"Why are you standing like this?" he shouts back at Falgyon. "We are all sitting down! We are listening to you. Why are you shouting?"

Then he launches into his own list of grievances, shouting back at Falgyon and a man named Wangchuk, an older soldier (whom everyone calls "Beda") and head of one of the three households in Pang Kumik, who has joined Falgyon in the corner. The three men trade jibes and accuse each other's candidates of corruption, of campaign law violations, of lying to and stealing from the people of Zanskar.

This goes on for a few minutes. The oldest village in Zanskar, the first to be destroyed, rending itself from within, acting out, or obviating, the curse of the *zbalu*, the angry *lha*, the tightening water vise. Rangdol is still shouting at Tsering Motup. Stobdan is shouting at Falgyon and Beda. The rest of us just gape, like rubberneckers at a car accident.

Beda, a large, dour guy, is literally vibrating with rage. He picks up the wooden *choktse* table. He shakes it at Stobdan. Is he threatening to bring it down on his head?

Around the room, eyes widen.

He holds it up there in the space between them for a very long moment—long enough that it seems to me that he's still making up his mind, right then and there, whether he will bludgeon his neighbor or just turn this into some kind of physical, rhetorical flourish.

Beda, perhaps subconsciously sensing the shock coursing through the hall, puts the *choktse* down.

(Later people will ask me over and over, in jokey tones that seek to downplay the whole thing, but still somehow manage to convey everyone's deep surprise and dismay: "Did you see when Beda lifted the *choktse* at Stobdan?" *Oh, that Beda!*)

The argument continues for a few more minutes at a slightly lower volume. The maneuver has drained the venom somehow, as though the men had edged up to the abyss and then backed away, chastened by what they had seen there.

Then quite suddenly, mysteriously (as many of the Kumikpas' unspoken signals are mysterious to me), the fracas breaks up. The antagonists have shot their bolt. Everyone gets up to leave. Tsewang Rigzin's kindly, long-faced Aunt Dolma gives me a mournful headshake as she walks past. We all filter out quietly into the star-speckled night.

Later I hear some Kumikpas whisper that some of the instigators of the confrontation have improperly benefited from preferential contracts that the incumbent has granted. Last year, Motup installed the wires that bring very sporadic power from the Karsha diesel generator (and sporadically from a new small solar plant near Kumik), while Falgyon was given the contract to build the new community hall. Another supporter's household was appointed to host the *anganwadi*, the village maternal and child nutrition center, a job that comes with a nice monthly stipend and none too demanding duties.

I point out in response that this kind of patronage—steering lucrative government contracts or jobs to your political supporters—is a common practice, not just in Zanskar but throughout India and, one might argue, the world. Someone's got to do those jobs, and to the victor go the spoils, right?

The difference in Zanskar and India is the scale and brazenness of the graft and how it filters down into every aspect of daily life. Everybody skims money and supplies. Certain funds designated for watershed development projects, for example, have tended to evaporate and condense into certain well-connected people's pockets, and then flow into new guesthouses and private vehicle purchases. It's debilitating and dispiriting, but people get resigned to it all, Over the past couple of decades, Kumikpas, like villagers throughout Zanskar and Ladakh, have become as cynical as any resident of a Chicago neighborhood controlled by a party machine. You're a fool if you're in power and don't take advantage while you can, or so the conventional wisdom has it. This sort of whispering—so-and-so "eats" the money—is now stock in trade and has been turned against just about everyone over the years, headman Tsewang Norboo and schoolteacher Tashi Stobdan included.*

The morning after the fracas Stobdan is a bit agitated. He is resigned

---

*Gara, the challenger, went on to win the councilor election, but his Congress Party would lose nationwide in a landslide almost two years later, largely due to corruption scandals.

to the storm he sees coming—some kind of impending freeze in relations between the two factions.

"This village goes in two parties in future. But I pray to god to never divide in two parties, because then is totally separate—even sheep is separated!"

Mostly he seems sad. "In the village we live," he says plaintively, to no one in particular. "We have to share happiness, everything! For instance, if someone dies, we need *phasphun*."

All that week, *phasphun* members had assembled from far and wide to handle every detail after the death of Zangmo's mother. It was not far from his mind.

"At the meeting I said, 'We must live together. For life. These guys come for one minute, one hour, and leave,' " he says. "But they are not listening."

He purses his lips and shakes his hand as though something has burned him. He is convinced that the other men cooked up the whole display in advance, that they are plotting to ask the CEC to remove him from his teaching post if he wins re-election. He wears the same distracted look I saw on the day his name was struck from the lineup at the Sadbhavana Mela, when the army gave Kumik its new generator.

I tell him that all seems a bit paranoid. But I am *chigyalpa*, so what do I know? I only know that I am out of my depth, aware that I have witnessed the emergence of subterranean currents that have built up some kind of invisible pressure through decades' worth of daily jostling and sharing and score-keeping in the shadow of these thirty-nine houses. From what I know about the methods of politicians in India, there is likely merit to both sides' accusations. From what I know of village life, there are many other dynamics at work. One could live here for thirty years and still not be wholly privy to the mycorrhizal network of nursed grievances and interlocking calculations in this remote village.

But what seems clear, even to an outsider like me, is that some powerful currents have upset the ancient equipoise between cooperation and competition in the "waterless village." The water and fire connections are being severed not just by external forces but from within.

• • •

The next morning Falgyon and Stobdan pass each other on the way to the spring to get water. They greet each other in neutral tones—*"Jullay." "Jullay."*—and move on.

Later in the day, Beda gives me a ride back from Padum in his Tata Mobile truck. Beda is his calm, collected, slightly sardonic self once again. He invites me into his home in Pang Kumik. His young daughter serves us tea, and we talk about the army, about the new house in Marthang he is finally getting around to building, the dimensions of which he and I had laid out together with string to face and harness the sun two years ago.

He recently got a promotion in the army, and it comes with a salary increase. The whole village turned out a couple of weeks ago to celebrate this good fortune. Stobdan and Falgyon and I had sat together at the same *choktse*, laughing and drinking *chang* into the wee hours, toasting Beda, toasting Kumik, toasting each other. Mindful of the approaching morning and the need to pace myself, I had kept trying to dilute mine by pouring water from a pitcher into my cup. Falgyon playfully snatched the pitcher away.

"Don't you know there is a drought in Kumik?" he said. "Drink *chang*, not water!"

We all had a good laugh at that one.

Beda and I look out the window at the setting sun. He has a load of newly built windows and doors in the back of his truck for his new solar house. We discuss the price of wood these days—expensive! Neither of us mentions the raised *choktse*.

In 2000, James Hansen, the well-known NASA scientist, published a paper with some colleagues in the *Proceedings of the National Academy of Sciences*. In it he proposed that reducing emissions of methane and other short-lived greenhouse gases, "combined with a reduction of black carbon emissions and plausible success in slowing CO_2 emissions . . . could lead to a decline in the rate of global warming, reducing the danger of dramatic climate change. Such a focus on air pollution has practical benefits that unite the interests of developed and developing countries."

This modest-seeming proposal (effectively saying, "There seem to be even more good reasons than we thought for taking a hard look at accelerating efforts to clean up air pollution") caused quite a fracas. Hansen had been one of the earliest and most vocal scientists to urge Congress to act to slow down emissions of greenhouse gases, especially carbon dioxide. Environmental groups accused the nation's leading climate scientist of everything from selling them out to needlessly muddying the waters by introducing an unwelcome distraction from the need to focus on carbon dioxide, everyone's agreed-upon global warming bête noire. Republican politicians opposed to curbing carbon dioxide emissions seized on the paper, distorting it to argue that climate scientists were now backtracking on earlier claims. In the waning seconds of one of their debates, George W. Bush seemed about to cite Hansen's work to Al Gore, in order to downplay the significance of the threat from global warming, before he was cut off by the moderator: "Some of the scientists, I believe, Mr. Vice President, haven't they been changing their opinion a little on global warming?"

Much new information has emerged since 2000 about the extent of black carbon's impacts on regional and global climate, thanks to the efforts of scientists like Tami Bond, Veerabhadran Ramanathan, Mark Jacobson, and many others. Uncertainties remain, but their best estimates suggest that black carbon contributes 60 percent of the radiative forcing that carbon dioxide does, making it the second biggest contributor to climate change.

The upshot of all that work: black carbon's influence is much worse than we thought, and there's more of it in the atmosphere than we realized.

But the fear remains in certain climate advocacy circles that focusing on black carbon runs the risk of taking policy makers' collective eye off the ball—and, further, that it may be used in a game of political bait-and-switch by disingenuous people who want to torpedo the next attempt to pass urgently needed legislation to put a cap or a price on carbon dioxide emitted by economic activity. Some warn that these climate obstructionists will say that tackling soot gets us off the hook, that we don't have to worry about the harder task of decarbonizing our entire economy. This concern assumes a finite attention span among those decision makers who are open to action on climate change and a cynical opportunism among the skeptics and deniers. While these are not unreasonable assumptions, they are beside the point.

The case of Senator James Inhofe is instructive. The Republican senator from Oklahoma is the most prominent climate change skeptic in Congress. He frequently calls global warming a "hoax" and taunts Al Gore whenever snow falls in Washington, DC. But he also cosponsored a bill in 2009, along with Democratic senators Barbara Boxer, John Kerry, and Tom Carper, requiring the EPA to conduct a thorough analysis on black carbon's sources and impacts and to recommend avenues for mitigating its emissions. He has since repeatedly called hearings about the pollutant and called for action to clean it up because he is concerned about its health impacts around the world. He is, paradoxically, simultaneously a flagrant climate denier and a black carbon hawk.

At one subcommittee hearing on black carbon, Inhofe said, "First, I should point out that black carbon has nothing to do with global warming or carbon dioxide." Totally false. But see what he's doing there? If not, his next sentence makes it clear: "It is, however, an important topic, especially on the continent of Africa." Very true. Black carbon and the other co-emitted pollutants that make up household air pollution steal more healthy years of life from African women than any other environmental risk factor, including lack of access to improved sanitation.

Inhofe's rhetorical move illustrates the reason some advocates for strong action to slow down climate change are nervous about shouting too loudly about the threat posed by black carbon emissions: it will serve the interests of those powerful political (and corporate) constituencies, they say, who are happy to keep us from talking about, let alone fighting, the elephant in the room. Which in the long term, as everybody knows, is carbon dioxide.*

---

*The great chemist Michael Faraday, who demonstrated in 1860 how to prevent soot formation, ended that same series of lectures by marveling at the power and abundance of "carbonic acid" (aka carbon dioxide): "As much as 5,000,000 pounds, or 548 tons, of

The laws of physics, of course, don't bend to political concerns. While this kind of gamesmanship is fretted over at conference panels and think tank meetings, those tiny little grapelike chains of black carbon keep doing their thing, soaking up rays and cranking up the temperature of the air over Greenland, over the Himalaya, even over the western United States, whether politicians in Washington like it or not.

As does carbon dioxide. Their one-two punch ensures that the Rongbuk Glacier will keep shrinking and the intrepid David Breashears will be there to document its death throes for the rest of us. The geophysicist Henry Pollack perhaps put it best when he wrote: "Ice asks no questions, presents no arguments, reads no newspapers, listens to no debates. It is not burdened by ideology and carries no political baggage as it changes from solid to liquid. It just melts."

For their part, Bond and Ramanathan, two of the leading scientific voices drawing attention to the "bad actor" qualities of black carbon, are always at pains to explain to decision makers that cleaning up soot is not a substitute for reducing carbon dioxide emissions. Based on the best current science, it is urgent, they say, repeatedly and loudly, to do *both*. "Reducing black carbon and ozone in the atmosphere is like applying an emergency brake in a car out of control," Bond testified to Congress in 2010. "It will slow the vehicle quickly and give you a little time to think. But the problem will continue if you don't take your foot off the gas pedal—that is, if CO_2 emissions are maintained."

"If you want to avoid near-term impacts which are accumulating and are going to get worse and worse, you've got to do *both* of these things," says Durwood Zaelke, the environmental law expert. "The people who talk about distraction and trade-off are not keeping up with the science and they're also misunderstanding policy. If you do aggressive mitigation on that [short-lived climate forcer] side using known technology, usually with existing laws and institutions, you can avoid

---

carbonic acid is formed by respiration in London alone in twenty-four hours. And where does all this go? Up into the air. . . . As charcoal burns, it becomes a vapor and passes off into the atmosphere, which is the great vehicle, the great carrier for conveying it away to other places." It would be another thirty-six years before Svante Arrhenius made his early calculations showing how all that accumulating "vapor" would turn Earth's atmosphere into a hothouse.

1.1 degrees of warming by the end of the century. If you do a 440 parts per million CO_2 mitigation—say you can do it—you avoid 1.1 Celsius by the end of the century. So they're exactly the same at the end of the century. At the midpoint of the century, if you cut the short-lived pollutants, you avoid 0.6 deg Celsius. If you cut CO_2 under 440 ppm, you avoid 0.1 degree Celsius by 2050. It doesn't mean you shouldn't do CO_2, because CO_2 over the long run will become an increasingly larger fraction. But what about between now and 2050? It's six times more effective. What about between now and 2100? Equally."

There is also the possibility that we delay the onset of perilous feedback loops in certain tipping elements—like methane escaping from the Arctic tundra—if we can slow down warming in the next couple decades. "We need to be thinking of the emergency lever," Zaelke concludes, "because we may find soon that these feedbacks tip us into some very dangerous territory that is demonstrable, and political will is galvanized"—and then leaders will want to know what fast-acting options are available for ramping down warming quickly.

The idea that attention to black carbon will detract from the effort to reduce CO_2 seems to stem from a "can't walk and chew gum at the same time" attitude. (As many glum scientists pointed out to me, after all, it's not encouraging that we've been at it this long and haven't gotten anywhere with reducing carbon dioxide emissions.) But in many, many cases, it's a moot point: actions to reduce black carbon *also* reduce carbon dioxide emissions. They often stream out together, needlessly, from the very same dirty fires.

In China, the world's fastest-growing economy and biggest carbon dioxide emitter, a lot of the black carbon and other carbonaceous PM2.5 comes from coal combustion for both industry and domestic heating, and from diesel. Any crash program—a scaled-up version of what the Chinese government is planning for Beijing—would crack down on the particles streaming out of those pipes *and* reduce the consumption of fossil fuels. A recent study estimates that a relatively modest carbon tax in China, in addition to reducing carbon dioxide emissions, would have the added benefit of preventing 89,000 premature deaths per year from exposure to air pollution and increase crop yields at the same time. Likewise, new regulations limiting carbon dioxide emissions from existing power plants, announced by the U.S. EPA in June 2014, are projected to prevent 4,000 premature deaths and 100,000 asthma attacks in their first year of implementation—because the caps will also reduce levels of fine particulate matter and other dangerous co-emitted pollutants. Win-win-win-win, as Sonam Wangchuk would say.

But there is an even thornier geopolitical source of wariness about black carbon, one that both frames and extends beyond the political dynamics around climate change in the U.S. capital.

It goes like this. The developing world is responsible for the bulk of global black carbon emissions. The United States' current share of global black carbon emissions is about 8 percent; emissions there and in Europe are projected to keep declining through 2030 and beyond. (Though that decline, too, could be greatly accelerated, and many more lives saved, with smarter, more aggressive programs.)

Meanwhile, China's black carbon output doubled between 2000 and 2006. The developed world—the United States, Europe, and other industrialized countries—are of course responsible for the bulk of historical carbon dioxide emissions. As of 2010, the United States was contributing over 16 percent of global carbon dioxide emissions, with less than 5 percent of the world's population.

Understandably, some people in India and China have expressed concern that black carbon will be used to shift the blame for climate change onto them, to obfuscate this larger warming ledger, to change the conversation at international negotiations and in the global media from carbon dioxide to the climate impacts of black carbon—most of which comes from fires lit to survive in the developing world.

"Now if you put it in a climate context, that's particularly sensitive," Durwood Zaelke says. "Because the way the G-77 [a block of developing countries in international climate negotiations] has positioned themselves on climate, they say, 'You in the West are the bad guys because historically you've taken up most of the air space, leaving us very little for our development. You're the bad guys, we're the good guys.' That's shifting because China is emitting more than anyone else in the world, and in a few years China's historical carbon dioxide emissions will overtake the United States'. When you look at black carbon, it flips it around: 'China and India, you're the bad guys. We've cleaned up our black carbon in the United States and Europe. Not completely, but we've done a good job.' "

Anumita Roychowdhury, executive director of the Centre for Science and Environment in New Delhi, acknowledges that there is indeed a fear among developing countries that the emerging black carbon narrative will shift the blame, and concomitant burden of taking action, onto India, China, and the sub-Saharan African nations—those countries least able to pay for upgrading their hearths and transport systems and power plants.

"It is that which is creating this whole suspicion, in the whole debate: 'What is it that you're up to?' " she says. "It's only because the black carbon debate is getting hijacked by the climate negotiation team. That has put the developing world a little bit on the guard. Equating the CO_2 with a bunch of other warming gases, and resolving it within the CO_2 framework—all that is creating a lot of confusion. Especially because the carbon dioxide debate is about apportioning responsibility globally. And you [in the West] have not yet been able to deal with that, address that equally, and now you're bringing in a bunch of other things in that platform. This whole blame, shifting of responsibility, it's now getting mixed up with that politics."

It does seem brazenly unjust and hypocritical: the West had its chance to develop, to improve its standard of living, through the use of fossil fuels. It polluted its way to prosperity. Now it's Asia and Africa's turn to lift people out of poverty using the Industrialization 2.0 playbook, and the West cries foul—"Look at all the soot they are producing!"—pointing the finger at people lighting fires just to survive and now, yes, finally enjoying the benefits of technology, like coal-fired electricity and diesel-powered transport. Just like those in the wealthier nations, who have conveniently forgotten their own Londons and Pittsburghs, their own sooty bootstrapping phase. It's kind of grotesque.

One recent study (led by an Indian researcher who is based in the United States) even analyzed the black and brown carbon produced by the ancient practice of cremating the dead on wooden funeral pyres throughout South Asia, noting that it adds up to a soot burden equivalent to a whopping 23 percent of the amount of carbonaceous aerosols produced by burning fossil fuels throughout India. Over 7 million corpses are burned on such pyres every year in India and Nepal. After a life spent burning just to survive, even their death is marked by one last sooty plume, which may escape into the beyond—but doesn't escape the new climate accounting.

Some officials in countries like India and China are thus loathe to admit black carbon into the conversation over climate change. The average carbon dioxide emissions of an Indian citizen, after all, are less than one-tenth of those of an American. And if black carbon is indeed used in such a cynical way, as a bait-and-switch to divert attention from industrialized countries' responsibility for the current mess we all find ourselves in—that would indeed be a travesty. Consider: about 17 percent of China's black carbon emissions in 2006 (and much of its carbon dioxide emissions, too) were generated in producing goods for export to the U.S. and European nations and other wealthy countries full of insatiate consumers. So if, as recent research suggests, black carbon and other aerosol pollution from Asia sweeps across the Pacific and powerfully disrupts global weather patterns, making storms deeper and stronger and more intense for the rest of the world downwind—well, for those of us in the United States, at least, it's a literal case of what goes around coming around.

In an interview in the lead-up to a round of climate talks in 2013, India's main negotiator, Jayanthi Natarajan said cagily, "As far as black carbon is concerned in the Himalayan region, we have done considerable studies, we have done modeling and other studies and we have come to our own conclusions on how far it is a climate forcer and how far it can be taken forward and its importance. I would emphasize that the country's economy as a developing nation and the strategic context of development of economy is absolutely vital and foundational in taking this discussion further."

Translation: We're too busy pulling our people out of poverty, and we can't afford to take our eyes off that ball for a second.

"That is a mistake we are making," says Roychowdhury. She advocates for strong action to reduce black carbon emissions, but in the "approaches and ways of dealing with it, we can't cast everything in that traditional mode of dealing with climate change."

"You cannot have this one-size-fits-all kind of approach to everything." She echoes other advocates and scientists in calling for a separate framework to deal with black carbon and other short-lived warming agents, one that takes into account the luxury-versus-survival character of the emissions coming out of Asia and other parts of the world still struggling with energy poverty. "You have to interpret your larger benefit, within the co-benefit context." The overwhelming reason to act to clean up black carbon, she argues, as with other sources of dangerous ultra-fine particles, is to protect people's health.

Indeed, the fears among Indian and Chinese leaders of confronting black carbon's full climate impact overlooks a crucial fact about the modern water-fire connection: precisely because the damage it causes is magnified *close to where black carbon is emitted*, the benefits of cleaning up black carbon redound overwhelmingly to the people who live there.

Because soot is what settles. It stays, for the most part, close to home. Cleaning up black carbon is a central battle in the larger fight to alleviate poverty in places like Uttar Pradesh, precisely because of the enormous costs—health, economic, time—it imposes on those who can least afford to pay them. This fact, combined with the emerging clarity of our understanding of soot's extensive damage, actually offers us all a tremendously cost-effective opportunity to accelerate *human development* by lightening the health burdens from Harlem to Harbin.

"Today anything you decide to do, you cannot have a single action that you want to take—we have to be judged from the perspective of co-benefits," Roychowdhury says. We can't afford to fall into the trap of making trade-offs between climate and health, she argues. "Diesel is actually an example of that trade-off. Say I come from a climate change [perspective], and I'm going to say diesel is a damn good thing to do because it's going to increase energy efficiency, and I'm not going to look at anything else. So I ignore the public health [impact] of this. Any strategy that you are pushing for must be tested on all the benchmarks."

Nowhere is the opportunity greater than in places like India and China and Africa. Cleaning up the air over the Himalaya is an issue of global interest and consequence, but the stakes are by far the highest for the people of the region. People like Stobdan and Zangmo and Tashi and Chorol and Urgain and Chondol and their children. And people in the villages and cities downstream, like Kaharan Purwa. As Joel Schwartz, the pioneering environmental health researcher at Harvard, has put the matter, "You only get the benefit of the health effects of reduced exposure to black carbon if you are the one who reduces the exposure, because these things occur locally. So China and India are the ones that are going to reap the health benefits of controlling black carbon in the future, and I think that has great prospects for helping us to convince them that it is time to act now."

Ramanathan, whose Project Surya is introducing clean cookstoves in Uttar Pradesh and measuring the attendant effects on health and climate, understands the reluctance of some of his countrymen to focus on the climate impacts of black carbon. But he's not interested in that particular political debate anymore. When we met in Lucknow, as he returned from a visit to some of India's poorest, most smoke-enshrouded villages, he pointed out that we now know enough to act on black carbon "on many fronts."

"Health alone is the winning argument," Ramanathan said. "There is an impact on agriculture, nobody denies that, glacier melting, monsoon. It's going to take some time to bring everybody onboard. On the CO_2, we have been at it for a hundred years, and half of America is [still] divided. What hope do we have to get 1.2 billion agreeing on the monsoon? It's not gonna happen."

"So," I asked, "what do you make of the reluctance to discuss black carbon's climate impact here in India?"

"Even that doesn't surprise me. You tell me why in the United States we have air pollution laws. Because of climate? Because of ecosystem? No! The whole western European clean air act is [about] health. I don't think Indians are different."

Roychowdhury makes the same point: U.S. EPA regulations on particulate matter pollution weren't motivated by climate considerations. "The driver has always been public health, right?" As I. H. Rehman, of TERI in New Delhi and a collaborator of Ramanathan's on Project Surya, puts it, increasing access to clean energy options in poor households "is a development imperative. It has huge implications in terms of women's health, in terms of drudgery, in terms of issues related to children.

"The Indian government is [interested] in the environmental benefits of it" as well, he says, "the co-benefits. But even if the environmental angle was not there, we would still need to do this.

"The thermal cooking energy needs, particularly of communities which are hugely dependent on traditional biomass, have not been attractive to the global community for various reasons," he observes a bit testily, "until it got into the politics of global climate change. Only then did you have people from the developed countries sitting up and saying, 'Aha, this is an important area!'

"I'm not talking of this politically," Rehman continues, "but from an implementer point of view. The immediate problem that a household needs to be concerned with is what? It's the problem he or she is having with the technology, their options, it's the drudgery, their health. That is the immediate concern, which is not to belittle or ignore the environmental concern—"

"It's just more abstract and distant and hard to imagine."

"Exactly. And it is human for a housewife or for a family to get concerned about their health first. But where the politics should not come into conflict is the fact that all of us are working in the same direction!"

All this speculation about political maneuvering and jockeying looks past the fact that we have an epidemic on our hands *right now*: as a major constituent of both household and ambient air pollution, black carbon and its co-emitted pollutants from inefficient fires contribute to close to 7 million premature deaths every year. These pollutants are killing people now, as a part of "atmospheric brown clouds" hovering over hotspots around the world, from Bakersfield, California, to Nairobi, like smaller versions of the vast one I glimpsed from my flight blanketing the Gangetic Plain.

As Tami Bond and her coauthors wrote in their definitive treatise on black carbon's climate impact: "Regardless of net climate forcing or other climatic effects, all BC mitigation options bring health benefits through reduced particulate matter exposure." Indeed, the United Nations Environment Program estimates that pursuing just seven black carbon technology and policy measures would deliver a value of $5 trillion in avoided damage to human health by 2030—*every year.* "The most substantial benefit will be felt in or in the vicinity of regions where action is taken to reduce emissions," a UNEP study notes. "The greatest health and crop benefits would be expected to occur in Asia."*

In 2012, UNEP, in concert with the U.S. State Department under then secretary of state Hillary Clinton, launched the Climate and Clean Air Coalition. The CCAC is a voluntary initiative; it is now made up of over sixty countries and institutional partners and works to accelerate efforts to reduce emissions of black carbon, methane, HFCs, and other short-lived climate pollutants through a mix of "highlighting and bolstering existing efforts" and spurring new research and action. "About half of the near-term temperature benefit could be achieved through measures that would result in net cost savings in the long run, even without accounting for their health and ecosystems benefits," another UNEP report notes.

This is one of the most striking conclusions from the UNEP analysis: some measures to mitigate black carbon—replacing conventional stoves with clean-burning stoves; replacing conventional brick kilns with vertical-shaft brick kilns—offer a *net cost savings* to society, even when the staggering health benefits and avoided crop losses aren't tallied up. The effective cost of the cookstove intervention alone is *negative* $6 per ton of CO_2 equivalent—meaning there is no net cost to society, but a net economic return in climate benefits alone.

"But prevailing short-term profit expectations make these measures less profitable for private investors," the UNEP authors note. "Hence, it is unlikely that the measures would be implemented solely through market forces." That's where national-level programs and multilateral efforts like the CCAC come into the picture. The CCAC, for example, has already approved programs targeting emissions from heavy-duty diesel engines and brick kilns in member countries, as well as from residential heating and cooking.

Those who oppose such efforts to clean up black carbon, on the

---

*Black carbon not only blocks sunlight needed by crops for photosynthesis; the incomplete combustion that produces it also emits significant amounts of carbon monoxide, volatile organic compounds, and nitrogen oxides, which produce ozone that damages crops.

grounds that it is a far lesser climate evil when compared to carbon dioxide, have to answer a simple question: what about the health impacts?

Let's say that aggressive investment and deployment of a suite of black carbon–reducing technologies and policies were pursued—vertical shaft brick kilns, diesel particulate filters, forced-draft gasifier cookstoves, and biogas systems. If, in fifteen years, new science emerged to suggest that the benefits to the climate weren't quite as substantial people thought in 2014, would we really regret the extremely cost-effective investment in saving all those lives and rescuing all that human potential?

Leading to the next obvious question: why aren't we making an all-out push to clean up 500 million smoke-filled kitchens, phase out and upgrade the dirtiest diesel engines, switch transport fleets over to clean fuels, help small industry operators build cleaner-burning kilns, ban open burning, and funnel agricultural waste into clean power generation and other innovative uses?

Regulations removing particulates from the air we breathe have been some of the most cost-effective public policy in modern history. According to the EPA, reductions in air pollution (including PM, ozone, sulfur dioxide, and nitrogen oxides) mandated by the Clean Air Act of 1970 prevented 160,000 premature deaths, 130,000 heart attacks, 13 million lost work days, and 1.7 million asthma attacks—*in 2010 alone*. The law's economic impact has been just as strong. An independent study commissioned by the Small Business Majority in 2010 concluded the ratio of financial benefits to the law's costs was 40 to 1. The EPA's own peer-reviewed study found that benefits exceeded costs by up to 90 to 1. Even as the act cut "criteria" air pollutants by 68 percent between 1970 and 2011, the American economy grew by 212 percent. The EPA estimates that the total cost to the economy of meeting its tighter soot standards will range from $53 million to $350 million, while the benefits will range from $4 billion to $9 billion each year in avoided premature deaths, illness, and hospital visits.

As is often the case, the state of California serves as a bellwether. Veerabhadran Ramanathan and some colleagues recently concluded a three-year study for the California Air Resources Board. They found that black carbon concentrations in California's air were reduced by 90 percent over a forty-five-year period beginning in 1967, when the CARB was set up. It's the largest "natural experiment" conducted to date on the climate benefits of black carbon reductions. Over that same forty-five-year period, diesel fuel consumption increased by a factor of 5. The economy churned ahead, but soot levels in the air went down. How?

In the 1970s, in response to mounting public concern over its notorious smog in urban areas, California instituted stringent emissions controls on diesel vehicles, the largest source of the state's airborne particulate matter. The decline was largely due to these diesel controls. But controls were also implemented for emissions from industrial activities, waste incineration, and residential wood burning.

Few if any Californians were inconvenienced in their day-to-day lives. There was no measurable drag on economic growth, no interruption of the steady rise in Californians' standard of living. The only noticeable difference was a steady clearing of the air since the state's own "crazy bad" days of the 1960s and 1970s. The study didn't even examine the health savings, but focused on the implications for global warming. Ramanathan summed up the lessons from California's experience thus: "If California's efforts in reducing black carbon can be replicated globally, we can slow down global warming in the coming decades by about 15 percent, in addition to protecting people's lives. It is a win-win solution if we also mitigate CO_2 emissions simultaneously." A coauthor, Phil Rasch, said that getting rid of black carbon "might lead to a cooling of the planet by half a degree to a degree Celsius." (Recall: we've already warmed the planet by 0.8 degrees, and going over 2 degrees of warming is widely regarded to be a serious threat to the continuation of human civilization.)

California's experience shows that reducing black carbon does not have to reduce economic growth, and as Tami Bond observed, that the transition can be accelerated through smart policies.

"Despite the fact that black carbon is unwanted, removing it is not free," Bond told Congress back in 2007. California required industry to meet certain targets. DPFs—diesel particulate filters—today raise the price of new diesel trucks—but they are extremely cost-effective in terms of quantifiable climate damage averted and, even more immediately and concretely, in terms of asthma attacks, hospitalizations, and premature deaths avoided.*

To reinvent fire and clean up both skies and kitchens, we don't need to reinvent many wheels. A large chunk of the reductions in black carbon (and methane and other short-lived pollutants) can be tackled through existing mechanisms, institutions, and treaties.

National-level initiatives, such as the one recently launched by India, can target black carbon emissions, if they have real teeth and real funding. Sector-focused multilateral initiatives, like the Global Alliance for Clean Cookstoves and the Climate and Clean Air Coalition, can promote public-private partnerships, sponsor research and development, and connect enterprises with interested funders. In Europe, efforts have also begun to address black carbon through existing mechanisms, such as the thirty-five-year-old Convention on Long-range Transboundary Air Pollution (CLRTAP), the first international treaty to "act on the link between air pollution and climate change." Likewise the Arctic Council, made up of representatives from the governments of the eight Arctic nations, is looking at ways to collaborate to "enhance efforts to reduce emissions of black carbon from the Arctic States." The International Maritime Organization is "currently considering whether to control black carbon emissions from ships." The Global Alliance for Clean Cookstoves, though it doesn't target black carbon specifically, is also focused on accelerating the transition away from the inefficient cooking methods that produce it.

"What I predict and what I'm advocating," says Zaelke, "is that we

---

*And lest we forget, it's not just India that still has its work cut out for it. Even in parts of some countries that have traveled the full arc of burning, just taking a breath can still be a very dangerous thing to do. California, for example, is home to the most counties of any state in the United States with soot levels exceeding the new EPA yearly average PM2.5 limits of 12 micrograms per cubic meter. "Although I take my medication and follow my treatment plan, I still worry about air pollution caused by smoke and truck exhaust because it triggers my asthma symptoms," ten-year-old Jaxin Woodward of Vallejo, California, testified at a 2013 public hearing in Sacramento on the proposed new EPA soot rule. Already a Junior Olympian runner, she aspired to compete in track and field in high school and college. But her severe asthma required her to take costly medications like Flovent and keep a detailed treatment plan based on where she travels to compete. "Getting a full breath of air—clean air—is really hard for me."

disaggregate climate, recognizing that it's not one problem but a package of problems. And we take pieces out and put them in the right venues, existing venues like Montreal, regional venues like the Gothenburg Protocol, CLRTAP, the Asian haze agreement, the Arctic movement. And we need to develop tailored governance systems for other pieces. Don't try to solve everything in one agreement."

There are policy and business "vehicles" out there already waiting to be accelerated, infused with more public and private support. These existing frameworks can be leveraged to tackle most of the black carbon burden in our skies and kitchens. There's no need for new treaties or wide-ranging agreements. Much of the work to be done is at the national or subnational level. Each country will have to devise its own response to its soot crisis. Air quality regulations, vehicle emissions controls, open-fire bans, incentive programs, public transit planning, vehicle buyback and retrofit programs, national cookstove programs—each has a place, and each will have to be tailored to air quality and economic conditions in specific regions, cities, and localities.

In many cases, fighting black carbon will require the same kind of pincer-movement approach that Sonam Wangchuk perfected in Ladakh, combining grassroots mobilization and energy and outrage—demanding cleaner skies—with a push to get leaders and decision makers onboard with concrete financial and institutional support. In New York City, it will be accelerating the effort, via expanded public incentive programs, to wean each and every building off No. 4 and No. 6 heating oil and cleaning up the private fleets of diesel trucks and buses that ply the streets near playgrounds where asthmatic children frolic. In California, it will be more stringent monitoring and enforcement of recent regulations requiring private truck operators to install diesel particulate filters. In Punjab, it will be diverting millions of tons of rice straw into uses like power generation and banning postharvest burning in the fields. In Asia and Africa and South America and beyond, it will be scaling up successful pilot efforts to build and maintain small clean-burning biogas systems for cooking and lighting in schools and family farms. It will mean supporting efforts big and small, public and private, to "leapfrog" over old, dirty technologies like kerosene lamps and diesel generators and coal-fired industry by using innovative approaches like Mera

Gao Power's solar-powered strategy of combating the darkness in rural Uttar Pradesh. It will mean incubating thousands of promising ventures struggling to grow, like nascent fires starved for the oxygen of technical, financial, and moral support.

It won't be easy—but it will be worth doing. Just ask Ratan Tata, India's leading corporate citizen. He was the chairman of the Tata Group, the country's foremost industrial conglomerate, for over two decades until he stepped down in 2012; he now chairs the group's charitable foundations, the Tata Trusts. The Tata Group is kind of like General Electric, General Motors, and Microsoft rolled into one. Its products are ubiquitous—you won't travel far on any road in India before you encounter the name Tata on everything from freight-carrying trucks to solar panels to cell phones. The Tata Group owns steel mills and coal power plants, Tetley Tea and the storied carmaker Jaguar Land Rover. Because his company (founded in 1868 by his great-grandfather Jamsetji Tata) is deeply involved in almost every sector of India's economy, Ratan Tata has given much thought to questions of pollution, growth, and clean development—and at the end of a long career guiding everything from car making to steel production to power generation to financial and consulting services, he laments the lack of seriousness with which his country's bureaucratic and business circles are tackling the central challenge of clean development.

Over tea in a suite in the Pierre Hotel just off Central Park in Manhattan (owned by Tata's Taj Hotels group) he tells me that he was once asked to support an "X prize" innovation competition for the development of clean cookstoves in India. He readily agreed to fund the entire prize out of his own pocket—equivalent to $1 million. And then, he says, "nothing happened. And I got a terse little note saying the project is dropped, we don't need the prize anymore. No explanation."

He shrugs and points out that innovation is important, but "thereafter there's a long journey to be followed to make it *happen*." Tata knows better than most people just how hard it is to successfully develop and then bring to market products that target low-income consumers. As the head of Tata Motors, he led the long, torturous (and controversial, given that India has 1.2 billion potential drivers) effort to develop the Tata Nano, a basic car designed for the common man, priced at 100,000 rupees (about $2,500 at the time it debuted).

Take electricity, for example. "The government is deceitful enough," Tata says, "to say that they've electrified a village, and 'electrified a village' in their parlance means one bulb in that village. You want to bring electricity to that village, we need to look at lower-cost forms of power generation. And on a clean sheet of paper, how do we do it? Today a generator is 2 lakh rupees, 3 lakh rupees, how do we make it 1 lakh? What do we do to make that happen? Unless industry is allocating their energies to doing that kind of thing, nothing is in fact going to take place.

"I've come to the conclusion that our whole power plan in India is kind of wrong, because it goes on ultra-megawatt systems, and maybe what we need are a bunch of microgrids. Small power-generation units that may power a village and would be in a grid that could, somewhat like cell phone base stations, instead of talking to each other, be wired with each other and could draw and give power. And when you go to the smaller ones suddenly the alternatives for fuel open up enormously. It's different than the big stations. The small ones can run on a series of alternate things. So you don't need four thousand megawatts in one location, with everything else that goes with it in terms of storage of coal, the harbor to bring in vessels, and so on. The infrastructure you create by that scale maybe is not the right thing for India, not for all parts of India."

I mention Husk Power, a start-up company that builds microgrids using generators fueled by rice husks to produce electricity for over 300 villages in Bihar, one of India's poorest states.

"Which is terrific! Maybe India will end up having four or five different solutions to this. I'm just saying we should have those solutions, rather than have four hundred megawatts or nothing, which is where we are today."

Take another huge, and growing, source of soot in India's skies and lungs: diesel passenger cars and trucks (of which Tata Motors is one of the leading producers). Tata points out, "If you look at diesel vehicles in Europe, it's not the smoke-belching, noisy, rough-running engine that there was. It's today a very clean engine because it's turbocharged, it's got particulate filters, and the fuel has improved. And can improve even further. They're talking of clean diesel now that's like water you can look through. If we use that, and maybe a more expensive fuel, if we insist on that, the diesel will be preferred to gasoline in many, many ways." The owners of the trucks that bear his name all over India adjust the fuel injection system to save fuel. "And they don't upkeep the vehicle, so they belch smoke—and they really do belch smoke!" he concedes. But the technology is there to solve this problem if governments enforce tight standards and industry rises to the challenge to make new, cleaner systems more affordable. "The truth is, the new diesel and gasoline engines have an electronic engine management system, so it would take care of this."

If India—which stands to lose more from the twin health and climate impacts of soot than almost any other nation—is serious about tackling black carbon, because of its many varied sources, Tata says, "it has to be a total awareness. We have to undertake this on a war footing if we want it to work. And not lip service on it as it should be."

To be effective, both the private and public sectors need to be all hands on deck, with a focus on all the critical sources. He gives the example of the millions of black carbon–belching diesel generators used in backyards and small businesses for power, which today are unregulated (and which are actually subsidized via the fuel supports). Not only should there be implementation of stricter emissions standards, in line with the current Euro 6 emissions controls in Europe, which aggressively reduce fine particulate and other pollutants, industry must also be held accountable. Tata says that most large firms have flouted India's pollution controls, and "they reap huge benefits, because the lack of enforcement allows them to do that. So one of the issues to get better control of emissions is to strengthen or make more powerful [enforcement] . . . and reduce corruption in the enforcement system."

But in addition to government agencies, says Tata, his fellow industrialists also have a long way to go. He tells me he was part of a group of forty businesspeople involved in putting together a presentation on sustainable development at the first Earth Summit in Rio de Janeiro in 1992. After that experience, he then funded a five-year-long exercise on sustainable development in India for the Confederation of Indian Industry. "And we got zero support from Indian industries. Zero meaning zero! It just wasn't on anybody's radar screen, they didn't want to put it on their screen, they had no interest in it, and then later, [also] in terms of even global warming."

"I'm sad to say that most of the carmakers are behind the scenes working on reducing [India's emissions] standards," he says. "They are saying, 'Well, since the EU will be in conformity with Euro 5 or 6 as the case might be, the rest of the world might be satisfied with Euro 4, and will stay there,' and they will meet that because they can't afford to produce cars that meet Euro 6 standards in Africa, for example. They're loving having another set of standards for the rest of the world.

"And that's going to prevail unless we can get our individual governments to work on something, bearing down on the fact that if we don't do something, we will be responsible for the world suffering sixty, eighty, a hundred years from now—irreparably."

Or much sooner. Things can change pretty fast, after all.

But not only in the pants-soiling, Greenland-melting, Arctic death spiral feedback mechanism pileup kind of way, like when Shi Yafeng spread soot on Gansu's glaciers and got immediate slushy results. Sudden changes happen in the human realm, too, and these have striking ramifications in the far-flung physical world. Researchers who manage air monitors in the Arctic have found that black carbon in the region started to sharply decline in 1992 because of the collapse of the Soviet Union and its ensuing slowdown in industrial activity. Ice cores from Greenland show clear evidence of the success of the U.S. Clean Air Act passed in 1970: layers of snow deposited after that year contain dramatically lower levels of nitrate (a marker for acid rain caused by coal power plants).

We forget how powerful we are, and how quickly we can upend the status quo and its "inevitable" march forward. Under sustained pressure from ordinary citizens, New York City's MTA switched its entire fleet of public buses—the world's largest—over to clean hybrid or DPF-fitted diesels within a few years. During the 2008 Beijing Olympics, Chinese officials restricted private cars to driving on alternate days and cut down on new vehicle registrations, and the skies noticeably cleared within days; researchers later found that if such programs were continued city residents' risk of developing lung cancer would be halved. Indonesia switched most of its population from burning kerosene to using LPG in less than four years. In 2003, officials in New Delhi finally responded to years of mounting public pressure and court orders, and within months, all auto-rickshaws and city buses were converted to cleaner-burning compressed natural gas. Levels of fine particles in the city came down quickly and dramatically.

"It gives you enormous confidence that if you see action, you will also see results," says Anumita Roychowdhury of CSE, a Delhi resident herself. "The media began quoting common people in the city who were saying, 'Now we can see the stars in the sky!' " Politicians had initially resisted the CNG conversion program, so "it was amazing to see how the political parties were then fighting each other to take credit for that program."*

It all shows the power of people coming together and demanding clean air. The skies clear quickly, and more light comes through, in its own kind of subtle, burden-lifting feedback mechanism.

What if California's rules on heavy trucks and buses, requiring DPF retrofits or newer trucks, were embraced by other states? If India adopted the stringent Euro 6 vehicle emissions standards and developed targeted, well-funded incentive programs for kiln owners to switch to vertical-shaft brick kilns? Or if mega-cities instituted congestion pricing, like London's program, or switched their public bus fleets over to compressed natural gas, as Delhi has done, and hybrid-electric, like the MTA in New York? What if laws against diesel trucks idling in New

*New Delhi's sheer growth in population has since overwhelmed such gains. The city gains a few thousand new vehicles a day, and its boundaries have spread outward to envelop the heavily polluting industrial estates that officials had relocated away from residential areas a decade or two ago. Roychowdhury speaks wistfully of the kind of "war footing" approach she sees materializing in China, where the pollution might be crazy bad, but not even as bad as the air in her own city. If they want to shed its current mantle as the world's most polluted city, New Delhi's leaders will have to redouble efforts to clean up its air.

York were actually enforced? If India rerouted subsidies and incentives so that instead of encouraging diesel and kerosene consumption, it supported the rapid spread of soot-free solar power? Or if the world's multilateral agencies, like the World Bank, decided to help turn the "counterfactual" level of 7 µg/m3 black carbon and other PM2.5 in the air of the kitchens of the poor into the *factual*, the new reality?

What if hundreds of thousands joined the 362 people who made small loans to Mera Gao Power, using Sunfunder, an online peer-to-peer crowd-funding platform, to bring solar lighting to (and banish kerosene from) 1,250 households in Uttar Pradesh and supported fast-growing, finance-hungry ventures like it from Kenya to Nepal? How many lives might be saved, and unburdened, for a relative pittance?

And what will happen if we do nothing . . . if we continue burning the way we do now, business as usual? Veerabhadran Ramanathan and Durwood Zaelke have given this a lot of thought.

"If you don't start now trying to reduce black carbon," Zaelke says, "you don't save those lives, you don't save the glaciers, and you don't save the climate if Ramanathan and Jacobson and others are right. Because you can't wait for scientific certainty before you start a policy. It's too late. And time is our enemy here. Things are getting worse so fast, we are just about to break the 400 ppm barrier for CO_2.* If we think we're going to have any chance for staying below 2 degrees [Celsius warming], it's going to require heroic efforts on short-lived pollutants. And we know we can do that. It's within our political grasp."

I climbed to the top of Sultan Largo, the 18,300-foot mountain above Kumik, with Lobzhang Tashi, Urgain's nephew, one day in late August.

The night before the *lha* had brought a hailstorm down on us, then lightning and snow. I had sat out in the open in a low swale on a foam pad as hail stones piled up around my legs. We weren't after its *nilim*, I shouted to the *lha*, just a view.

---

*The 400 ppm threshold (as measured over the northern hemisphere for an entire month) was passed in April 2014.

The next morning dawned clear and bright, sun blazing in an azure sky. We climbed up the soft surface of the small, sloped glacier to the summit. From there we could see a long way: down the other side, where a larger glacier stretches toward Stongde, way off to the northwest in Pakistan, all the way to K2, the world's second-highest mountain. We could see the Karakoram, the Great Himalaya, the Zanskar Range.

Turning around, we could see where Meme Paldan and his team made a shallow, futile canal many decades ago in the scree at the foot of the glacier on the saddle between the two peaks. We could see, a few dozen yards away, the trench that Stobdan and his companions had dug for the same purpose in 1997. It's more clearly defined than its predecessor, but it runs parallel and in the same direction, toward the lip of the precipice and the spot where the wind pushed it back as useless spray in Stobdan's face.

We could see that both were dry.

On those heights, we could see something to support every Kumikpa's theory. We could see that the thrust of the old proverb is basically right—the water does flow mostly down the other sides of the mountain, toward Shade, toward those water-soaked bastards down in Shila. We could see the fragments of permanent snowfields, isolated from the main field and from each other, rendered more vulnerable to the dark, light-absorbing soil all around them, not long for this world. We could see the face of the small glacier, facing north, streaked and pitted with dust, and we could hear the water percolating underneath as we stood on its surface.

We could see the water connection. We could see most of Zanskar, and could see how most of Zanskar's settlements were nestled at the foot of debouching streams pouring down from swaths of snow and ice. We could see how those streams feed the rivers Stod and Lungnak, and how they in turn united to become the Zanskar, which flows north to join the Indus on its journey to the Arabian Sea. We could see Marthang, the red place, and the new canal. We could just make out the buildings way down there, so exposed, and yet right in the middle of everything.

We could see alternative futures branching out from that summit, standing there on top of a doomed glacier: the top of the world laid bare—or a world without soot.

I tried for a moment to picture a world without its alpine ice—all sere browns, purple rock, russet slopes. From the dark triangular mass of Mount Everest to the snowless serrated peaks of Montana's Glacier National Park (which will need a new name within a few decades). I tried to see the day, far in the future, after all the floods and surges, when those spent frozen reservoirs would send their final trickle into the Lungnak, the Stod, the Zanskar, the Indus, and the hard times to come for the billions downstream when those rivers seasonally run dry.

Looking up from the top of Sultan Largo, what I couldn't see was any black carbon—just sun and sky. What streaked the dying glacier was instead dirt and mud—not fresh soot, which had done its dark hot work and sighed into its frozen belly. What little was streaming up from Zanskari stoves and Tata Mobiles below was overwhelmed and absorbed by the blue dome. Most of it was south beyond the mountains, lapping at their feet, the monsoon rains scavenging them from the air.

The sunny crime scene still seemed incongruous. It took an act of imagination to picture their nano-scale assault on these white sentinels, to reconcile how small Kumik seemed, far below, with the immense scale of the Himalayan peaks stretching out to the horizon. But I knew those dark particles were there.

Given what I had discovered, it was easy to imagine all that soot settling out of the sky after a week and all those women and children breathing easier, from Harlem to Haku Besi. This view of the world, downstream and downwind, with branching possibilities, seemed to offer an implicit test of our civilization's maturity, which is to say, of our collective imagination and gumption: to see the writing on the wall, how it might all end, and to see a path forward, to a different future. Toward a greener Marthang and clearer skies.

Despite P. Namgyal's dire prognosis, I couldn't picture a world without its 20,000 Zanskaris, there among those mountains, singing as they tossed rocks across a desert plain.

"They'll never do worth a damn as long as they've got two choices," an old neighbor once told the writer and farmer Wendell Berry, in reference to some common acquaintances, I suppose. I'm not sure I fully understood the sentiment until I watched the Kumikpas make their slow, resolute turn toward Marthang.

In 1740, Benjamin Franklin founded the Union Fire Company in Philadelphia, perhaps the first formal volunteer fire brigade in America. Like many of the ventures that sprouted out of meetings of the Junto, his mutual-aid-cum-ale-drinking society, there was a civic-minded element to the plan. But, as usual with Franklin's schemes, alongside the thought of the greater good there was also some self-interest at work: each member lived in a flammable house.

Plus Franklin chafed at the inefficiencies he saw, the stupidity of the waste: a house would burn down just because there were holes in their leather buckets or the hose was kept too far away. With a little planning, some discipline and forethought and training and investment, he argued, we could have a proper fire brigade, and our success rate will improve dramatically. It was so successful that he had to turn away prospective new members of his fire company. Go form fire companies of your own in your neighborhoods, he told them.

The idea of a "global citizen" has always been somewhat incoherent. Nobody lives globally. Everybody lives in a particular place, surrounded by particular people—neighbors who will call you on your bullshit, put you in your place, strike your name from the list of those who will shake the general's hand at the Farmer's Mela. Neighbors who will carry embers to your hearth if your fire goes out, or leather buckets if that same fire gets out of control. Neighbors who might occasionally think about breaking a wooden *choktse* over your head.

We don't have the cognitive or emotional bandwidth, as carbon-based creatures of limited ken, to take onboard a meaningful connection to 7.2 billion people. Social psychologists think our limit is about 200, actually. So if you're not moved by the plight of someone halfway around the world, I get it: Zangmo is *my* friend; you've never met her. Why should you care that she might be breathing insane quantities of soot every day? But chances are, you're breathing some of it, too (remember, there is no known safe threshold for exposure to black carbon and other PM2.5), and that someone you know will be adversely affected by it. The whole world is burning and hotspots abound. We all live downwind of someone else's fire.

To make that soot-free world I glimpsed from the summit of Sultan Largo a reality, information, whether in the form of powerful epidemiological studies or climate model runs or the rigorous cost-benefit analyses they engender, is not enough. It never has been. The knowledge that India's kerosene subsidies, for example, waste vast amounts of taxpayer money even as they endanger the people they're meant to benefit will not be sufficient to change the status quo. To lever that money into supporting truly clean lighting and cooking technologies and fuels requires the patient, frustrating work of politics, where there are winners and losers and structural impediments to overcome. (To wit: when the Indian government lowered its subsidy for diesel fuel in September 2012, the ensuing protests almost toppled the government.) The knowledge of "attentive Minds," in the words of Ben Franklin, requires some pragmatic follow-through: "What signifies Philosophy that does not apply to some Use?" It requires the relentlessness of someone hacking a canal into a cliff or a mountaintop with a pickaxe.

But that first step of seeing black carbon can help us perceive and reengineer the *goals* of our opaque, modern, "progress"-seeking systems. These dark particles offer us a new risk management framework—a kind of unified field theory of climate change mitigation, resilience building, poverty alleviation, and truly clean development.

The concept of *lamdon*—the "lamp that lights the way"—is an old, venerated trope in Zanskari and Ladakhi legend. The *lamdon* shows us the way forward in unfamiliar terrain, illuminating what's important. We now know that black carbon *is* the flame—that the *lamdon* is, in fact, the *shremok*. Those energized particles, heated to the point of incandescence, illuminate everything around them. The light they throw is even more revealing in the clear mountain heights around Kumik, where the ligaments of survival are laid bare as in few places on Earth.

Like its alter ego, the *lamdon,* black carbon shines a light on a possible path forward. As *shremok*, black carbon can help us see what really matters, what's urgent out there in the dark: the fanged threats of GLOFs, rising seas, disappearing snows, the life-shattering plagues of pneumonia and COPD and all those burdened breaths we take every day. Black carbon is thus the flashing red light on civilization's dashboard, signaling that it's time for some maintenance, for another concerted effort of Promethean bootstrapping. We get to choose what we see in it: potent carcinogen or india ink, airborne assassin or *terra preta*, obstacle or opportunity, despair or hope.

Even the robust conclusions of science—or of common sense—can only take us so far. Beyond their limits, bounded as they are by uncertainty, we must engage in some calculations of risk. One example: balance bracing for impact with efforts to ensure that the impact isn't as destructive as it might otherwise be, in the absence of any action today.

"What I think what we ought to be preventing is chaos—chaos in the social system," Tami Bond told me, "displacement, injustice, all that stuff."

That means working now to prevent runaway climate change from making much of our world unlivable (think feedback mechanisms, think Mad Max)—but it also means working now to meet the basic needs (water, energy, shelter, light, livelihoods) of the huge part of the human family living without, or with merely intermittent, access to them—those still in the smoking section. "One of the ways to do that is by paying attention to the bottom of the spectrum and providing a reasonable, sustained path for people to move up and become more resilient in the resources they have," Bond continued. "The nice thing about the black carbon question is that it's forced people to look at the lowest levels of society and say, all this stuff we've been doing hasn't touched a major issue."

Black carbon thus trains our gaze onto what we've been missing all along: the enormous overlooked health epidemic caused by breathing in soot and smoke; the impacts of the dark haze we re-create every day on particular, vulnerable, and hugely important "secret Cabinets" of the planet, like the Arctic and the Himalaya and the few blocks around the George Washington Bridge bus station on 178th Street in Washington Heights, New York.

This approach to understanding the urgent set of intertwined risks facing us in the twenty-first century—climate, health, poverty, water, and fire all bleeding into the political—aligns uncannily with the traditional risk management framework employed by Zanskaris and Ladakhis, embodied in the concept of *chu len me len*, the water connection and fire connection. The imperative at the heart of it was the prevention of social chaos that would imperil the efficient and equitable and reciprocal use of scarce resources like water and draft power and even hot embers, and thereby threaten *everybody's* survival. The other, deeper insight at the heart of it is simply that, when we are cut off from our neighbors, life becomes a little less worth living. No water and a cold hearth are daunting problems, but not nearly as scary if your neighbors have your back.

This kind of appeal to solidarity, to collective interest, can seem a little naïve, especially in the context of the diverging national interests that seem to keep countries like China and India and the United States pointed perennially in different directions, at international climate negotiations and other forums. Political scientists understandably roll their eyes at such talk. But it's instructive to dwell on the root of the Zanskari-Ladakhi risk management framework: kinship groups, solidarity groups, collective labor groups, and the principle of *chu len me len*—the water connection and fire connection.

The connection refers to the sharing of those primal forces or entities that keep us alive on this lonely spinning globe: fuel and fire, flowing water. But the phrase is almost exclusively used in the context of *severing* those connections. It is the ultimate punishment, a kind of village-scale nuclear option. There is nothing kumbaya about it—a social boycott, a death-in-life. It's incredibly ruthless and effective. This is what many visitors to Zanskar and other remote parts of the Tibetan cultural region get wrong.

The outward attitude of philosophical cheerfulness, the hardiness and warmth and hospitable qualities of the denizens, leads many to conclude that it comes from some indwelling spiritual wisdom. Recall one gloss on Zanskar's name, "the good white land," invoking the purity of its residents; it is said to be a revealed land, the *chos yul*, "the land of religion." Some chalk it up to their Buddhism, though you will find the same qualities in spades in Ladakhi and Zanskari Muslims, like my friend Mohammed Amin. Perhaps faith can explain some of it.

I attribute it instead to clarity. Zanskaris are the ultimate pragmatists. Up there in their mountain settlements, it's easier to perceive all the ties that bind us—to each other and to the land, the skies, the fire, and the water. Easier, too, to discern the costs of our headlong race to our supposed Shangri-la, as measured by all that burning we do to get by and get ahead.

These linkages are a bit harder to discern for the rest of us who dwell at less vaunted heights. In lieu of a view from Sultan Largo, we must rely on snapshots like the one captured by NASA's Terra satellite of the western United States on January 18, 2014. It shows the brown wrinkles of the Sierra Nevada, denuded of snow except at their very tips. At the time the snowpack was around 12 percent of historically normal levels. And just to the west, at the foot of the mountains, lay the Central Valley, the United States' "food palace," normally verdant with strawberries, almonds, cotton, lettuce, broccoli, you name it—now a dull, parched, sandy brown hue. A Marthang of sorts. When I saw the photo, it reminded me of the silent fields that fill the space between Kumik and Stongde, beneath the long-vanished, magic-person-spit-dissolved glacier. They were once green, too. ("Under such conditions, California may be prone to water shortages, crop loss and the loss of farm jobs, and increased wildfires," observed the note appending the images, quoting the governor's emergency declaration the day before. "The waters which are from heaven . . . may those divine waters protect me here," sang the Vedic farmers of ancient India.)

Meme Ishay Paldan does not require a satellite to read the writing on the wall. I would submit that this transparency and immediacy is a partial explanation of his and his neighbors' remarkable pluck. Kumikpas can be as venal and petty and shortsighted, and as compassionate and hardworking and self-sacrificing and breathtakingly generous, as any other people, anywhere. They have no special claim on deep insight into the workings of visible and invisible worlds, the atmosphere, the soil, or the human heart. They just happen to live in a place that is harder than most, and where the contours of survival are exceedingly legible. They are inheritors of traditions born of painful trial and error over many centuries that have evolved to mitigate the considerable risks of life at 12,000 feet, in an arid, raw landscape.

One of the co-benefits of these deadly serious reciprocal arrangements is plain old fun. Mirth and joy and impromptu song-and-dance-and-mud-brick-making routines tend to sprout out of their fertile ground. Much has been written on the extraordinary sense of community that emerges in times of disaster, crisis, upheaval—from the aftermath of Hurricane Sandy to the meals that friends and neighbors bring when you're in the hospital. This feeling fades fast, but in Zanskar, this is institutionalized in the *phasphun*, which handles everything for the principal players during funerals (and births and weddings, too). The *phasphun*, the *bes* groups, even the ritualized sharing of *chang*, are strategic: these overlapping structures bind people closer, in a joy-filled dance of both celebration and support, crisis management and crisis aversion. The *phasphun*, for instance, may have started in the distant past as a kinship group. The members share the same *pha lha*, or god of the hearth. "Now it's beyond kin, people are spread widely," Stobdan says. Gara and Tashi Dawa, his close neighbors, are in his *phasphun*, but so are some "Baltis," Ladakhi Muslims who live hundreds of kilometers away. "Even in Leh, anywhere, the same *pha lha*." When someone dies, when there's a wedding or a birth, the far-flung *phasphun* members come together and help the household in need.

"We say, *ruspa chik-chik*. Blood changes, but the bones are the same."

It all adds up to a sophisticated, elaborate enmeshing, a built-in risk aversion and mitigation mechanism. Spend enough time in a place like Kumik and you start to see it manifesting everywhere. When Zangmo gets up at 4:00 a.m. to work in her neighbors' fields. When Stenzin Thinlas helps Stobdan trek to the pastures late at night to rescue his sick cow and drag home the carcass of another cow that has died so the meat won't go to waste. When the whole village gathers to celebrate the first birthday of Mingyur's son (in a place where survival to that age remains less certain than in most parts of the world)—with, yes, lots of *chang*.

For most of us, it requires a heroic act of imagination to perceive these constraints, these binds, the paradoxical source of our greatest strength. Our societies are too diffuse, the connections too opaque. Not so in Zanskar! The resources available to a farmer like Gara can be gauged by glancing out his window: at the snows on the mountain above, at his neat fields, at the numbers and attitudes of his neighbors just across the dung-strewn way. The system boundaries of survival are painfully apparent.

Extrapolating that lesson to a global scale is fraught, and somewhat quixotic. Climate change is among the most cosmopolitan threats we have ever faced in the sense that it reminds us we are all at the mercy of systems beyond our control, all in the same spherical lifeboat among the stars. The *lha* can strike in the form of Sandy, or Katrina, or Haiyan in the Philippines.

But the world's connective tissue isn't transparent in the way that Zanskar's snows and streams and sown fields are. It seems too much to ask of people: we are not hardwired to see threats that abstract and diffuse in time and space, and few people have the time or energy to engage in a direct way. Like Kunzang in Langtang, like Zangmo in Kumik, they are too busy gathering fuel and hauling water, too busy getting by, to worry about something like climate change or the death of a glacier half a century hence.

And yet, if we don't recognize the global *chu len me len*, and how we are breaking the threads of these connections each day, we are unquestionably as doomed as the kingdom of Guge, as the Anasazi, as the ancient Harappan civilization whose comfortable monsoon-fed existence suddenly dried up. Doing nothing, many have pointed out, removes all uncertainty. Imja Tso *will* burst. The Arctic sea ice *will* disappear. The Sierras *will* be mostly bare of snow. Seven million will die, year after year, just from breathing, for not very good reasons. For all of these reasons, the *chu len me len* is the most powerful, most hardnosed and pragmatic risk management framework I have ever come across. It is even a little terrifying.

Framing matters. The stories we tell ourselves shape our decisions. Researchers at Stanford once ran an experiment with some volunteers, a version of the classic prisoner's dilemma, in which two people being questioned in separate rooms will both go free if they cooperate and protect each other but will both get harsher sentences if they rat each other out. The researchers found that if they told the participants they were playing something called the "Wall Street game," just one-third would choose to cooperate in the scenario. If they said they were playing a "community game," two-thirds of the participants cooperated. The Kumikpas are expert, self-aware practitioners of the community game. They're savvy to the improved odds of surviving and thriving that playing this game offers, both to the collective and to the individual.

So I would propose taking a page from the Zanskari playbook and approaching the scourge of black carbon and incomplete combustion as a community game. We have a hard time imagining the risks until they happen, hitting us in the face. And we have a hard time imagining the possibilities, too.

Remember what killed the character in Jack London's story: it wasn't bad luck or nasty weather or spiteful nature gods. It was a failure of imagination. The sooty markings on the walls of the cave, that night after my dunk in the Zanskar River, made it clear. What can kill us—what *is* killing us, almost 7 million of us, every year—is that sooty stew streaming out of a billion fires. What could save us? Each other: Thinles pulling me from the river, Urgain making the fire that restored my circulation. People carrying embers from hearth to hearth.

Serious-minded people who think about climate change policy and energy futures for a living often say, "There's no silver bullet. Only silver buckshot." It's become a ubiquitous cliché, but like most ubiquitous clichés it happens to be true. In terms of technology, we've got to throw everything and the kitchen sink at this one. I just would offer one amendment: the closest thing we have to a silver bullet is captured in that phrase—*yato meta?*—and its dark twin—*chu len me len chaden*.

It's the only silver bullet there is: the Prometheus next door. The fire brigade, if you like. The idea behind the fire connection in the old days in Ladakh and Zanskar is pretty much the same idea animating new institutions like the Global Alliance for Clean Cookstoves and the Climate and Clean Air Coalition. Carry embers to The Boss and the young upstarts behind BioLite, and they will carry them back to you with their clever, clean-burning devices. With its energy leapfrogging, should it choose to pursue that brighter path, India could carry embers of a different sort to the world, by showing the rest of the world that a newer, cleaner, smarter, more equitable path of low-carbon "development" (in terms of both lower emissions of black carbon, *and* of carbon dioxide)—one that skips the dark, airborne excesses of the original Industrial Revolution—is still possible.

And the small, waterless village of Kumik, Zanskar, India, can do the same, providing the rest of us with an object lesson in how to both mitigate and adapt. How to take responsibility, and then, how to take focused, patient, strategic action. An ember of hope in a time of dark and imminent danger.

The morning after the "*choktse* incident," most of Kumik walks down to Marthang. One or two people from every household stroll along the perimeter of the metal fence that encloses the new village, which was installed this time around not by vengeful *zbalu* but by government contractors, a couple of years back. They walk the full length, pausing in places to twist the metal links back in place or shore up fence posts that animals or mischief makers have pushed over.

During a break, Stenzin Thinlas, the head of Gonpapa, the oldest house in Zanskar, calls me over. He wants advice on the design of the new house he is building. "A simple house," he says, "just three rooms. Warm is most important."

I have drawn him a simple solar design. We pace it out on the baked earth and drag our heels in the dust to mark the orientation toward the sun, the location of the insulated "jacket" walls. And I think back to five years earlier, on my first day in Marthang, when I had come to see Stobdan's foundation and check his solar orientation—when I had first learned about a waterless village and, puzzled, listened to a cryptic joke.

It's late in the morning, time for a break. Everyone sits down in a circle in front of Tashi's new house, men and women, young and old, all mixed together, no sign of *bral* (the traditional seniority seating hierarchy, based on proximity to the stove or hearth). Tashi and Chorol, Lower Kumik's first year-round residents, are pleased and proud to be hosting. They run in and out of their freshly cement-plastered house with their finest cups and bottles of buttermilk and *chang*. They pour tea for forty-six of their neighbors and one toddler.

As they eat and drink, the Kumikpas hash out the previous night's arguments again. Soon Tsewang Norboo and Mingyur are shouting, not so much in anger, as to be heard. Murmurs, nervous laughs, tense jokes run through the group.

I watch from a distance. It occurs to me, for the hundredth time, how in the center of the action they are down here. ("Zanskar seems remote, but to a Zanskari it's the center of the world," an anthropologist once observed.) One really can see all of central Zanskar from here—the new *kun mik*—and all the weather systems—dust storms, dark cloud banks—that approach from both north and south. The villagers call me over to join them and drink tea. I sit down on the edge of the group and a cup materializes from someone's bag.

It's an overcast, windy day. While they argue, Nepali laborers keep building up the walls of the house next door. A dog keeps barking, punctuating a steady hum of debate. Chapatis, rice, bread, barley flour and butter, yogurt and buttermilk, get passed around and shared by all. Zangmo mouths silent prayers. Two tractors loaded with stone rumble by. Rangdol cracks a joke. Dust enshrouds us. A light rain starts falling. Thinlas's younger brother (and Tashi's older brother), Nyimbum, leans over and tells me that tomorrow he will plant apricot trees. "Ten, fifteen years before, it was too cold," he says with a shrug. "But now it's getting warmer. Maybe they will grow."

Stenzin Thinlas suddenly makes an announcement. He passes around a yellow bag lent by a monk, and each person tosses in an item—a bracelet, keys, coins, some things that their neighbors all recognize as theirs. One by one, Thinlas and Tsewang Norboo pull those keys, coins, rings, and strings out of the bag and call out their owners' names. For the next sixty days households will be matched in pairs. Congress supporters and anti-Congress, big houses and little houses, those with income and those without, they will all send members to take turns patrolling together, to keep cattle from eating the newly planted saplings of Marthang. Making teams for the protection of Lower Kumik. And with each pairing I see the cord of the *chu len me len* getting rebraided a bit. Some seasonal maintenance.

People seem to be in better spirits today. Have they scheduled this point of business anyway? Just a prosaic session of pragmatic Marthang planning? Or was the meeting skillfully put together to cool down the flames of the "*choktse* incident"?

Zanskaris, with their poker-face grins, would never tell. Either way, you've got to keep the cows from eating the saplings.

A few days later, as I walk one last time through Kumik before I head home, I pass a three-year-old girl. She is hauling water, taking careful mincing steps toward home with a little jerry can she has filled up from the spring. I meet Mingyur on the path, and we go to visit a spell with Ishay Paldan, who is out there alone, hunched over in his fields, cutting grass with a sickle. Then I run into Tashi, who tells me that the son of one of his neighbors got into an argument—about politics—with his uncle, near the road-building camp outside of Padum. He threw the older man to the ground twice. I think of what Tashi's wife Chorol said—"When people live too close, they quarrel. Privacy and space is better"—while listing all the virtues of life in Marthang. Everyone down there is putting up their own stone wall around their three acres of land. In lieu of *zbalu*, they hire Nepali teenagers and Eicher tractors to build their walls. And I think of Frost: you know, good fences making good neighbors.

It's a windless, Technicolor noonday. I fill my bottle with spring water from the pipe. On days like this, with everyone out in the fields, one gets a glimpse of Kumik's future as a ghost town. "Things reveal themselves passing away," the poet William Butler Yeats once remarked. Today, under an unsparing sun, Kumik and its materials are laid bare, like the snowfields above stripped of a fresh coat of snow. From the prayer wheel over the stream at the heart of the village I can see it all. *Kun mik*.

The empty silent *gompa*. The weathered guardian *chortens*. The alluvial fan north of Karsha tapering to a distant, needled ridge. The low-slung electric wires, strung up by Tsering Motup in 2012, now a source of bitterness. The clumps of poplars like punctuation marks dotting the slope below. The hidden stream flowing north. The southwest-facing pueblo-like core of Kumik. The piles of stone laid up for building here and there, like a promise or threads tied around a finger. The old scars and grooves on the land, footpaths and dusty dry water courses. The new cut in the hillside to lay pipe to Pang Kumik (the resolution of the argument that took place the year before), the new switchback road, the old jeep track. Gonpapa's three ancient *chortens* and two more down the ridge, like a reticle for sighting Marthang. Piles of drying peas and fodder grasses. The solar photovoltaic station, a hasty piece of government patronage already malfunctioning after less than one year. The neat rows of harvested barley in Pang Kumik, two vehicles parked nearby, with some bright new wooden doors destined for Marthang. The stand of trees like some last witness of the mountain's liquid largesse. The roofs' motley burden: solar panels, dung discs, drying hay, *burtse*, gnarled sticks, prayer flags, Dish TV, *tsepo* baskets and sheepskins, drying clothes, larger chunks of twisted wood in case of a really cold winter.

Signs of habitation still abound. But how will some sights strike a future visitor to the ghost town of Kumik? The small willow grove surrounded by a high stone wall marked with the words "two-wheeler garage," where Zangmo makes *arak* and *chang* and Stobdan parks his rickety motorcycle. The *zbalu*'s wall recently cleaved, after a thousand years, to let Tata Mobiles rumble on through. The verdant medallion of Stongde, like Marthang's looking-glass twin. The stepless pathways and house perimeters, culmination of a long series of semi-conscious cost-benefit calculations (better to spend time making stone steps or drinking *chang* together, cementing social ties?).

It's all still alive, still there and still loved. But for how long? I think back to a day from the previous summer, when Stobdan and I rode up the rough link road to Kumik in the back of a Tata Mobile, with a weak calf that had been lost for a couple of days lowing pitifully next to us. We rolled into the central square, greeted by shouting boys, familiar faces, and gently waving willow fronds. "Some are saying they don't want to leave," he said, "because 'This was my father's father's father's *meme*'s *meme*'s *meme*'s.' " He mimicked these neighbors in a tone suggesting he was too modern for that kind of narrow nostalgia. But Stobdan—the great booster of Marthang!—admitted to sometimes wanting to stay in the old village. He, too, feels the pull of his lifelong home, his ancestors' gods, his small green valley, the only world that he's ever known, now slowly falling apart. "Bees are drawn to the flowers," he told me with a shrug, gazing at the poplars throbbing in the breeze. "It's like this."

Lundup, a young soldier on leave, walks by with a kid on his shoulders. "*Jullay,* Mr. Jon! How are you?" Bundles-with-legs trundle up hill and up stepladders onto mud roofs: Kumikpas carrying fuel for their animals. Here comes Rigzin's father now with one. With his knit cap and sweater, plaid shirt, corduroys, jacket, and boots, and kindly laconic ways Rigzin's father could be one of my Vermont neighbors.

Cloud masses scudder over the Himalaya and then fracture. A shout issues from field to house: *Chu ma-yongs!* The water hasn't come! And then a figure dips his head back down below the heads of barley. On days like this, the trickling sound of the stream fills the narrow valley and sounds bigger than it really is.

As much as the loss of the water stored in the dying ice and shrinking snow, one laments the passing of a certain beauty from the world. Kumik's children wave to me from the wall of the willow grove. Today they are washing clothes. How wondrous, how miraculous!

Later, back in the cool kitchen of Kharpa, I look around. There is a life insurance policy laid out on a *choktse*. On another are Stobdan's books, for his college correspondence courses. *Indira Gandhi National University B.A. Foundation Course. Humanities and Social Sciences. Some Concepts for Communication in English. ESO-14 Society and Stratification, Introducing Social Stratification. Government and Politics in India: Historical Background. Section 2.3: The World of the Peasantry.* Flowers on the sill in old cans. Zangmo's prayer beads. A prayer book. A copy of the newspaper *Rangyul* ("Oxygen Bar to Come Up at Khardung La"). A lottery ticket from Sani Nasjal. The Jugnu Solar Home Lighting system charge controller 12 volt, 10 amps sits idle; something's broken. Bronze ladles hang above it. An empty bottle of *arak* that we emptied last night is on the floor. Under the shelves that hold the dowry plates and bronzed pots sits Stobdan's hard green suitcase, full of ID cards, birth certificates, and various important documents. It also contains the detailed registers that Stobdan has kept of every project—the canal to Marthang, the journal of his climb to Sultan Largo, the construction of the *lhakhang*.

The solar *lhakhang* now has walls, windows, a roof, and a statue of a serene Buddha looking south toward the Great Himalaya Range. It's a humble, simple little building, with two large windows that open to the south, framing a grand view of the glaciers above Padum. The double walls have cavities full of straw, wood shavings, and sawdust for insulation. The main room has been plastered. The entryway has framed photos of all the villagers who worked to help build it, swinging picks and shovels and packing mud into brick forms. Stobdan's father, Rigzin Tantar, visits twice daily to light the butter lamps, and the young soldiers and their wives, the pioneers of Marthang, sometimes congregate in the unplastered kitchen, just to be together. It's started to become what everyone intended—a village-scale hearth, a hub for what one day may become a "model village."

But two years after work began, five years after my first visit to Kumik, it's not quite finished. So, as "strong committee" members, Stobdan and I sit in the summer kitchen, going over accounts and the work yet to be done. We need glass for the double-paned windows, more insulation for the roof, and a verandah where people can sit in the winter sun. The walls need another coat of plaster; the temporary stone steps need to become permanent. Stobdan brightens when he shows me the little yellow donation books he had printed up in Leh to provide receipts for people who want to give—20 rupees, 50 rupees, whatever—to help finish the solar *lhakhang* of Lower Kumik.

In his logbook Stobdan has recorded every expenditure, from a few rupees for nails to 2,500 rupees for an Eicher tractor hired for the day, so that the project is transparent to anyone who wants to know how donated money—from villagers, from other Zanskaris, from foundations and generous individuals in *chigyal*—has been spent. Every household, and the name of every member of the Kumik family who has participated, is written in its pages, next to the number of days they have worked or the rupees they have given instead if they couldn't come to work one day. I flip through the pages and read the names of every

Kumikpa therein. Norzom Palmo. Nawang Phunchok. Sonam Dawa. Stanzin Choszom. Stenzin Dorjay. Tsewang Norboo. Tsering Motup. Tsetan Angchuk. Sonam Yangzon. Tashi Dawa. I think of the setbacks: The time we showed up and water overflowing from our little mixing pond had destroyed two days' worth of mud bricks. The time a wind and dust storm forced forty of us to shelter behind the half-built walls. Or when we formed a fire brigade of twenty pairs of hands, passing rocks to each other and singing at the top of our lungs as we tossed them into the foundation. Yes, it's a humble little building, still in progress—but it sure took a lot of work. And a lot of shouting, laughing, singing.

I thumb back to the first page, where Stobdan has written:

DETAILED EXPENDITURE REGISTER OF LAKHANG GOMPA LOWER KUMIK ZANSKAR

> Kumik is a village consisting 40 household and is known to be first settled village in the Zanskar. Due to failure of snowfall the people couldn't harvest even blade of grass and it is also known to be most draughtful village in the Zanskar. And consequently they had to sale their yak, cows and etc at very nominal prizes. So the people of Kumik held a meeting in 2001 and decided to construct a cannal up to lower Kumik from the river of Lungnak. Now the dreams of Kumikpa is come to near the get water at the barren land. Also they construct more than 25 house at the lower Kumik.

He is confident that people will continue to give, not just money but of their time and skill. And then they will come, right to the heart of Zanskar, to gather together within the *lhakhang*'s walls, insulated by straw from Kumik and wood shavings from Padum and heated by the sun.

Signs of human care now abound in Marthang. Chorol's little garden of flowers and potatoes is thriving. Stobdan has a sizable field of wheat, and one of peas, that he'll plow back under to add nitrogen and organic matter to the soil. Nearby, Tsewang Norboo's house has healthy willows in its courtyard. Beda's house is going up quickly, pointed due south. Thinlas Samphel is proud of his new root cellar. After five years, it's all come a long way.

There's still a long way to go; the Kumikpas, thirteen years after their decision to build this community, are really just getting started. It's certainly no Shangri-la—it's still a "red, dusty place." Perhaps, though, it is one of the *beyul*—a hidden land that becomes revealed to those who make a strong effort to find it, to see it.* A place where the things that matter most, things that are most foundational to human welfare, are made more apparent.

Once people see what's possible, they say, more will come and we'll all be together. ("Then life here will be *demo*," Chorol told me. Beautiful.) And once they imagine the life they want—if they want to breathe cleaner air and keep snowmelt flowing through their new canal for decades to come—Kumikpas will face some other choices. So will those of us in *chigyal*. Whether clearer skies and better days come to Marthang is up to them—and how they decide to cook their meals, light and heat their homes, get from place to place—and up to the rest of us living downstream and upwind of the roof of the world, as well.

"But it will take time," Stobdan says. He takes the account book, flips to the last page, and hands it back to me. There, in his blocky script, Stobdan has written: "Coming together is a beginning. Keeping together is progress. Working together is success."

And the clouds part a little, and I finally understand that he has known all along something as easy to overlook, to take for granted, as the sooty particles that stream up quietly from a bright flame: He is

---

*According to legend, Guru Rinpoche designated these hidden valleys throughout the Himalaya. They would be places of refuge during times of global strife. Those trying to reach *beyul* inevitably encounter hardship and peril. Like the heroic yogic masters meditating in the mountains, seeking to overcome their own desires and passions, they too must overcome great obstacles—storms, floods, their own despair—to even be able to *see* these hidden places, let alone reach them. Stories abound of seekers walking right over the ground of *beyul* and right past it, unable to perceive its superb qualities. To get to *beyul*, the path must be illuminated, and then the place would be revealed at the right time. Zanskar, the *chos yul*, as some have called it, is now one of the revealed lands.

not alone. He has a whole lot of *yato*. That knowledge can make the impossible—like building a new home in an uncertain world—seem almost inevitable. He leans in with a conspiratorial grin, and claps a hand on my shoulder.

"We will be successful one day, yeah?"

Then he goes out to help Zangmo in the fields.

Epilogue: Carrying Embers

In the Zangskari's world there is no place for chance; events are essentially of a moral nature and everything is the expression of a cause-effect relationship in which personal motivation and awareness have major contributions to make. An interdependence between the moral and the natural forms the basis of all attributions of causality.

—JOHN CROOK, *HIMALAYAN BUDDHIST VILLAGES*

I had simply wanted to know what was drying up Kumik.

Every scientist I would talk to, over the years since my first day in the doomed village, emphasized the uncertainty, the unknowability, the need for more research—in the same way that identifying, with complete confidence, the hidden hand that took away Tsewang Rigzin's infant siblings is fundamentally impossible. When we met in his office near the shores of Kashmir's Dal Lake, Professor Shakeel Romshoo, who has studied the snow and ice of Zanskar for over twenty years, sounded this need for caution.

"If you look at the Zanskar glaciers," he told me, "they are in the cold desert region, a completely different type of topographic and climate regime there. But even within that regime, Zanskar, Suru, even within one basin, there may be three hundred or four hundred glaciers, and they behave differently—because of geomorphological indicators, because of microclimates, even in the same topographic regime. If a glacier is there, or if a glacier is here, it makes an impact, because temperature there is cooler, it is hotter here." The same goes for snowfields. Even after all this, Kumik's glacier up on Sultan Largo could indeed be the victim of dust deposition, or some weird micro-climatic fluctuations in its vicinity, a butterfly flapping its wings in Chengdu—or some combination thereof.

But with the evidence before us, the answer seems very likely traceable to some modern variant of the *chu len me len*: Kumik's tough situation is a consequence of all the inefficient fires we light to survive and thrive every day. Some close to home—in Zanskar, in Kashmir, in Punjab—and some far, far away.

Jammu and Kashmir has lost 20 percent of its total glacier mass in the past six decades. "So the recession we have seen in these glaciers," Romshoo said, "how much of it is because of natural variability and how much is because of enhanced temperature, how much is because of black carbon—that is a big research question, and we haven't worked on all these aspects. But we know this is very simple: we have the trends of the temperature increase here, we have one hundred years [of] data, and this is very simple science. When the temperature increases, melting will be more. There is nothing to prove in this.

"And if you look at the state of Jammu and Kashmir, the indicators of climate change are very clear and loud, if you see the recession of glaciers, the decrease in precipitation, especially the winter precipitation, the increase of temperatures, the availability of water in different basins—that is also showing a decline. There is no doubt that the climate change is happening and it is impacting various sectors of the economy in this state of Jammu and Kashmir."

Oliver Wendell Holmes once wrote, "The only simplicity for which I would give a straw is that which is on the other side of the complex—not that which never has divined it." Simplicity on the far side of complexity. Common sense vindicated, compounded, and amplified by a whole lot of rigorous, inventive, peer-reviewed science. Dark stuff is warm. Warmer air melts snow and ice. Breathing smoke is bad for you. Better burning means less smoke. We're all in this thing together. "Build the dam before the flood comes."

The revelation of black carbon's role in all of this, its dark maneuverings on the roof of the world, led, like bread crumbs through a forest, like snowmelt cascading down some ravine, to a series of other sobering revelations about the extent of its damage to human life in the here and now, and to human possibilities in the not-so-distant future. I found much more than I bargained for.

I still keep coming back to certain epiphanies I found following black carbon's dark trail downstream, beyond the crippling blow it dealt the glaciers of the Alps and now seems to be meting out to the Himalayan cryosphere. Namely: that 3 billion people still burn solid fuels; that the soot-laden smoke from our indoor and outdoor fires kills almost 7 million people each year—nearly 1 in every 1,000 people on the planet; that exposure to sooty smoke from hearth fires is implicated in half of all child pneumonia cases (and that pneumonia is the leading cause of death of children under five); that we've long known how to clean this mess up. (And: that acting on black carbon without acting quickly and aggressively to reduce carbon dioxide emissions is simply delaying slightly the inevitable day of reckoning.)

I keep hearing Kunzang's reply when I asked her if all that soot bothered her and her family (what must have seemed like the dumbest question she had ever heard). Cradling her baby daughter there in the shadow of the receding Langtang glacier, her eyes flashing, she said: "Of course it does! But what to do? We have to eat!"

I keep hearing Kirk Smith, on the morning of the release of the 2010 Global Burden of Disease report telling us that household air pollution kills more people than any other environmental factor: "Enough already! We know this combustion is bad. Let's get rid of this stuff!"

I keep hearing Sonam Wangchuk at the kitchen table in Kathmandu, lamenting our habitual neglect of these opportunities to preserve so much human potential and harness so much abundant energy all around us at the same time: "It seems a crime not to do it!"

And I keep hearing Sandeep Pandey, dodging jackals and monsoon rains with his merry band of "lamp-men" as he brought light to the Darkness: "I don't know who has said, but nothing is impossible."

I had also wondered, in a twist on James Crowden's question of winter 1976: how on earth would these people survive in Marthang?

How would they start over, leaving the Holocene that is old Kumik and building a new home in Marthang, their own little version of the wild and risk-filled Anthropocene? From what I could see, the answer seemed clear enough: *yato!*

When called upon, whether to dig a *yura* or move a mountain or build a solar *lhakhang*, even as modern economic forces and old rivalries tug them apart, like the cosmic dark matter pushing the stars away from each other, the Kumikpas still reflexively put Ben Franklin's old credo in action:—"The good men may do separately is small compared with what they may do collectively." And with a little push from the pioneers and visionaries among them—the Stobdans and Tashis and Chorols—to see what might be possible in their uncertain future out there in the wind and dust, they shape and choose their own destiny. Even in their quarrels, the Kumikpas are united, bound together by both peril—the prospect of the coals in their hearth going cold, of the water vanishing—and by promise—of a neighbor carrying embers through the dawn, of a hand-dug canal watering a few saplings and the "red place" slowly becoming green.

And finally, there was my original nagging question: why was Stobdan laughing there by the sign, with its tale of doom?

That one's a bit harder to solve. Part of him certainly welcomes the step into the unknown, with the chance to reinvent himself that it brings—the chance to become a bit closer to the rest of the world and its constant flux of goods, ideas, contradictions, razzmatazz. To become, for better or for worse, a little more "modern." (*Chandigarh!* The *chadar* road! Cooking with gas!) And he is especially keen for Zangmo to work a little less and rest more. But part of Stobdan still feels the gravity of a thousand years of good living in the "place where everything grows," in Kharpa, the house of his ancestors.

Ultimately, I think that what explains his deep confidence is that same sense of not being alone, his knowledge that he has the ultimate safety net: his neighbors there in the lifeboat with him, sailing into rough waters together, toward the light of some distant port. On my last visit to Marthang, I watched as Stobdan bustled about his yard setting up a low-cost, gravity-powered sprinkler, and then watched as his neighbor Beda crouched down next to him, and the two men earnestly discussed the most efficient way to water their precious saplings. Even Falgyon and Beda, when push comes to shove, are his *yato*.

But there was an even deeper note humming through that laugh by the doleful sign. This is the balance of the answer that is harder to articulate.

People often ask me why I keep going back to Ladakh and Zanskar, time and again, after all these years. I have to admit that, given what I now know of the accumulating consequences, it gets harder and harder to justify the carbon emissions of my flight to and from Delhi and the soot-laced trail left by the Tata Sumo ride from Leh or Srinagar to Padum. But still I go. Why?

"The people," I reply.

They are magnetic. They are good company. They know how to have a fine time, even when—especially when—the chips are down. They understand the value of each other's company, and they are attuned to that old question: "How long will this life last?" They aren't necessarily any more wise or virtuous or farsighted than people anywhere else. In fact, they remind me a lot of some people back home in the United States. Meaning: they are just people.

But truth be told, I also keep going back for more lessons in the Zanskari art of risk management, in the great clarity of those mountain heights. I gain hope and strength when I hear my friend Urgain Dorjay whistle brightly up at the leaden, threatening *chadar* sky as he pulls on his frozen boots.

And I think that I'd like to learn how to whistle like that.

Acknowledgments

This book would not be possible without the generosity of the people of Kumik, who are too many to list here. Over the past several years, they have welcomed me into their homes and shared their hopes and fears for the future, their daily joys and struggles. I have learned from each and every one of them, since that first day I sipped butter tea with Ishay Paldan. I am especially indebted to Tashi Stobdan, Tsewang Zangmo, and Tsewang Rigzin for their tireless help and friendship.

This story started with an invitation to work together to build a new village in the "red place"—a journey that is far from over. Many people helped build the humble little *lhakhang* that now sits at the heart of Marthang—Kumikpas of all ages, masons from Nepal, carpenters from nearby Stongde, the monks of Stongde Monastery, and many generous supporters from "outside the kingdom." I'd especially like to thank John Parsons and the American Alpine Club for the Zach Martin Breaking Barriers grant that helped initiate the project, and the Hans Saari Memorial Foundation, Rencontres au Bout du Monde (RBM), and the Goggio Family Foundation for providing critical financial support as well.

Over the course of several trips to Zanskar discussed in these pages, I've had the good fortune to travel and work with Gregg Smith, Thinlas Chorol, Choszang Namgyal, Vincent Stauffer, Ashley Morton, Tsewang Thinles, Lobzhang Gyatso, Rigzin Yangdol, Zach Lamb, Carey

Clouse, Mohammed Amin Zanskari, Phuntsog "P." Namgyal, Lobzhang Tashi, and Ben "Shangku" Stephenson—who invited me to join the Zanskar Ski School for some glacier skiing in the summer of 2003, and got this whole ball rolling, in a way. Over the past decade, Urgain Dorjay and Sonam Chondol and their family have opened their hearth and home to me again and again. Our many conversations helped open a window onto their land and culture; I am forever in their debt, and forever grateful for our friendship. For similar reasons, for their generous advice, and for so many wonderful lessons in compassion, courage, and pragmatic problem-solving, I am happily beholden to Sonam Wangchuk, Rebecca Norman, Konchok Norgay, and the students and staff of SECMOL in Ladakh.

The seed of this book was a writing project for a course taught by Michael Pollan at UC Berkeley; his early encouragement and guidance were instrumental in its evolution into this broader tale. It was a great pleasure to work closely, from afar, with Laird Townsend of Project Word to then develop that seed into a story published in the *Boston Globe*, and with Nicolas Villaume, on a related photo-essay for display at the Smithsonian Museum of the American Indian. Both of them came to the story with great respect for the people at the heart of it and great instincts for how to share it, with accuracy and nuance, with a wider audience.

Thanks also go to friends, classmates, and faculty mentors at the Energy and Resources Group of UC Berkeley, where I first started exploring the inextricably linked issues of household energy, resource poverty, air pollution, and climate change under the expert guidance of Dan Kammen, Isha Ray, and Ashok Gadgil, among others. Conversations with fellow "ERGies" Christian Casillas, Mary Louise Gifford, Stacy Jackson, Sam Borgeson, Merrian Fuller Borgeson, and Josh Apte have also been helpful at different stages of researching and writing. I'm indebted to the Wildacres Residency Program for the gift of a week of doing nothing but writing in a peaceful setting, high in the Blue Ridge Mountains. Christopher Shaw, Bill McKibben, and the redoubtable fellows of the Middlebury Fellowship in Environmental Journalism program gave—and continue to give—me confidence that this whole line of work is worth pursuing. Shahan Mufti and Tom Powers gave helpful (and frank) advice on the book-writing process.

I am indebted to many others for giving liberally of their counsel and encouragement—and occasionally shelter and sustenance—in the course of my reporting and writing, from Vermont to New York to India and beyond: Curtis and Ed Koren, Kathy and David Hooke, Preeti and Rati Puri, Viraj Puri, the Gangba family, Sam and Merrian Borgeson, Andreas Stavropoulos, Ivan Cestero, Cam Pittelkow and Sarah Stewart, Maggie Doyne and the residents of Kopila Valley, and Megan Shull. The unstinting confidence and support of my sister Christina and my brother-in-law Jeff (and, in their own way, my nephews Reid and Timothy) helped me cross the finish line. I am grateful to my editor, George Witte, for his skillful shepherding of both the story and this first-time author, and to my agent, Nicholas Ellison, for believing in the project and in my ability to pull it off. Sydelle Kramer, Joe Spieler, Sydney Parker, and Nicholas Ellison provided helpful feedback on my early proposals, and Minh Le offered detailed, incisive comments on early drafts. Liza Cochran gave me careful edits and constant encouragement—my gratitude for her support, for her faith in me and in this story's possibilities, and for her many other gifts, goes beyond words.

Rigzin Yangdol, Urgain Dorjay, Lobzhang Tashi, Tashi Stobdan, and Tsewang Rigzin provided help with translation of certain conversations in Zanskar. K. K. Rai helped with translation in Uttar Pradesh, Suraj Kumar Sharma in Nepal, and Yin Fang in Yunnan, China. Rebecca Norman provided invaluable guidance on the meaning and usage of certain Zanskari and Ladakhi words and translated the folk song "Wakhai Drakbu."

Many scientists, researchers, entrepreneurs, and policy experts gave generously of their time and insights. I hope I have accurately described their important work—responsibility for errors of fact or interpretation rests entirely with me. I am especially grateful for illuminating conversations with Kirk Smith, Tami Bond, Veerabhadran Ramanathan, Thomas Painter, Jason Box, I. H. Rehman, Anumita Roychowdhury, Veena Joshi, Jonathan Cedar, Nikhil Jaisinghani, Sandeep Pandey, Zeng Juhong, Suraj

Kumar Sharma, Shakeel Romshoo, Leslie Cordes, Catherine Witherspoon, Jacob Moss, Susan Anenberg, Durwood Zaelke, Steven Chillrud, Patrick Kinney, Tica Novakov, Thomas Kirchstetter, Dr. Stenzin Namgyal, Dr. Tashi Norboo, Dr. Nawang Chosdan, Peggy Shepard, Carlos Jusino, Sarath Guttikunda, Arnico Pandey, Pradeep Mool, Johannes Lehmann, Surabi Menon, Seb Mankelow, David Breashears, and Ratan Tata.

When I was a boy, my late grandfather James B. O'Neil gave me a subscription to *National Geographic*; years later, when he inquired why I was going to India, I reminded him of this fact and puckishly replied: "Because of you!" ("You got me!" he said with a pleased chuckle and a twinkle in his eye.) He gave me endless encouragement and helped instill in me an appreciation for the marvels of this world, and the speed with which it changes, that infuses this work.

Finally, I don't know how to properly thank my parents, James and Barbara, who have supported me in everything I've done, who have been my teachers, my friends, my wise counselors, my compass points. I can only hope this story expresses some small measure of my bottomless gratitude, reflects some tiny gleam of all the light they throw around them, captures and shares a few of the many, many lessons I've learned, and continue to learn, from them.

Glossary of Zanskari-Ladakhi Words

Zanskari is generally regarded as a dialect of the Ladakhi language; the two spoken languages are very similar, varying only in some vocabulary and pronunciation. Certain words borrowed from Hindi and/or Urdu are frequently used by speakers of Zanskari and Ladakhi. (Hindi words are denoted below in parentheses.)

NOTE ON TRANSLATION: *Most quotations from Urgain Dorjay, Tsewang Rigzin, and Tashi Stobdan are in English (their third or fourth language). All other quotations of Zanskaris or Ladakhis are rendered into English via my own or others' translation of the speakers' original Zanskari or Ladakhi. All quotations of Sonam Wangchuk are in English.*

abi-ley: grandmother (*-ley* is a suffix added to names that denotes respect)
ache: older sister
acho: older brother
ama tsogspa: Women's Alliance
anganwadi: literally, a courtyard shelter; also the name of an Indian government program of small centers in villages supporting child health and nutrition (Hindi)
arak: a strong liquor made from distilling barley wine
azhang: maternal uncle; also, term of address for men of middle age or elder than the speaker
bao: cave
bati-wallah: lamp-man or light man (Hindi)
bes: reciprocal labor-sharing arrangements between a group of households
beyul: literally "revealed land"; remote

valleys in the Himalayan region believed to be places of refuge, visible only to certain advanced practitioners of Tibetan Buddhism

bodyik: the written, Tibetan-derived script of Ladakhi and Zanskari

bokhari: a metal stove used for heating a room (Hindi)

bral: the traditional ritualized seating arrangement in which elders and honored guests sit closest to the hearth and others sit in decreasing proximity according to their social status

butpa: bellows used to fan flames

chadar: also sometimes rendered as *chadur* or *tchadur*; literally, the "ice way," a trekking route that forms for six to eight weeks each winter on the frozen surface of the Zanskar River, offering the only way in or out of Zanskar

chang: a kind of barley wine

changnet: hangover

Chi choen: "What to do?"

chigyal: outside the kingdom; foreign lands

chigyalpa: foreigner; literally: one from *chigyal*

choktse: small, carved wooden table

chorten: a rectangular Buddhist monument

chos yul: literally, "land of religion" (a description traditionally applied to Zanskar)

chosphun: a ceremony in which two people are randomly chosen by monks to be spiritual friends for life

chotkhang: rooftop shrine room found in every Buddhist house

Chu len me len chaden: "The water connection and fire connection is cut"; a social boycott in which irrigation supply and embers for relighting hearth fires are cut off to the offending household

chu tangches: to "give water," to irrigate fields

chulha: traditional cooking stove (Hindi)

chumchar: waterfall

chumik: a spring; literally, "water-eye"

churpon: "lord of the water," a rotating position in villages of Ladakh's Indus Valley for adjudicating disputes over water allocation

chu-rut: flood

crore: 10 million (Hindi)

daman: drums

dhaba: a small, cheap eatery (Hindi)

doksa: the area around the simple stone shepherds' huts in the high pastures

dzo: the hybrid offspring of a yak and a cow, prized for their draft power

dzong: a type of castle fortress built in Bhutan

gara: blacksmith; also a nickname given to infants whose older siblings died in infancy, derived from a tradition in which a blacksmith gives the child milk from a special spoon

Garshung: "little Gara," Tsewang Rigzin's nickname

genti: pickaxe (Hindi)

geshe: a learned monk, one who has completed twelve years of study of Buddhist scriptures and theory

goba: village headman

gompa: monastery
goncha: a woolen robe; the traditional dress of Ladakh and Zanskar
Gonpapa: the oldest, original house of Kumik, located on the ridge above the *lhakhang*
Hala: "Isn't it so?"
hamo: difficult
hariyali: green place or plantation (Hindi)
homche-rak: to be thirsty
hup alahes: exclamation used upon rising to one's feet with a heavy load
jullay: all-purpose word used for greetings, requests, thank-yous, and farewells (analogous to Hawaiian "aloha")
kangri: literally, "ice mountain"; used to describe both glaciers and ice- and snow-covered peaks
Kangyur: the central texts of the Tibetan Buddhist canon, consisting of 108 volumes, including translations of the sutras, or words of the Buddha
Karu chachey: informal Zanskari phrase meaning "Where are you going?'
Kha Kumik, chu Shila: "The snow falls above Kumik, but the water goes to Shila"; Zanskari proverb used to say something along the lines of "isn't that ironic?"
kha tangches: to snow
khangchen: the "big house"; the main family ancestral home typically inherited by the oldest son
khangchung: the "little house"; an offshoot house built by a younger brother, into which parents move after ceding the "big house" to the oldest son (also *khangbu*)
Kharpa: Tashi Stobdan's ancestral house in the center of Kumik
kholak: a common traditional dish made from mixing barley flour and butter tea or buttermilk
khuyus: threshing
Kumik: a village in central Zanskar, Kargil District, Jammu, and Kashmir, India; the name has three different possible derivations and interpretations: from *kun min*, meaning "the place where everything grows"; from *kun mik*, meaning "the place where one sees everything"; from *gu mi*, meaning "man from Guge," an ancient kingdom in western Tibet
Kumikpa: a resident of the village of Kumik
Ladakh: a mountainous region in the northwest Himalaya, in the far northern part of India
lamdon: lamp or torch; a "guiding light"
lchangma: tree or trees
lche: dried dung used for fuel
lha: god or spirit
lhakhang: a small temple or prayer meeting hall, found in most villages
lhande: hungry ghost or demon; monster
lhato: a stone shrine on the edge of the village or on a house
lhu: underground spirits
lorapa: the rotating village officer responsible for keeping animals out of agricultural fields
lungspo: wind
mani: stones carved with Buddhist

mantras; also, a long wall of such stones

Mani Ringmo: the main bazaar of Padum, Zanskar's largest town (literally "long *mani* wall")

mantra: a phrase, word, or sound with religious significance and sacred power meant to be repeated during prayer

Marthang: literally, the "red place," the no-man's land north of Padum in central Zanskar where the new village of Lower Kumik is being built

meme: grandfather

mera gao: "my village" (Hindi)

naktsik: a marking of soot placed on infants' foreheads to ward off the evil eye

nang: room; also used to describe smaller subsections of a field

nilim: sapphire (Hindi)

numbirdar: a village-level official in India's local governance system (Hindi)

nyampo: together

onpo: a village astrologer who performs exorcisms and religious rites

paba: a traditional dish consisting of boiled dough made from mixing barley, wheat, and pea flours

palimentok: a small yellow wildflower

panch: a local leader in India's village self-government system (Hindi)

pha lha: the god of the hearth

phasphun: a kinship group of households that share the same *pha lha*, called into action to give support during births, weddings, and funerals

phu: the high mountain pastures where cattle and yaks graze in the summer

puja: prayer (Hindi)

pukka: literally, "ripe"; used to suggest a building that is solid and durable or, generally, first-class, high-quality, or modern; also specifically used in reference to structures made with concrete (Hindi)

ranthaks: water-powered stone mills used to grind roasted barley into flour

rarzepa: the shepherd who grazes sheep and goats beyond the village (a post that rotates from household to household each day in summer)

ridaks: a catchall term referring to wild sheep and mountain goats, such as ibex, blue sheep, urial, Argali, etc.

rigpa: a bricklike chunk of dried sheep and cow dung, burned for space heating

romkhang: cremation ground

salwar kameez: traditional north Indian clothing, consisting of loose-fitting pants and a tunic-like shirt (Hindi)

sati: the forbidden Hindu practice in which widows jump onto their husbands' funeral pyres (Hindi)

sermo: a variety of barley with larger grain heads, grown in certain villages with abundant water in Zanskar

sheep sheep: "slowly but continuously"; often used in reference to drinking *chang*

shremok: dry soot

skitpo: happy, fun

skora: clockwise circumambulation of religious monuments or spaces
song: literally, "it went"; in Zanskar, colloquial meaning is "agreed" or "understood"
spang: grassy place, often spring-fed
stanches: to show; the way something looks
sunches: to be lonely or bored
surya: sun (Hindi)
tagi khambir: a yogurt-infused bread
tarlok: a term used on the *chadar* to describe the continuous "turning over" or changing of the ice
tehsildar: government officer who supervises property holdings and tax revenue (Hindi)
tenten talu: ritual characters in skeleton costumes at monastery festivals, usually played by child monks
thukpa: a noodle soup
trangmo: cold
tsangma chik choste: "having all come together as one"
tsepo: a woven basket used for carrying dung and fodder grasses
tsokpo: dirty, bad
tsot-tsot: traditional Zanskari word games and riddles told on winter evenings
tutkhung: the hole in the roof of a kitchen to evacuate smoke from hearth fires
tutpa: smoke
tutsi* or *tutpa chakse: forms of soot mixed with condensation to form a tarry mixture that coats ceiling beams
yampa: a long, flat stone used for baking bread
yato: help; friends
yato meta: literally, "Are you without help?"; colloquially, "Where are your friends?" or "What are you crazy, to travel alone?"
yokhang: the slightly sunken central room or "winter kitchen" of traditional Zanskari homes
yura: irrigation canal
Zanskar: also sometimes rendered as Zangskar; a remote valley and former Buddhist kingdom in the northwest Himalaya, in the Indian state of Jammu and Kashmir; the name has three possible meanings or readings: *zan khar*, "food palace"; *zangs kar,* "land of white copper"; and *zang khar*, "the good white place"
Zanskari: a person from Zanskar; *Zanskarpa*, in their own language
zbalu: a small, fairy-like, unpredictable spirit only visible to some individuals
zhing mos: spring plowing and planting
zing: a storage pond for irrigation
zumo: illness or disease

Bibliography

1: THE CURSE

Books

Crook, John, and Henry Osmaston, eds. *Himalayan Buddhist Villages: Environment, Resources, Society and Religious Life in Zangskar, Ladakh*. Bristol, UK: University of Bristol, 1994.

Heraclitus. *Fragments: The Collected Wisdom of Heraclitus*. Translated by Brooks Haxton. New York: Viking Penguin, 2001.

Hesiod. *The Works and Days; Theogony; The Shield of Herakles*. Translated by Richmond Lattimore. Ann Arbor: University of Michigan Press, 1959.

Muller, Friedrich Max. *Rig-Veda-Sanhita: The Sacred Hymns of the Brahmans*. London: Trubner, 1869.

Rizvi, Janet. *Ladakh: Crossroads of High Asia*. New Delhi: Oxford University Press, 1996.

Articles and Reports

Bajracharya, Samjwal Ratna, and Basanta Shrestha. *The Status of Glaciers in the Hindu Kush-Himalayan Region*. Kathmandu: International Centre for Integrated Mountain Development, November 2011. http://lib.icimod.org/record/9419.

Bhutiyani, M. R., V. S. Kale, and N. J. Pawar. "Long-Term Trends in Maximum, Minimum and Mean Annual Air Temperatures Across the Northwestern Himalaya During the Twentieth Century." *Climatic Change* 85 (2007): 159–77.

Bond, Tami C., et al. "Bounding the Role of Black Carbon in the Climate System: A Scientific Assessment." *Journal of Geophysical Research: Atmospheres* 118, no. 11 (June 16, 2013): 5380–552.

Borenstein, Seth. "Arctic Sea Ice Melts to All-Time Record Low." Associated Press, September 20, 2012.

Carrington, Damian. "More Than 1,000 New Coal Plants Planned Worldwide, Figures Show." *Guardian*, November 19, 2012.

Chao, Julie. "Black Carbon a Significant Factor in Melting of Himalayan Glaciers." *Berkeley Lab News Center*, February 3, 2010. http://newscenter.lbl.gov/feature-stories/2010/02/03/black-carbon-himalayan-glaciers/.

"Cloak Protects Glacier from Sun," *BBC News,* May 10, 2005. http://news.bbc.co.uk/2/hi/europe/4533945.stm.

Cohen, J. B., and C. Wang. "Estimating Global Black Carbon Emissions Using a Top-Down Kalman Filter Approach." *Journal of Geophysical Research: Atmospheres* 119, no. 1 (2014): 307–23.

Cook-Anderson, Gretchen. "New Study Turns Up the Heat on Soot's Role in Himalayan Warming." *NASA Earth Science News,* December 14, 2009. http://www.nasa.gov/topics/earth/features/himalayan-warming_prt.htm.

Gautam, Mahesh R., Govinda R. Timilsina, and Kumud Acharya. *Climate Change in the Himalayas: Current State of Knowledge*. Policy Research Working Paper no. 2013. WPS 6516. Washington, DC: World Bank. http://documents.worldbank.org/curated/en/2013/06/17935389/climate-change-himalayas-current-state-knowledge.

Goldberg, Jeffrey. "Drowning Kiribati." *Bloomberg Businessweek*. November 21, 2013.

Gustafsson, Örjan, et al. "Brown Clouds over South Asia: Biomass or Fossil Fuel Combustion?" *Science* 323, no. 5913 (January 23, 2009): 495–98. doi: 10.1126/science.1164857.

Hansen, James, and Larissa Nazarenko. "Soot Climate Forcing via Snow and Ice Albedos." *Proceedings of the National Academies of Sciences* 101, no. 2 (January 13, 2004): 423–28.

Harrabin, Roger. "China Building More Power Plants." *BBC News,* June 19, 2007.

Immerzeel, Walter W., Ludovicus P. H. van Beek, and Mark F. P. Bierkens. "Climate Change Will Affect the Asian Water Towers." *Science* 328 (June 11, 2010): 1382–85. doi:10.1126/science.1183188.

Intergovernmental Panel on Climate Change. "Summary for Policymakers." In *Climate Change 2007: The Physical Science Basis. Contribution of Working Group I to the Fourth Assessment Report of the Intergovernmental Panel on Climate Change*. D. Qin and M. Manning, eds. Cambridge: Cambridge University Press, 2007.

Kehrwald, Natalie M., et al. "Mass Loss on Himalayan Glacier Endangers

Water Resources." *Geophysical Research Letters* 35, no. 22 (2008): 503, doi:10.1029/2008gl035556.

Mann, Michael E. "Little Ice Age." In *Encyclopedia of Global Environmental Change,* vol. 1, *The Earth System: Physical and Chemical Dimensions of Global Environmental Change,* Michael C. MacCracken and John S. Perry, eds. Chichester: John Wiley, 2002, 504–09.

Menon, S., D. Koch, et al. "Black Carbon Aerosols and the Third Polar Ice Cap." *Atmospheric Chemistry and Physics Discussions* 9 (2009): 26593–625.

Mingle, Jonathan. "When the Glacier Left." *Boston Globe*, November 29, 2009.

NASA. "NASA Finds 2011 Ninth-Warmest Year on Record," January 19, 2012. http://www.nasa.gov/topics/earth/features/2011-temps.html.

National Research Council. *Himalayan Glaciers: Climate Change, Water Resources, and* Water Security. Washington, DC: National Academies Press, 2012.

Painter, Thomas H., et al. "End of the Little Ice Age in the Alps Forced by Industrial Black Carbon." *Proceedings of the National Academy of Sciences* 110, no. 38 (September 3, 2013): 15216–221. doi: 10.1073/pnas.1302570110.

Panday, Arnico. *Black Carbon in the Hindu Kush-Himalayan Region*. Kathmandu: International Centre for Integrated Mountain Development, 2011. http://lib.icimod.org/record/26925.

Pontifical Academy of Sciences. "Fate of Mountain Glaciers in the Anthropocene," May 5, 2011. https://jifresse.ucla.edu/events/PAS_Glacier_050511_final-sr-klw.pdf.

Qui, Jane. "China: The Third Pole." *Nature* 454 (2008): 393–96. doi:10.1038/454393a.

Rai, Sandeep Chamling, ed. *An Overview of Glaciers, Glacier Retreat, and Subsequent Impacts in Nepal, India and China*. WWF Nepal Program, 2005.

Ramanathan, V., and G. Carmichael. "Global and Regional Climate Changes due to Black Carbon." *Nature Geoscience* 1 (April 2008): 221–27. doi:10.1038/ngeo156.

———, and Y. Feng. "On Avoiding Dangerous Anthropogenic Interference with the Climate System: Formidable Challenges Ahead." *Proceedings of the National Academies of Sciences* 105, no. 38 (September 23, 2008): 14245–250.

Sackur, Stephen. "The Alaskan Village Set to Disappear Under Water in a Decade." *BBC News Magazine*, July 29, 2013.

Steffen, Will, et al. "The Anthropocene: From Global Change to Planetary Stewardship." *Ambio,* October 12, 2011. doi 10.1007/s13280-011-0185-x.

U.S. Environmental Protection Agency. *Report to Congress on Black Carbon.* EPA-450/R-12-001. March 2012. http://www.epa.gov/blackcarbon.

World Bank and International Cryosphere Climate Initiative. *On Thin Ice: How Cutting Pollution Can Slow Warming and Save Lives*. Washington, DC, October 2013.

World Health Organization. *Ambient (Outdoor) Air Pollution in Cities Database 2014*. Geneva. May 2014.

Yang, Ailun, and Yiyun Cui. "Global Coal Risk Assessment: Data Analysis and Market Research." Washington, DC: World Resources Institute, November 2012. http://www.wri.org/publication/global-coal-risk-assessment.

2: OUR DARK MATERIALS

Books

Archer, David. The *Long Thaw: How Humans Are Changing the Next 100,000 Years of Earth's Climate*. Princeton, NJ: Princeton University Press, 2009.

Crowley, John E. *The Invention of Comfort: Sensibilities and Design in Early Modern Britain and Early America*. Baltimore: Johns Hopkins University Press, 2000.

Edwards, Phillip, ed. *Hamlet, Prince of Denmark*. New Cambridge Shakespeare Series. Cambridge: Cambridge University Press, 1985.

Evelyn, John. *Fumifugium. Or, the Inconvenience of the Aer and Smoake of London Dissipated: Together with Some Remedies Humbly Proposed by J. E. Esq. to His Sacred Majestie, and to the Parliament Now Assembled*. Elmsford, NY: Maxwell Reprint Co., 1969.

Faraday, Michael. *The Chemical History of a Candle*. New York: Crowell. 1957.

Franklin, Benjamin. Letter to Mary Stevenson, November 1760 or September 1761. *The Papers of Benjamin Franklin*, vol. 9. Edited by Leonard W. Labaree. New Haven, CT: Yale University Press, 1966.

Gottsegen, Mark D. *The Painter's Handbook: A Complete Reference*. New York: Watson-Guptill, 2006.

Grann, David. *The Lost City of Z: A Tale of Deadly Obsession in the Amazon*. New York: Doubleday, 2009.

London, Jack. "To Build a Fire." *The Jack London Reader*. Philadelphia, PA: Courage Books, 1994.

Peissel, Michel. *Zanskar: The Hidden Kingdom*. New York: Dutton, 1979.

Randall, Willard Sterne. *Ethan Allen: His Life and Times*. New York: Norton, 2011.

Warwick, Hugh, and Alison Doig. *Smoke: The Killer in the Kitchen: Indoor Air Pollution in Developing Countries*. London: ITDG Publishing, 2004.

Articles and Reports

American Geophysical Union. "Scientists Find Extensive Glacial Retreat in Mount Everest Region." American Geophysical Union Newsroom. May 13, 2013.

http://news.agu.org/press-release/scientists-find-extensive-glacial-retreat-in-mount-everest-region/.

Balakrishnan, Kalpana, Aaron Cohen, and Kirk R. Smith. "Addressing the Burden of Disease Attributable to Air Pollution in India: The Need to Integrate Across Household and Ambient Air Pollution Exposures." *Environmental Health Perspectives* 122, no. 1 (January 2014). http://dx.doi.org/10.1289/ehp.1307822.

Bond, Tami C. Testimony for the House Select Committee on Energy Independence and Global Warming. *Clearing the Smoke: Understanding the Impacts of Black Carbon Pollution*. 111th Cong., 2nd sess., March 16, 2010.

———, et al. "Bounding the Role of Black Carbon in the Climate System: A Scientific Assessment." *Journal of Geophysical Research: Atmospheres* 118, no. 11 (June 16, 2013): 5380–552.

Breashears, David. "Tracking the Himalaya's Melting Glaciers." Yale Environment 360. http://e360.yale.edu/slideshow/the_himalayas_melting_glaciers/8/1/.

Bruce, N., et al. The *Health Effects of Indoor Air Pollution Exposure in Developing Countries*. World Health Organization. 2002. http://www.who.int/peh/air/indoor/oeh02.5.pdf.

Flanner, M. G., C. S. Zender, J. T. Randerson, and P. J. Rasch. 2007. "Present-Day Climate Forcing and Response from Black Carbon in Snow." *Journal of Geophysical Research: Atmospheres* 112, no. D11 (June 16, 2007): 202.

———, N. Hess, M. Mahowald, T. H. Painter, V. Ramanathan, and P. J. Rasch. "Springtime Warming and Reduced Snow Cover from Carbonaceous Particles." *Atmospheric Chemistry and Physics* 9 (2009): 2481–97.

House Committee on Oversight and Government Reform. *Black Carbon and Global Warming*. Testimony for Joel Schwartz. 110th Cong., 1st sess., October 18, 2007.

Janssen, Nicole A. H., et al. *Health Effects of Black Carbon*. Geneva: World Health Organization, 2012. http://www.euro.who.int/en/health-topics/environment-and-health/air-quality/publications/2012/health-effects-of-black-carbon.

Janssen, N. A. H., G. Hoek, M. Simic-Lawson, P. Fischer, L. van Bree, H. ten Brink, et al. "Black carbon as an additional indicator of the adverse health effects of airborne particles compared with PM10 and PM2.5." *Environmental Health Perspectives* 119 (2011): 1691–99.

Jenkins, B. M., et al. "Combustion Properties of Biomass." *Fuel Processing Technology* 54 (1998): 17–46.

Kaspari, S. D., et al. "Recent Increase in Black Carbon Concentrations from a Mt. Everest Ice Core Spanning 1860–2000 AD." *Geophysical Research Letters* 38, no. 4 (2011): 703.

Lim, Stephen S., et al. "A Comparative Risk Assessment of Burden of Disease and Injury Attributable to 67 Risk Factors and Risk Factor Clusters in 21 Regions, 1990–2010: A Systematic Analysis for the Global Burden of Disease Study 2010." *Lancet* 380, no. 9859 (December 15, 2012): 2224–60.

Naeher, Luke P., et al. "Woodsmoke Health Effects: A Review." *Inhalation Toxicology* 19 (2007): 67–106. doi: 10.1080/08958370600985875.

Nussbaumer, Thomas. "Combustion and Co-Combustion of Biomass: Fundamentals, Technologies and Primary Measures for Emissions Reduction." *Energy & Fuels* 17 (2003): 1510–21.

Qian, Y., M. G. Flanner, L. R. Leung, and W. Wang. "Sensitivity Studies on the Impacts of Tibetan Plateau Snowpack Pollution on the Asian Hydrological Cycle and Monsoon Climate." *Atmospheric Chemistry and Physics* 11, no. 5 (2011): 1929. doi: 10.5194/acp-11-1929-2011.

Ramanathan, Veerabhadran. Testimony for the House Select Committee on Energy Independence and Global Warming. *Clearing the Smoke: Understanding the Impacts of Black Carbon Pollution*. 111th Cong., 2nd sess., March 16, 2010.

Robinson, Alexander, Reinhard Caloy, and Andrey Ganopolski. "Multistability and Critical Thresholds of the Greenland Ice Sheet." *Nature Climate Change* 2 (2012): 429–32. doi:10.1038/nclimate1449.

Schwartz, Joel. "Particulate Air Pollution and Daily Mortality: A Synthesis." *Public Health Reviews* 19 (1992): 39–60.

———, Francine Laden, and Antonella Zanobetti. "The Concentration-Response Relation Between PM2.5 and Daily Deaths." *Environmental Health Perspectives* 110, no. 10 (October 2002): 1025–29.

Smith, Kirk R. "Clean Cooking Fuels: What's the Big Deal?" Presentation to Clean Cookstoves Forum, Global Alliance for Clean Cookstoves, Phnom Penh, March 19, 2013. http://ehs.sph.berkeley.edu/krsmith/presentations/2013/gacc%20forum.pdf.

———, J. Zhang, R. Uma, V. V. N. Kishore, V. Joshi, and M. A. K. Khalil. "Greenhouse Implications of Household Fuels: An Analysis for India." *Annual Review of Energy and Environment* 25 (2000): 741–63.

Tollefson, Jeff. "Climate's Smoky Spectre." *Nature* 460 (July 2, 2009): 29–32.

U.S. Environmental Protection Agency. *Report to Congress on Black Carbon*. EPA-450/R-12-001. March 2012.

Valladas, Helene. "Direct Radiocarbon Dating of Prehistoric Cave Paintings by Accelerator Mass Spectrometry." *Measurement Science and Technology* 14, no. 9 (September 1, 2003): 1487–92. doi:10.1088/0957-0233/14/9/301.

Von Neumann, John. "Can We Survive Technology?" *Fortune*, June 1955.

Waldron, H. A. "A Brief History of Scrotal Cancer." *British Journal of Industrial Medicine* 40, no. 4 (1983): 390–401.

World Bank: "Studies Seek Paths to Clean Cooking Solutions." August 6, 2013. http://www.worldbank.org/en/news/feature/2013/08/06/studies-seek-paths-to-clean-cooking-solutions.

World Health Organization. *Fuel for Life: Household Energy and Health*. 2006. http://www.who.int/indoorair/publications/fuelforlife/en/.

Xu, Baiqing, et al. "Black Soot and the Survival of Tibetan Glaciers." *Proceedings of the National Academies of Sciences* 106, no. 52 (December 29, 2009): 2114–18.

Zhang, Juichen, and David R. Oldroyd. "'Red and Expert': Chinese Glaciology During the Mao Tse-Tung Period (1958–1976)." *Geological Society of London, Special Publications* 310, no. 1 (2009): 145–54. doi: 10.1144/sp310.17.

Zorich, Zach. "A Chauvet Primer." *Archaeology* 64, no. 2 (March-April 2011): 39.

3: WATER CONNECTION, FIRE CONNECTION

Books

Cullen, Heidi. The *Weather of the Future: Heat Waves, Extreme Storms, and Other Scenes from a Climate-Changed Planet.* New York: HarperCollins, 2010.

Duka, T. *Life and Works of Alexander Csoma de Koros.* New Delhi: Manjushri Publishing House, 1972.

Fox, Edward. *The Hungarian Who Walked to Heaven: Alexander Csoma de Koros, 1784–1842.* London: Short Lives, 2001.

Reisner, Marc. *Cadillac Desert.* New York: Penguin Books, 1993.

Articles and Reports

Ball, Keith, and Jonathan Elford. "Health in Zangskar." In *Himalayan Buddhist Villages: Environment, Resources, Society and Religious Life in Zangskar, Ladakh.* Edited by John Crook and Henry Osmaston, 405–32. Bristol, UK: University of Bristol, 1994.

California Department of Water Resources. *California State Water Project Overview.* 2014. http://www.water.ca.gov/swp/index.cfm.

Cayan, Dan, et al. *Scenarios of Climate Change in California: An Overview.* California Climate Change Center, February 2006. http://www.energy.ca.gov/2005publications/CEC-500-2005-186/CEC-500-2005-186-SF.pdf.

Crowden, James. "Development and Change in Zangskar, 1977–1989." In *Recent Research on Ladakh 4 & 5: Proceedings of the Fourth and Fifth International Colloquia on Ladakh, Leh 1995.* Edited by H. Osmaston and P. Denwood. New Delhi: Motilal Banarsidass Publishers.

———. "The Road to Padum: Its Effect on Zangskar." In *Recent Research on Ladakh 6, Proceedings of the Sixth International Colloquium on Ladakh, Leh 1993.* Edited by H. Osmaston & T. Tsering. New Delhi: Motilal Banarsidass Publishers, 1997, 53–66.

Fujii, Reed. "Drought Disaster Declared in S.J., Elsewhere." *Stockton Record*, August 31, 2013.

Gutschow, Kim, and Seb Mankelow. "Dry Winters, Dry Summers: Water Shortages in Zangskar." International Association for Ladakh Studies, *Ladakh Studies* 15, August 2001.

Hadley, O. L., et al. "Measured Black Carbon Deposition on the Sierra Nevada Snow Pack and Implication for Snow Pack Retreat." *Atmospheric Chemistry and Physics Discussions* 10 (2010): 10463–85.

Healy, Jack. "Thin Snowpack in West Signals Summer of Drought." *New York Times*, February 22, 2013.

House Committee on Oversight and Government Reform. *Black Carbon and Global Warming*. Testimony for Charles Zender. 110th Cong., 1st sess., October 18, 2007.

Koch, D., and J. Hansen. "Distant Origins of Arctic Black Carbon: A Goddard Institute for Space Studies ModelE Experiment." *Journal of Geophysical Research* 110, D04204 (February 25, 2005). doi:10.1029/2004jd005296.

McIntyre, Kim, ed. "Ice, Snow and Water: Impacts of Climate Change on California and Himalayan Asia." La Jolla, CA: UC San Diego Sustainability Solutions Institute, 2009.

Norboo, T., et al. "Domestic Pollution and Respiratory Illness in a Ladakhi Village." *International Journal of Epidemiology* 20, no. 3 (1991): 749–57.

Office of Governor Edmund G. Brown, Jr. "Governor Brown Declares Drought State of Emergency." January 17, 2014. http://gov.ca.gov/news.php?id=18368.

Onishi, Norimitsu, and Malia Wollan. "Severe Drought Grows Worse in California." *New York Times,* January 17, 2014.

Pirie, Fernanda. "The Fragile Web of Order: Conflict Avoidance and Dispute Resolution in Ladakh." PhD Thesis. Oxford: Wolfson College, Oxford University. 2002.

Ramanathan, V. "Mitigation and Black Carbon." Presentation to EPA Black Carbon Workshop, Chapel Hill, NC, March 4, 2010.

Reisewitz, Annie. "Black Carbon from Wildfires Shown to Warm Southern California Skies." Scripps News Website. December 16, 2009. https://scripps.ucsd.edu/news/2274.

Romm, Joe. "Leading Scientists Explain How Climate Change Is Worsening California's Epic Drought." *Thinkprogress.org*. January 31, 2014.

Sewall, Jacob O., and Lisa Cirbus Sloan. "Disappearing Arctic Sea Ice Reduces Available Water in the American West." *Geophysical Research Letters* 31, no. 6, March 2004. doi: 10.1029/2003gl019133.

Tankersley, Jim. "California Farms, Vineyards in Peril from Warming, U.S. Energy Secretary Warns." *Los Angeles Times,* February 4, 2009.

4: THE ROAD TO SHANGRI-LA

Books

Dewey, Scott Hamilton. *Don't Breathe the Air: Air Pollution and U.S. Environmental Politics, 1945–1970*. College Station: Texas A&M University Press, 2000.

Dickens, Charles. A *Christmas Carol*. Edited by Richard Kelly. New York: Broadview Press, 2003.

Harvey, Andrew. A *Journey in Ladakh*. Boston: Houghton Mifflin, 1983.

Heraclitus. *Fragments: The Collected Wisdom of Heraclitus*. Translated by Brooks Haxton. New York: Viking Penguin, 2001.

Hilton, James. *Lost Horizon*. New York: Morrow, 1936.

Lorant, Stefan. *Pittsburgh: The Story of an American City*, 5th ed. Pittsburgh: Esselmont Books, 1999.

Nanamoli, Bhikkhu. The *Life of the Buddha: According to the Pali Canon*. Seattle: BPS Pariyatti Editions, 2001.

Nitske, W. Robert, and Charles Morrow Wilson. *Rudolf Diesel: Pioneer of the Age of Power*. Norman: University of Oklahoma Press, 1965.

Norberg-Hodge, Helena. *Ancient Futures: Learning from Ladakh*. San Francisco: Sierra Club Books, 1991.

Peissel, Michel. *Zanskar: The Hidden Kingdom*. New York: Dutton, 1979.

Rizvi, Janet. *Ladakh: Crossroads of High Asia*. New Delhi: Oxford University Press, 1996.

Tarr, Joel A. *Devastation and Renewal: An Environmental History of Pittsburgh and Its Region*. Pittsburgh, PA: University of Pittsburgh Press, 2003.

Wells, H. G. *The Outline of History: Being a Plain History of Life and Mankind*. New York: Macmillan, 1921.

Articles and Reports

Bacha, John, et al. *Diesel Fuels Technical Review*. Chevron Corporation, 2007. https://www.chevronwithtechron.com/products/documents/diesel_fuel_tech_review.pdf.

Back, Aaron, and Josh Chin. "Chinese Premier Wen Calls for Action on Air Pollution." *Wall Street Journal*, January 29, 2013.

Bell, Michelle L., Devra L. Davis, and Tony Fletcher. "A Retrospective Assessment of Mortality from the London Smog Episode of 1952: The Role of Influenza and Pollution." *Environmental Health Perspectives* 112, no. 1 (January 2004): 6–8.

Bergstrom, Robert W., et al. "On the Wavelength Dependence of the Absorption of Black Carbon Particles: Predictions and Results from the Tarfox Experiment and Implications for the Aerosol Single Scattering Albedo." *Journal of Atmospheric Sciences*, Global Aerosol Climatology Special Issue, December 2000.

Bond, Tami C. Testimony for the House Select Committee on Energy Independence and Global Warming. *Clearing the Smoke: Understanding the Impacts of Black Carbon Pollution*. 111th Cong., 2nd sess., March 16, 2010.

———, et al. "Historical Emissions of Black and Organic Carbon Aerosol from Energy-Related Combustion, 1850–2000." *Global Biogeochemical Cycles* 21, no. 2 (June 2007). doi: 10.1029/2006gb002840.

———, et al. "Bounding the Role of Black Carbon in the Climate System: A Scientific Assessment." *Journal of Geophysical Research: Atmospheres* 118, no. 11 (June 16, 2013): 5380–552.

Chatterjee, Sumeet, and Nidhi Verma. "Diesel Price Hike May Boost Refiners, Divestment Drive." *Reuters*, January 18, 2013. http://in.reuters.com/article/2013/01/18/india-diesel-sharesale-idindee90h08920130118.

Chen, Yuyu, et al. "Evidence on the Impact of Sustained Exposure to Air Pollution on Life Expectancy from China's Huai River Policy." *Proceedings of the National Academy of Sciences* 110, no. 32 (2013). doi: 10.1073/pnas.1300018110.

Cohen, J. B., and C. Wang. "Estimating Global Black Carbon Emissions Using a Top-Down Kalman Filter Approach." *Journal of Geophysical Research: Atmospheres* 119 (2014): 1–17. doi: 10.1002/2013jd019912.

Cornell, Alexandra G., et al. "Domestic Airborne Black Carbon and Exhaled Nitric Oxide in Children in NYC." *Journal of Exposure Science and Environmental Epidemiology* 22, no. 3 (May-June 2012): 258–66.

Crook, John. "Zangskari Attitudes." In *Himalayan Buddhist Villages: Environment, Resources, Society and Religious Life in Zangskar, Ladakh*. Edited by John Crook and Henry Osmaston, 405–32. Bristol: University of Bristol, 1994, 533–49.

Crowden, J. "The Road to Padum: Its Effect on Zangskar." In *Recent Research on Ladakh 6, Proceedings of the Sixth International Colloquium on Ladakh, Leh 1993*. Edited by H. Osmaston & T. Tsering. New Delhi: Motilal Banarsidass Publishers, 53–66.

Cusick, Daniel. "China's Soaring Coal Consumption Poses Climate Challenge." *Scientific American*, January 30, 2013.

Fang, Shona C., et al. "Residential Black Carbon Exposure and Circulating Markers of Systemic Inflammation in Elderly Males: The Normative Aging Study." *Environmental Health Perspectives* 120, no. 5 (2012): 674–80.

Forbes. "Without Much Fanfare, Apple Has Sold Its 500 Millionth Iphone." *Forbes.Com*, March 25, 2014. http://www.forbes.com/sites/markrogowsky/2014/03/25/without-much-fanfare-apple-has-sold-its-500-millionth-iphone/.

Jung, Kyung Hwa, et al. "Childhood Exposure to Fine Particulate Matter and Black Carbon and the Development of New Wheeze Between Ages 5 and 7 in an Urban Prospective Cohort." *Environment International* 45 (2012): 44–50.

Louchouarn, Patrick, et al. "Elemental and Molecular Evidence of Soot- and Char-Derived Black Carbon Inputs to New York City's Atmosphere During the Twentieth Century." *Environmental Science and Technology* 41, no. 1 (2007): 82–7.

Metropolitan Transportation Authority. "New York City Transit and the Environment." http://web.mta.info/ny.

Navarro, Mireya. "City Issues Rule to Ban Dirtiest Oils at Buildings." *New York Times*, April 21, 2011.

New York City Department of Health and Mental Hygiene. "New York City Trends in Air Pollution and Its Health Consequences." September 26, 2013.

Novakov, T., S. G. Chang, and A. B. Harker. "Sulfates as Pollution Particulates: Catalytic Formation on Carbon (Soot) Particles." *Science* 186, no 4160 (October 18, 1974): 259–61. doi: 10.1126/science.186.4160.259.

Novakov, Tica, and Hal Rosen. "The Black Carbon Story: Early History and New Perspectives." *Ambio*, April 5, 2013.

Pearson, Natalie Obiko, and Rakteem Katakey. "Diesel-Powered Vehicles Leave New Delhi's Air Worse than Beijing's." *Washington Post*, March 4, 2014.

Power, Melinda C., et al. "Traffic-Related Air Pollution and Cognitive Function in a Cohort of Older Men." *Environmental Health Perspectives* 119, no. 5 (2010): 682–87.

Ramanathan, V., and G. Carmichael. "Global and Regional Climate Changes due to Black Carbon." *Nature Geoscience* 1 (April 2008): 221–27. doi:10.1038/ngeo156.

Seamons, David, et al. *The Bottom of the Barrel: How the Dirtiest Heating Oil Pollutes Our Air and Harms Our Health*. Report prepared for the Environmental Defense Fund, December 16, 2009. http://www.edf.org/sites/default/files/10085_EDF_Heating_Oil_Report.pdf.

Shao, Heng. "Bizarre China Report: Five Benefits of a Smoggy China." *Forbes*, December 9, 2013. http://www.forbes.com/sites/hengshao/2013/12/09/bizarre-china-report-ii/.

Smoke Control Lantern Slide Collection, ca. 1940–1950, Ais.1978.22. Historic Pittsburgh Image collection, Archives Service Center, University of Pittsburgh. http://digital.library.pitt.edu/images/pittsburgh/smokecontrol.html.

Steward, R. C. "Industrial and Non-Industrial Melanism in the Peppered Moth *Biston Betularia* (L.)." *Ecological Entomology* 2, no. 3 (1977): 231–43. doi:10.1111/j.1365-2311.1977.tb00886.x.

U.S. Agency for International Development. "Black Carbon Emissions in Asia: Sources, Impacts and Abatement Opportunities." Bangkok, April 2010.

Watt, Louise. "Super Smog Hits North China City; Flights Cancelled." Associated

Press, October 21, 2013. http://bigstory.ap.org/article/heavy-smog-hits-north-china-city-flights-canceled.

Watts, Jonathan. "Twitter Gaffe: US Embassy Announces 'Crazy Bad' Beijing Air Pollution." *Guardian*, November 19, 2010.

Weir, Kirsten. "Smog in Our Brains." *Monitor on Psychology* 43, no. 7 (July-August 2012).

Wilson, Mark. "Drayage Truck Emissions at the Port of Oakland." *Environmental Energy Technologies Division News* 10, no. 3 (Winter 2012). http://eetd.lbl.gov/newsletter/nl38/eetd-nl38-4-oaklandport.html.

Wood, Michael. "Ancient History in Depth: Shangri-La." *BBC News*, February 17, 2007.

World Bank. "China Overview." April 1, 2014. http://www.worldbank.org/en/country/china/overview.

World Health Organization. "Diesel Engine Exhaust Carcinogenic." International Agency for Research on Cancer, Press Release no. 213, June 12, 2012.

5: THE BURDENS OF FIRE

Books

Brimblecombe, Peter. *The Big Smoke: A History of Air Pollution in London Since Medieval Times*. London: Methuen, 1987.

Fu Du. The *Selected Poems of Tu Fu*. Translated by David Hinton. New York: New Direction, 1989.

Isaacson, Walter. *Benjamin Franklin: An American Life*. New York: Simon and Schuster, 2004.

Articles and Reports

American Lung Association. *Sick of Soot: How the EPA Can Save Lives by Cleaning Up Fine Particle Air Pollution*. A Report by the Clean Air Task Force, Earthjustice, and the American Lung Association, 2011.

Eilperin, Juliet. "White House Weakened EPA Soot Proposal, Documents Show." *Washington Post*, July 18, 2012.

Elford, Jonathan. "Kumik: A Demographic Profile." In *Himalayan Buddhist Villages*, Edited by John Crook and Henry Osmaston, 331–60, Bristol, UK: University of Bristol, 1994.

Fang, Shona C., et al. "Residential Black Carbon Exposure and Circulating Markers of Systemic Inflammation in Elderly Males: The Normative Aging Study." *Environmental Health Perspectives* 120, no. 5 (2012): 674–80.

Franklin, Benjamin. "An Account of the New Invented Pennsylvanian Fire-Places." In *The Papers of Benjamin Franklin,* vol. 2, *January 1, 1735 through December 31, 1744,* 419. Edited by L. W. Labaree. New Haven, CT: Yale University Press, 1960.

Global Alliance for Clean Cookstoves. "Igniting Change: A Strategy for Universal Adoption of Clean Cookstoves and Fuels." November 2011.

Harris, Gardiner. "Beijing's Bad Air Would Be Step Up for Smoggy Delhi." *New York Times*, January 25, 2014.

Janssen, Nicole A. H., et al. *Health Effects of Black Carbon.* Geneva: World Health Organization, 2012. http://www.euro.who.int/en/health-topics/environment-and-health/air-quality/publications/2012/health-effects-of-black-carbon.

Laxmi, Vijay, et al. "Household Energy, Women's Hardship and Health Impacts in Rural Rajasthan, India: Need for Sustainable Energy Solutions." *Energy for Sustainable Development* 7, no. 1 (March 2003).

Lewis, Carla, and Jonathan Elford. "Social and Demographic Change in a Zangskari Village Between 1981 and 2005." Unpublished report of Durham University 2005 expedition to Kumik.

Lim, Stephen S., et al. "A Comparative Risk Assessment of Burden of Disease and Injury Attributable to 67 Risk Factors and Risk Factor Clusters in 21 Regions, 1990–2010: A Systematic Analysis for the Global Burden of Disease Study 2010." *Lancet* 380, no. 9859 (December 15, 2012): 2224–60.

Million Death Study Collaborators. "Causes of Neonatal and Child Mortality in India: A Nationally Representative Mortality Survey." *Lancet* 376, no. 9755 (2010): 1853–60. doi: 10.1016/s0140-6736(10)61461-4.

Montague, Peter. "Invisible Killers: Fine Particles." *Rachel's Hazardous Waste News* 373 (January 20, 1994).

Rock, Joseph. "Banishing the Devil of Disease Among the Nashi." *National Geographic* 46, no. 5 (November 1924).

Schwartz, Joel. Testimony for the House Committee on Oversight and Government Reform. *Black Carbon and Global Warming.* 110th Cong., 1st sess., October 18, 2007.

Smith, Kirk, Sumi Mehta, and Mirjam Maeusezahl-Feuz: "The Global Burden of Disease from Household Use of Solid Fuels: A Source of Indoor Air Pollution." In *Comparative Quantification of Health Risks: The Global Burden of Disease due to Selected Risk Factors.* Geneva: World Health Organization, 2004.

———, et al. "Effect of Reduction in Household Air Pollution on Childhood Pneumonia in Guatemala (Respire): A Randomised Controlled Trial." *Lancet* 378 (2011): 1717–26.

Smith-Sivertsen, T., et al. "Effect of Reducing Indoor Air Pollution on Women's Respiratory Symptoms and Lung Function: The Respire Randomized Trial." *American Journal of Epidemiology* 170, no. 2 (2009): 211–20.

UN Children's Fund. *Levels and Trends in Child Mortality: Estimates Developed by the UN Inter-Agency Group for Child Mortality Estimation.* New York, 2013. http://www.childinfo.org/files/Child_Mortality_Report_2013.pdf.

World Health Organization. "Air Quality and Health: Fact Sheet No. 313." Geneva: September 2011.

———. *Health Effects of Particulate Matter: Policy Implications for Countries in Eastern Europe, Caucasus and Central Asia.* 2013. http://www.euro.who.int/__data/assets/pdf_file/0006/189051/Health-effects-of-particulate-matter-final-Eng.pdf?ua=1.

———. "Metrics: Disability-Adjusted Life Year (DALY). Quantifying the Burden of Disease from mortality and morbidity." 2014. http://www.who.int/healthinfo/global_burden_disease/metrics_daly/en/.

Zongxing, Li. "Changes of Climate, Glaciers and Runoff in China's Monsoonal Temperate Glacier Region During the Last Several Decades." *Quaternary International* 218 (2010): 13–28

6: WATER TOWERS FALLING

Books

Archer, David. The *Long Thaw: How Humans Are Changing the Next 100,000 Years of Earth's Climate.* Princeton, NJ: Princeton University Press, 2009.

Carey, Mark. *In the Shadow of Melting Glaciers: Climate Change and Andean Society.* New York: Oxford University Press, 2010.

De Saint-Exupéry, Antoine. Translation by Lewis Galantiere. *Wind, Sand and Stars.* Harcourt Brace: New York. 1992.

Norman, Rebecca. A *Dictionary of the Language Spoken by Ladakhis.* In preparation, 2011.

Articles and Reports

American Geophysical Union. "What's Going On in the Arctic?" AGU Fall Meeting Press Briefing, December 5, 2012.

Bajracharya, S. R., and Basanta Shrestha, eds. *The Status of Glaciers in the Hindu Kush-Himalayan Region.* International Centre for Integrated Mountain Development, November 2011. http://lib.icimod.org/record/26925.

Bolch, Tobias, et al. "The State and Fate of Himalayan Glaciers." *Science* 336, no. 6079 (April 20, 2012): 310–14. doi: 10.1126/science.1215828.

Bond, Tami C. Testimony for the House Select Committee on Energy Independence and Global Warming. *Clearing the Smoke: Understanding the Impacts of Black Carbon Pollution*. 111th Cong., 2nd sess., March 16, 2010.

———, et al. "Bounding the Role of Black Carbon in the Climate System: A Scientific Assessment." *Journal of Geophysical Research: Atmospheres* 118, no. 11, (June 16, 2013): 5380–552.

Box, J. E. "Dark Snow Project First Science Results." Dark Snow Project Blog, January 8, 2014. http://darksnow.org/2014/01/.

———. "Examining the Role of Black Carbon and Microbial Abundance in Greenland Ice Sheet Albedo Feedback." Presentation to AGU fall meeting, December 12, 2013.

———, et al. "Greenland Ice Sheet Albedo Feedback: Thermodynamics and Atmospheric Drivers." *Cryosphere* 6 (2012): 821–39.

Buis, Alan. "Is a Sleeping Giant Stirring in the Arctic?" *NASA News and Features,* June 10, 2013. http://www.NASA.gov/topics/earth/features/earth20130610.html#.U231zyvrbjc.

Cogley, J. G., et al. "Tracking the Source of Glacier Misinformation." *Science* 327, no. 5965 (January 29, 2010): 522. doi: 10.1126/science.327.5965.522-a.

Cruz, R. V., et al. "Climate Change 2007: Impacts, Adaptation and Vulnerability." *Contribution of Working Group II to the Fourth Assessment Report of the Intergovernmental Panel on Climate Change*, 469–506. Edited by M. L. Parry et al. Cambridge: Cambridge University Press, 2007.

Dixit, Yama, David A. Hodell, and Cameron A. Petrie. "Abrupt Weakening of the Summer Monsoon in Northwest India ~ 4100 Years Ago." *Geology*, February 24, 2014. doi: 10.1130/G35236.1.

Fujita, Koji, and Takayuki Nuimura. "Spatially Heterogeneous Wastage of Himalayan Glaciers." *Proceedings of the National Academy of Sciences* 108, no. 34 (2011): 14011–14. http://www.pnas.org/content/108/34/14011.full.

Hays, J. D., J. Imbrie, and N. J. Shackleton. (1976). "Variations in the Earth's Orbit: Pacemaker of the Ice Ages." *Science* 194, no. 4270 (December 10, 1976): 1121–32. doi:10.1126/science.194.4270.1121.

"India Floods: Coming to Terms with Loss in Uttarakhand." *BBC News,* July 15, 2013. http://www.bbc.com/news/world-asia-india-23311298.

Indian Space Research Organization. "Snow and Glaciers of the Himalaya." May 2011.

Institute for Governance and Sustainable Development. "Primer on Short-Lived Climate Pollutants." Washington, DC. November 2012.

"Is Black Carbon Affecting the Asian Monsoon?" *BBC News,* July 8, 2011. http://www.bbc.co.uk/news/science-environment-14047815.

Ives, Jack D., Rajendra B. Shrestha, and Pradeep K. Mool. *Formation of Glacial Lakes in the Hindu Kush-Himalayas and GLOF Risk Assessment*. International

Centre for Integrated Mountain Development. May 2010. http://www.indiaenvironmentportal.org.in/files/ICIMODGLOF.pdf.

Jacobson, Mark Z. "Climate Response of Fossil Fuel and Biofuel Soot, Accounting for Soot's Feedback to Snow and Sea Ice Albedo and Emissivity." *Journal of Geophysical Research: Atmospheres* 109, no. D21 (2004).

———. "Short-Term Effects of Controlling Fossil-Fuel Soot, Biofuel Soot and Gases, and Methane on Climate, Arctic Ice, and Air Pollution Health." *Journal of Geophysical Research: Atmospheres* 115, no. D14 (2010).

———. Testimony for the House Committee on Oversight and Government Reform. *Black Carbon and Global Warming*. 110th Cong., 1st sess., October 18, 2007.

Kamp, Ulrich, Martin Byrne, and Tobias Bolch. "Glacier Fluctuations Between 1975 and 2008 in the Greater Himalaya Range of Zanskar, Southern Ladakh." *Journal of Mountain Science* 8 (2011): 374–89.

Khadka, Navin Singh. "Is Black Carbon Affecting the Asian Monsoon?" *BBC News,* July 8, 2011.

Lamsang, Tenzing. "Panic in Punakha." *Kuensel,* April 30, 2009.

Lau, K. M., M. K. Kim, and K. M. Kim. "Asian Summer Monsoon Anomalies Induced by Aerosol Direct Forcing: The Role of the Tibetan Plateau." *Climate Dynamics* 26 (2006): 855–64.

Lenton, Timothy M. "Arctic Climate Tipping Points." *Ambio* 41 (2012): 10–22

———, et al. "Tipping Elements in the Earth's Climate System." *Proceedings of the National Academies of Science* 105, no. 6 (February 12, 2008): 1786–93.

Osmaston, Henry. "The Geology, Geomorphology and Quaternary History of Zangskar." In *Himalayan Buddhist Villages: Environment, Resources, Society and Religious Life in Zangskar, Ladakh*. Edited by John Crook and Henry Osmaston, 1–36. Bristol, UK: University of Bristol, 1994.

Ramanathan, V., and G. Carmichael. "Global and Regional Climate Changes due to Black Carbon." *Nature Geoscience* 1 (April 2008): 221–27. doi:10.1038/ngeo156.

———, et al. "Atmospheric Brown Clouds: Impacts on South Asian Climate and Hydrological Cycle." *Proceedings of the National Academies of Science* 102 (2005): 5326–33.

Roberts, David. "Hope and Fellowship." *Grist,* August 30, 2013.

Royal Government of Bhutan. "National Report on Bhutan for World Conference on Disaster Reduction." Thimphu, Bhutan: Department of Local Governance, Ministry of Home and Cultural Affairs, 2005. http://www.unisdr.org/2005/mdgs-drr/national-reports/Bhutan-report.pdf.

Schuur, Edward A. G., and Benjamin Abbott. "Climate Change: High Risk of Permafrost Thaw." *Nature* 480 (December 1, 2011): 32–33.

Shindell, D., and G. Faluvegi. "Climate Response to Regional Radiative Forcing During the Twentieth Century." *Nature Geoscience* 2 (2009): 294–300. doi:10.1038/ngeo473.

Taylor, Adam, and Samuel Blackstone. "Why the Monsoon Is the Biggest Factor in the South Asian Economy." *Business Insider,* June 20, 2012.

Turner, Christi. "A Darker Shade of Snow." *Boulder Weekly,* January 9, 2014.

Voiland, Adam. "Aerosols May Drive a Significant Portion of Arctic Warming." *NASA News and Features,* April 8, 2009. http://www.nasa.gov/topics/earth/features/warming_aerosols.html.

Wangdi, Kencho. "24 Glacial Lakes in Bhutan Identified as 'Potentially Dangerous.' " *Kuensel*, May 14, 2010.

Watanabe, Teiji, and Daniel Rothacher. "The 1994 Lugge Tsho Glacial Lake Outburst Flood, Bhutan Himalaya." *Mountain Research and Development* 16, no. 1 (February 1996): 77–81.

Yao, Tandong, et al. "Recent Glacial Retreat in High Asia in China and Its Impact on Water Resource in Northwest China." *Science in China Series D Earth Sciences* 47, no. 12 (2004): 1065–75. doi:10.1360/03yd0256.

Zaelke, Durwood J., and Veerabhadran Ramanathan. "Going Beyond Carbon Dioxide." *New York Times*, December 6, 2012.

Zender, Charles. Testimony for the House Committee on Oversight and Government Reform. *Black Carbon and Global Warming.* 110th Cong., 1st sess., October 18, 2007.

7: IN SEARCH OF PHUNSUKH WANGDU

Books

Adiga, Aravind. The *White Tiger*. New York: Free Press, 2008.

Faraday, Michael. *The Chemical History of a Candle.* New York: Crowell, 1957.

Van Doren, Carl. *Benjamin Franklin.* New York: Viking Press, 1938.

Articles and Reports

Acharya, Jiwan, M. Sundar Bajgain, and Prem Sagar Subedi. "Scaling Up Biogas in Nepal: What Else Is Needed?" *Boiling Point*, no. 50 (2005).

Ashden Awards. "Biogas Sector Partnership, Nepal." Ashden Awards Case Study, December 2009. http://www.ashden.org/files/bsp%20case%20study%20full.pdf.

Bond, Tami C., et al. "Bounding the Role of Black Carbon in the Climate System: A Scientific Assessment." *Journal of Geophysical Research: Atmospheres* 118, no. 11 (June 16, 2013): 5380–552.

Franklin, Benjamin. "To Jan Ingenhousz," Letter, August 28, 1785 (unpublished). *The Papers of Benjamin Franklin,* Digital Archive, Yale University, Franklinpapers.org.

Goenka, Debi, and Sarath Guttikunda. "Coal Kills: An Assessment of Death and Disease Caused by India's Dirtiest Energy Source." Urban Emissions, with Conservation Action Trust and Greenpeace India. January 2013. http://www.greenpeace.org/india/Global/india/report/Coal_Kills.pdf.

Jacobson, Arne. "Case Study 14: Rural Electrification in Ladakh, India." in *Special Report on Methodological and Technical Issues in Technology Transfer*, Intergovernmental Panel on Climate Change (IPCC) Working Group II. 1999. http://www.grida.no/climate/ipcc/tectran/342.htm.

Joshi, S. K., and I. Dudani. "Environmental Health Effects of Brick Kilns in Kathmandu Valley." *Kathmandu University Medical Journal* 6, no. 1 (2008): 3–11. http://www.academia.edu/183945/Environmental_health_effects_of_brick_kilns_in_Kathmandu_valley.

Kumar, Arun, Geeta Vaidyanathan, and K. R. Lakshmikanthan. "Cleaner Brick Production in India: A Transectoral Initiative." *Industry and Environment* 21, no. 1-2 (January-June 1998): 77–80.

Lam, Nicholas, et al. "Household Light Makes Global Heat: High Black Carbon Emissions From Kerosene Wick Lamps." *Environmental Science and Technology* 46, no. 24 (2012): 13531–38.

Mingle, Jonathan. "Rewriting the Books in Ladakh." *Cultural Survival Quarterly* 27, no. 4 (Winter 2003).

Pokhrel, A. K., et al. "Tuberculosis and Indoor Biomass and Kerosene Use in Nepal: A Case-Control Study." *Environmental Health Perspectives* 118, no. 4 (April 2010): 558–64.

Pushkarna, Vijay. "Man of the Year: Master Touch." *The Week*, December 23, 2001.

Ramanathan, Veerabhadran, and David G. Victor. "To Fight Climate Change, Clear the Air." *New York Times*, November 27, 2010.

Shindell, Drew, et al. "Simultaneously Mitigating Near-Term Climate Change and Improving Human Health and Food Security." *Science* 335, no. 6065 (January 13, 2012): 183–89. doi: 10.1126/science.1210026.

Smith, Kirk R., and Evan Haigler. "Co-Benefits of Climate Mitigation and Health Protection in Energy Systems: Scoping Methods." *Annual Review of Public Health* 29 (2008): 11–25.

Sonam, Wangchuk. "Are Bicycles a Win-Win Solution for Ladakh? I Don't Think So!" *Reach Ladakh*, January 18, 2014. http://www.reachladakh.com/print_news-details.php?pID=2154.

Sood, Jyotika. "NASA Draws Attention to Fires in Punjab Fields." *Down to Earth*, November 5, 2012. http://www.downtoearth.org.in/content/nasa-draws-attention-fires-punjab-fields.

Tierney, John. "Climate Proposal Puts Practicality Ahead of Sacrifice." *New York Times*, January 16, 2012.

World Bank. *On Thin Ice: How Cutting Pollution Can Slow Warming and Save*

Lives. A Joint Report of the World Bank and the International Cryosphere Climate Initiative. Washington, DC, October 2013.

Yee, Amy. "India Increases Effort to Harness Biomass Energy." *New York Times*, October 8, 2013.

Zaelke, Durwood J., and Veerabhadran Ramanathan. "Going Beyond Carbon Dioxide." *New York Times*, December 6, 2012.

8: NOW WE'RE COOKING WITH GAS!

Books

Parr, S. W. *Fuel, Gas, Water and Lubricants*. New York: McGraw-Hill, 1932.

Articles and Reports

American Gas Association Monthly, vol. 23, 1941.

American Pet Products Association. "U.S. Pet Industry Spending Figures and Future Outlook." http://www.americanpetproducts.org/press_industrytrends.asp.

Anenberg, S. C., et al. Global Air Quality and Health Co-Benefits of Mitigating Near-Term Climate Change Through Methane and Black Carbon Emission Controls. *Environmental Health Perspective* 120 (2012): 831–39.

Barron, James. "Storm Barrels Through Region, Leaving Destructive Path." *New York Times*, October 30, 2012.

BioLite. "HomeStove H4 Testing: Odisha, India." January 1, 2013. http://www.biolitestove.com/stories/#8.

Bond, Tami C. Testimony for House Committee on Oversight and Government Reform. *Black Carbon and Global Warming*. 110th Cong., 1st sess., October 18, 2007.

Budya, H., and M. Arofat. "Providing Cleaner Energy Access in Indonesia Through the Megaproject of Kerosene Conversion to LPG." *Energy Policy* 39, no. 12 (2011): 7575–86.

Clinton, Hillary. "Remarks on Global Alliance for Clean Cookstoves at Clinton Global Initiative." September 21, 2010. http://m.state.gov/md147500.htm.

Cordes, Leslie. *Igniting Change: A Strategy for Universal Adoption of Clean Cookstoves and Fuels*. Global Alliance for Clean Cookstoves, November 2011. http://www.cleancookstoves.org/resources/fact-sheets/igniting_change.pdf.

Franklin, Benjamin. "An Account of the New Invented Pennsylvanian Fire-Places." In *The Papers of Benjamin Franklin*, vol. 2, *January 1, 1735 through December 31, 1744*, 419. Edited by L. W. Labaree. New Haven, CT: Yale University Press, 1960.

Global Alliance for Clean Cookstoves Database. "Data and Statistics." http://www.cleancookstoves.org/resources/data-and-statistics/.

Household Energy Network. "An Interview with Professor Kirk R. Smith." *Boiling Point* no. 56: Liquid Fuels. http://www.hedon.info/bp56:interviewwithprofessorkirksmith?bl=y.

International Energy Agency. *World Energy Outlook: 2011*. Paris: International Energy Agency, 2011. http://www.iea.org/publications/freepublications/publication/weo2011_web.pdf.

Kar, Abhishek, et al. "Real-Time Assessment of Black Carbon Pollution in Indian Households due to Traditional and Improved Biomass Cookstoves." *Environmental Science and Technology* 46, no. 5 (2012): 2993–3000. doi: 10.1021/es203388g.

Lambe, Fiona, and Aaron Atteridge. "Putting the Cook Before the Stove: A User-Centred Approach to Understanding Household Energy Decision-Making. A Case Study of Haryana State, Northern India." Stockholm Environment Institute. May 2012.

Lim, Stephen S., et al. "A Comparative Risk Assessment of Burden of Disease and Injury Attributable to 67 Risk Factors and Risk Factor Clusters in 21 Regions, 1990–2010: A Systematic Analysis for the Global Burden of Disease Study 2010." *Lancet*, no. 9859 (December 15, 2012): 2224–60.

Mingle, Jonathan. "Burning Desire." *Dartmouth Alumni Magazine*, July-August 2012.

Mooney, Chris. "Why Greenland's Melting Could Be the Biggest Climate Disaster of All." *Climate Desk*. January 28, 2013. http://climatedesk.org/2013/01/why-greenlands-melting-could-be-the-biggest-climate-disaster-of-all/.

Mukhopadhyay, Rupak, et al. "Cooking Practices, Air Quality, and the Acceptability of Advanced Cookstoves in Haryana, India: An Exploratory Study to Inform Large-Scale Interventions." *Global Health Action* 5 (2012): 19016.

Ramanathan, V., and G. Carmichael. "Global and Regional Climate Changes due to Black Carbon." *Nature Geoscience* 1 (April 2008): 221–27. doi:10.1038/ngeo156.

Ransom, Diana. "Hurricane Sandy and Its Aftermath: By the Numbers." *Entrepreneur*. October 24, 2013. http://www.entrepreneur.com/article/229593.

Smith, Kirk R. "Comparative Risk Assessment of Household Air Pollution in the GBD 2010 Study." Presentation to Global Alliance for Clean Cookstoves, Washington, DC, December 17, 2012.

———. "In Praise of Petroleum." *Science* 298 (December 6, 2002): 1847.

———, and Karabi Dutta. "Cooking with Gas." *Energy for Sustainable Development* 15 (2011): 115–16.

Warwick, Hugh, and Alison Doig. *The Killer in the Kitchen: Indoor Air Pollution in Developing Countries*. London: ITDG Publishing, 2004.

Wilkinson, Paul, et al. "Public Health Benefits of Strategies to Reduce Greenhouse-

Gas Emissions: Household Energy." *Lancet* 374, no. 9705 (2009): 1917–25. doi:10.1016/s0140-6736(09)61713-X.

World Bank. *Household Cookstoves, Environment, Health and Climate Change: A New Look at an Old Problem*. Washington, DC, 2010. http://documents.worldbank.org/curated/en/2010/03/14600224/household-cook-stoves-environment-health-climate-change-new-look-old-problem.

———. "On Thin Ice: How Cutting Pollution Can Slow Warming and Save Lives." A Joint Report of the World Bank and the International Cryosphere Climate Initiative. Washington, DC, October 2013.

9: THE VIEW FROM SULTAN LARGO

Books

Howe, Mark Dewolf, ed. *Holmes-Pollock Letters: The Correspondence of Mr. Justice Holmes and Sir Frederick Pollock, 1874–1932,* 2d ed. Cambridge, MA: Harvard University Press, 1961.

Pollack, Henry. A *World Without Ice*. New York: Penguin, 2009.

Van Doren, Carl. *Benjamin Franklin*. New York: Viking Press, 1938.

Articles and Reports

Barboza, Tony. "Obama Administration Limits on Soot Pollution Upheld by Appeals Court." *Los Angeles Times*, May 9, 2014.

Boden, Tom, Bob Andres, and Gregg Marland. "Ranking of the World's Countries by 2010 Total CO_2 Emissions from Fossil-Fuel Burning, Cement Production, and Gas Flaring." Carbon Dioxide Information Analysis Center. http://cdiac.ornl.gov/trends/emis/top2010.tot.

Bond, Tami C. Testimony for the House Committee on Oversight and Government Reform. *Black Carbon and Global Warming*. 110th Cong., 1st sess., October 18, 2007.

———. Testimony for the House Select Committee on Energy Independence and Global Warming. *Clearing the Smoke: Understanding the Impacts of Black Carbon Pollution*. 111th Cong., 2nd sess., March 16, 2010.

Chakrabarty, Rajan K., et al. "Funeral Pyres in South Asia: Brown Carbon Aerosol Emissions and Climate Impacts." *Environmental Science and Technology Letters* 1, no. 1 (2014): 44–48. doi: 10.1021/ez4000669.

Dunbar, R. I. M. "Neocortex Size as a Constraint on Group Size in Primates." *Journal of Human Evolution* 22, no. 6 (1992): 469–93. doi:10.1016/0047-2484(92)90081-J.

Feng Y., V. Ramanathan, and V. R. Kotamarthi. "Brown Carbon: A Significant Atmospheric Absorber of Solar Radiation?" *Atmospheric Chemistry and Physics* 13 (2013): 8607–21.

Hansen, James, et al. "Global Warming in the Twenty-First Century: An Alternative Scenario." *Proceedings of the National Academies of Science* 97, no. 18 (August 29, 2000): 9875–80.

Hooker, Jake. "Beijing Announces Traffic Plan for Olympics." *New York Times,* June 21, 2008.

Institute for Governance and Sustainable Development. "Primer on Short-Lived Climate Pollutants." Washington, DC: November 2012.

Kar, Abhishek, et al. "Real-Time Assessment of Black Carbon Pollution in Indian Households due to Traditional and Improved Biomass Cookstoves." *Environmental Science and Technology* 46, no. 5 (2012): 2993–3000. doi: 10.1021/es203388g.

Mankelow, J. S. "Watershed Development in Central Zangskar." *Ladakh Studies* no. 19 (2003): 49–58.

Molina, Mario, et al. "Reducing Abrupt Climate Change Risk Using the Montreal Protocol and Other Regulatory Actions to Complement Cuts in CO2 Emissions." *Proceedings of the National Academy of Sciences* 106, no. 49 (December 8, 2009): 20616–21.

Monastersky, Richard. "The Storm at the Center of Climate Science." *Chronicle of Higher Education,* November 10, 2000.

NASA Earth Observatory. "All Dry on the Western Front." Image of the Day, January 23, 2014. http://earthobservatory.Nasa.Gov/Iotd/View.php?Id=82910.

Neumaier, Eva K. "Historiography and the Narration of Cultural Identity in Zanskar." In *Tibet: Past and Present; Proceedings of the Ninth Seminar of the International Association for Tibetan Studies, 2000,* 327–40. Edited by Henk Blezer and A. Zadoks. Leiden, Netherlands: Koninklijke Brill NV, 2002.

Nielsen, Chris P., and Mun S. Ho, "Clearing the Air in China." *New York Times,* October 25, 2013.

Ramanathan, V., et al. "Black Carbon and the Regional Climate of California." Report to the California Air Resources Board. January 22, 2013.

Schwartz, Joel. Testimony for House Committee on Oversight and Government Reform. *Black Carbon and Global Warming.* 110th Cong., 1st sess., October 18, 2007.

Sethi, Nitin. "'India Is Not a Nay-Sayer on Climate Change.'" *The Hindu,* November 7, 2013.

U.S. Environmental Protection Agency. "Public Hearing for Proposed Rules—National Ambient Air Quality Standards for Particulate Matter." Sacramento, CA. July 19, 2012.

———. "The Benefits and Costs of the Clean Air Act from 1990 to 2020." Office of Air and Radiation. March 2011.

UN Environment Programme. "Near-Term Climate Protection and Clean Air Benefits: Actions for Controlling Short-Lived Climate Forcers: A UNEP Synthesis Report." Nairobi, Kenya. November 2011.

———. "Towards An Action Plan for Near-Term Climate Protection and Clean Air Benefits." UNEP Science-Policy Brief. 2011.

Van Atten, Christopher, and Lily Hoffman-Andrews. "The Clean Air Act's Economic Benefits: Past, Present and Future." Report for the Small Business Majority and Main Street Alliance. October 2010.

Voorhees, Josh. "Forget Polar Bears." *Slate*. June 3, 2014.

World Bank. "On Thin Ice: How Cutting Pollution Can Slow Warming and Save Lives." Joint Report of the World Bank and the International Cryosphere Climate Initiative, Washington, DC, October 2013.

Index